Digging Deeper: Inside Africa's Agricultural, Food and Nutrition Dynamics

African Dynamics

VOLUME 13

The titles published in this series are listed at *brill.com/ad*

Digging Deeper: Inside Africa's Agricultural, Food and Nutrition Dynamics

Edited by

Akinyinka Akinyoade
Wijnand Klaver
Sebastiaan Soeters
Dick Foeken

BRILL

LEIDEN | BOSTON

Cover illustration: (top left and right) Photographs by Fernando Sanchez Bueno, Kirkissoye Neighbourhood, close to the Niger River in Niamey (Niger), (down left) Seeds for sale in market place.

Library of Congress Cataloging-in-Publication Data

Digging deeper : inside Africa's agricultural, food and nutrition dynamics / edited by Akinyinka Akinyoade, Wijnand Klaver, Sebastiaan Soeters, Dick Foeken.
pages cm. -- (African dynamics ; volume 13)
ISBN 978-90-04-28268-1 (pbk. : alk. paper) -- ISBN 978-90-04-28269-8 (e-book) 1. Agricultural productivity--Africa. 2. Food security--Africa. 3. Food consumption--Africa. 4. Agriculture--Africa. I. Akinyoade, Akinyinka. II. Title: Inside Africa's agricultural, food and nutrition dynamics. III. Series: African dynamics ; 13.
S472.A1D54 2014
338.1'6096--dc23

2014032453

This publication has been typeset in the multilingual "Brill" typeface. With over 5,100 characters covering Latin, IPA, Greek, and Cyrillic, this typeface is especially suitable for use in the humanities. For more information, please see www.brill.com/brill-typeface.

ISSN 1568-1777
ISBN 978-90-04-28268-1 (paperback)
ISBN 978-90-04-28269-8 (e-book)

Contents

List of Figures, Maps, Photos and Tables

Figures

Maps

Photos

Tables

List of Acronyms and Abbreviations

AB	Aktiebolag (Swedish for limited company)
ACET	African Center for Economic Transformation
ACGS	Agricultural Credit Guarantee Scheme
ACPC	African Climate Policy Centre
ADBG	African Development Bank Group
ADC	Agricultural Development Corporation
ADER	Average Dietary Energy Requirement
AEZ	Agro-Ecological Zoning
AFC	Agricultural Finance Corporation
AFCA	Agro-Food Clusters in Africa
AfDB	African Development Bank
AfricaRice	Africa Rice Center (CGIAR)
AFSUN	African Food Security Urban Network
AGRA	Alliance for a Green Revolution in Africa
AGRF	Africa Green Revolution Forum
AHITI	Animal Health Training Institute
AKAM	Aga Khan Agency for Microfinance
APPP	Africa Power and Politics Project
ARCN	Agricultural Research Council of Nigeria
ASARECA	Association for Strengthening Agricultural Research in East and Central Africa
ASC	African Studies Centre
ASCU	Agricultural Sector Coordination Unit
AU	African Union
BMI	Body Mass Index
BMR	Basal Metabolic Rate
C4C	Capacity for Competitiveness Programme
CAADP	Comprehensive Africa Agriculture Development Program
cap	capita
CBN	Central Bank of Nigeria
CCARDESA	Centre for Coordination of Agricultural Resources and Development for Southern Africa
CDKN	Climate and Development Knowledge Network
CDM	Clean Development Mechanism
CEE	Central and East European Countries
CERES	Research School for Resource Studies for Development
CFS	Committee on World Food Security

CGE	Computable General Equilibrium
CGIAR	Consultative Group on International Agricultural Research
CIAT	Centro Internacional de Agricultura Tropical/International Center of Tropical Agriculture (CGIAR)
CIP	Centro Internacional de la Papa/International Potato Center (CGIAR)
CIS	Commonwealth of Independent States
CLIMAFRICA	EU FP7 Research programme on Climate Change Predictions in Sub-Saharan Africa
CORAF/WECARD	West and Central African Council for Agricultural Development
DCW	Digital Chart of the World
DES	Dietary Energy Supply
DFID	Department for International Development
DHS	Demographic and Health Surveys
DISI	Development Information Services International
DRA	Developmental Regimes in Africa project
DSTV	Digital Satellite television
DTI	Dairy Training Institute
DVC	Dairy Value Chain
ECA	Economic Commission for Africa
ECOWAS	Economic Community of West African States
EGU	Egerton University
EIA	Environmental Impact Assessment
EMBRAPA	Empresa Brasileira de Pesquisa Agropecuária (Brazilian Agricultural Research Corporation)
EPZ	Export Processing Zone
ETB	Ethiopian Birr
EU	European Union
EU FP6	The EU 6th Framework Programme for Research and Technological Development (FP6)
EU FP7	The EU 7th Framework Programme for Research and Technological Development (FP7)
EWCA	Ethiopian Wild Life Conservation Authority
FAIS	(see INTERFAIS)
FANTA	Food and Nutrition Technical Assistance
FAO	Food and Agriculture Organization of the United Nations
FAOSTAT	Statistical Database of FAO
FARA	Forum for Agricultural Research in Africa
Faulu Kenya	Faulu Microfinance Bank Ltd
FBS	Food Balance Sheet

FCA	Fadama Community Association
Fe	Iron
FMANR	Federal Ministry of Agriculture and Natural Resources (Nigeria)
FOODMETRES	EU FP7 Research programme on Food Planning and Innovation for Sustainable Metropolitan Regions
FPRE	Farmer-Participatory Research and Extension
FUG	Fadama User Group
GDP	Gross Domestic Product
GES	Growth Enhancement Scheme (Nigeria)
GHG	Greenhouse gas
GTP	Growth & Transformation Plan
ha	Hectare
HIV/AIDS	Human Immunodeficiency Virus infection / Acquired Immuno Deficiency Syndrome
IAC	Inter-Academy Council
IATP	Institute for Agriculture and Trade Policy
ICAR	Indian Council of Agricultural Research
ICF Macro	Macro International market research firm
ICN	1992 International Conference on Nutrition
ICN + 21	2014 International Conference on Nutrition 21 years later
ICRISAT	International Crops Research Institute for the Semi-Arid Tropics (CGIAR)
IDA	International Development Association
IDC	International Data Corporation
IDRC	International Development Research Centre
IFAD	International Fund for Agricultural Development
IFPRI	International Food Policy Research Institute (CGIAR)
IITA	International Institute of Tropical Agriculture (CGIAR)
ILRI	International Livestock Research Institute (CGIAR)
INSTAPA	EU FP7 Research programme on Improved Nutrition through Staple Foods in Africa
INTERFAIS	International Food Aid Information System
IPCC	Intergovernmental Panel on Climate Change
IPP	Intellectual Property Protection
IPR	Intellectual Property Rights
IT	Information Technology
IWMI	International Water Management Institute (CGIAR)
JKUAT	Jomo Kenyatta University of Agriculture and Technology
JRC	European Joint Research Centre
KARI	Kenya Agricultural Research Institute

kcal	Kilocalorie
KCB	Kenya Commercial Bank
KDB	Kenya Dairy Board
KDPA	Kenya Dairy Producers Association
KEBS	Kenya Bureau of Standards
KIPPRA	Kenya Institute of Public Policy Research and Advocacy
KITLV	Royal Netherlands Institute of South East Asian and Caribbean Studies
KNOMAD	Global Knowledge Partnership on Migration and Development
K-REP	Kenya Rural Enterprise Programme Bank
KUFSALCC	Kampala Urban Food Security, Agriculture and Livestock Coordinating Committee
KWFT	Kenya Women Finance Trust
LANDac	Netherlands Academy for Land Governance for Equitable and Sustainable Development
LDP	Local Development Planning
LGA	Local Government Area
M&E	Monitoring & evaluation
MDER	Minimum Dietary Energy Requirement
MDG	Millennium Development Goal
MICS	Multi Indicator Cluster Survey
MoA	Ministry of Agriculture
MoFED	Ministry of Finance and Economic Development (Ethiopia)
MSM	Maastricht School of Management
MT	Metric Ton = tonne = 1000 kg
MW	Megawatt
N	Naira (Nigerian currency)
N/A	Not applicable
NACB	Nigerian Agricultural and Cooperative Bank
NARES	National Agricultural Research and Extension System
NARI	National Agricultural Research Institute
NARO	National Agricultural Research Organisation
NARS	National Agricultural Research System
NBS	National Bureau of Statistics
NCHS	National Centre for Health Statistics
NEFSALF	Nairobi and Environs Food Security, Agriculture and Livestock Forum
NEPAD	New Partnership for Africa's Development (AU)
NES	Not elsewhere specified
NGO	Non-governmental organization

NKCC	New Kenya Cooperative Creameries
NPC	National Planning Commission
NSF	Networking Support Function
NUFFIC	Netherlands Universities Foundation for International Cooperation
ODI	Overseas Development Institute
PAL	Physical Activity Level
PBL	Netherlands Environmental Assessment Agency
PG	Population growth
PHDR	Policy and Human Resource Development Fund (Japan)
PLC	Private Limited Company (incorporated business form in Ethiopia)
PoU	Prevalence of Undernourishment
R4D	Research for Development
RTA	Research Associate/Assistant
RUAF	Resource Centres on Urban Agriculture and Food Security
SAA	Sasakawa Africa Association
SAM	Social Accounting Matrix
SAP	Structural Adjustment Policy
SD	standard deviation
SG2000	Sasakawa Global 2000
SOW-VU	Centre for World Food Studies (Dutch acronym SOW-VU for "Stichting Onderzoek Wereldvoedselvoorziening van de Vrije Universiteit")
SRI	Stanford Research Institute
SRO	Sub-regional Research Organization
SSA	Sub-Saharan Africa
SUN	Scaling Up Nutrition
TAMA	Tamale Metropolitan Area
TFP	Total Factor Productivity
TGD	Technology Generation and Dissemination
UN	United Nations
UNDP	United Nations Development Programme
UNECA	United Nations Economic Commission for Africa
UNHabitat	United Nations Human Settlements Programme
UNHCR	United Nations High Commissioner for Refugees (UN Refugee Agency)
UNICEF	United Nations International Children's Emergency Fund
UNIDO	United Nations Industrial Development Organization
UNU-EHS	United Nations University Institute for Environment and Human Security
UPA	Urban and Peri-urban Agriculture
USAID	United States Agency for International Development
VU	VU University Amsterdam (Dutch acronym VU for "Vrije Universiteit")
WAAPP	West Africa Agriculture Productivity Programme

WARDA	West Africa Rice Development Association (former name of Africa Rice Center)
WFP	World Food Programme of the United Nations
WHO	World Health Organization of the United Nations
ZARI	Zambian Agricultural Research Institute
Zn	Zinc

List of Contributors

Sheu-Usman Akanbi

is involved in research on agricultural production and farm management at the University of Ilorin. He was the project coordinator of the Fadama Development Project in Kwara State, Nigeria from 2005 to 2013. This was a rural agricultural development programme that was financed by the World Bank and the African Development Bank and was designed to address the production and productivity constraints of Nigerian farmers. E-mail: dipoakanbi@gmail.com.

Akinyinka Akinyoade

is a demographer and a senior researcher at the African Studies Centre in Leiden and is the convener of its Agro-Food Clusters in Africa collaborative research group. His primary research interest is food and nutrition security, with special reference to the interplay between Africa's population growth and the continent's ability to feed itself. He is also involved in research on health and development, and fertility dynamics and family planning in West Africa; migration, human trafficking and forced labour (mainly in Nigeria); and decentralization, including public-service delivery in the education and health sectors in Nigeria, Cameroon and Tanzania. He played a major role in the ASC's recent Tracking Development project that compared the development trajectories of various African and South East Asian countries. E-mail: aakinyoade@ascleiden.nl.

Maru Shete

is a PhD fellow at the Netherlands Academy for Land Governance for Equitable and Sustainable Development (LANDac), where he is studying the impact of large-scale land acquisition on equitable and sustainable development in Ethiopia. He has many years of research and teaching experience and has published numerous articles on land tenure and land deals in commercial agriculture, impact and livelihood studies, poverty and food security, and rural development in Africa. In 2010, he won a research fellowship from the Future Agricultures Consortium Early Career Fellowship Program to work on issues surrounding land deals in Ethiopia. He was a research director and assistant professor at St Mary's University College in Ethiopia until he started his PhD in August 2011. E-mail: marushet@yahoo.com.

Joost Beuving

is a social anthropologist and is currently teaching anthropology and development studies at Radboud University Nijmegen, where he is also the coordinator of the Nuffic-Ethiopia project. His research interests focus on three main areas: everyday economic life in Africa with particular reference to globalizing markets; the ethnographic study of entrepreneurship; and qualitative/naturalistic inquiry. He has undertaken anthropological field research in various parts of Africa and Europe, and has published extensively in leading academic journals specializing in anthropology, African studies, development studies and social geography. His latest book, which he co-authored with G. de Vries, is entitled *The Craft of Naturalistic Inquiry* (Amsterdam University Press, 2014). E-mail: j.beuving@maw.ru.nl.

Inge Brouwer
is an associate professor in the Human Nutrition Division at Wageningen University. Her research focuses on the improvement of micronutrient intake among women and young children through food-based approaches like dietary diversification and (bio)fortification. This has provided research projects and (training) opportunities for numerous (inter)national MSc students. She has been awarded extensive research funding from the food industry, national funding programmes as well as EU FP6 and FP7 Framework Programmes. She coordinated a large FP7 INSTAPA research programme that aimed to improve nutrition through enhancing staple crops (sorghum, millet, maize and cassava) in Africa through biofortification, fortification and post-harvest processing involving over forty scientists from European and African countries. E-mail: inge.brouwer@wur.nl.

Diederik de Boer
has a Masters in public administration and an MBA from Bradford University (UK)/Tilburg University (NL) with a minor in small and micro-credit schemes from Harvard University (US). He was awarded a PhD at Erasmus University in Rotterdam in 2013 for his work on local economic development. He has worked for Euro-Consult in Bolivia, *Médecins sans Frontières* in Liberia and for the United Nations (WFP and UNDP) and the Netherlands Ministry of Foreign Affairs in Cambodia, Mongolia, Korea and the Palestinian Territories. He joined the Maastricht School of Management (MSM) as a project consultant in 1996 and was the director of its Sustainable Development Centre for seven years. He is currently coordinating the Round Table Africa programme and is a senior project consultant and researcher at MSM where his research focuses on local economic development and cluster competitiveness in Asia and Africa. E-mail: ddeboer@msm.nl.

Ton Dietz
was appointed Director of the African Studies Centre in Leiden and Professor of the Study of African Development at Leiden University in May 2010. Prior to this, he was Professor of Human Geography at the University of Amsterdam, a

(part-time) Professor of Social Sciences at Utrecht University (2002–2007) and the Scientific Director of CERES, the Research School for Resource Studies for Development. His research focuses on human geography, political environmental geography and the political geography of Africa, poverty analysis, rural development, food security and agrohubs, pastoralism, land and water conflicts, dry-land developments, the impact of climate change and the political ecology of forest areas, development aid, the participatory assessment of development assistance and change, civil society, NGOs, and geopolitics and Africa in the multi-polar world. E-mail: dietzaj@ascleiden.nl.

Abraham Falola

is a PhD student in agricultural economics at the University of Ilorin, Nigeria, where he was awarded an MSc in agricultural economics in 2011. He is currently a lecturer in the Department of Agricultural Economics and Farm Management and has published widely in the field of agricultural research. He is a member of the Nigerian Association of Agricultural Economists, the Agricultural Society of Nigeria and the Farm Management Association of Nigeria and has participated in numerous field and community development activities. E-mail: falolaabraham@yahoo.com.

Dick Foeken

is a human geographer whose main research interests are urban poverty, urban water supply and urban agriculture in Africa. He retired from the African Studies Centre in May 2014 but is still active in the field of urban water supply. He has published in many leading journals including *Political Geography*, *The Journal of Peasant Studies*, *Food and Nutrition Bulletin*, *Ecology of Food and Nutrition*, *Habitat International*, *Geoforum* and *GeoJournal* and co-edited three volumes of the ASC's African Dynamics series. He has been involved in recent projects on water-sector reforms in urban Kenya; the gender aspects of urban agriculture in Eldoret, Kenya; integrated biodiversity conservation in West Africa's Sahel region; and food security and the clustering of metropolitan food chains. E-mail: dfoeken@ascleiden.nl.

Wijnand Klaver

works at the African Studies Centre in Leiden as a senior researcher on food and nutrition security. His particular research interest lies in the analysis of household surveys on health, nutrition and agriculture, including urban agriculture and school feeding. He has recently undertaken an analysis of trends over the past fifty years in the area of food and nutrition in relation to demographic development, food production and poverty alleviation, using data on child growth, nutritional requirements, and food composition and consumption. This will act as a stepping stone to incorporate nutrition concerns in scenario studies of food security in African countries and cities towards 2030 and 2050. E-mail: klaver@ascleiden.nl.

Jackson Langat
did his PhD in agribusiness management and his MSc in agricultural and applied economics in East, Central and South Africa at Egerton University in Kenya. He has been teaching in its Department of Agricultural Economics and Agribusiness Management at the Faculty of Agriculture since 2008. Before this, he worked as the programme accountant for the Aga Khan Agency for Microfinance (AKAM) where he was responsible for Kenya and Tanzania. He was also involved as programme coordinator in the Village Bank programme at the K-Rep development agency. His interests lie in a system-thinking approach to research and he is currently involved in cluster analysis and value chain studies. He has participated in conferences on value chains, bioenergy and extension services in Ethiopia, Kenya and the Netherlands. E-mail: jlangat@gmail.com.

Diana Lee-Smith
co-founded the Mazingira Institute in 1985. It studies Kenyan urban agriculture and now hosts the Nairobi and Environs Food Security, Agriculture and Livestock Forum (NEFSALF). From 2002 to 2005, she was the African regional coordinator for Urban Harvest, a programme organized by the Consultative Group on International Agricultural Research (CGIAR). Dr Lee-Smith was Visiting Professor in Public Health at the University of Toronto from 2006 to 2009 where she taught and edited two books: *Healthy City Harvests* (CIP-IDRC, 2008) and *African Urban Harvest* (Springer NY-IDRC, 2010). She is currently based at the Mazingira Institute in Nairobi. E-mail: diana.leesmith@gmail.com.

Melle Leenstra
studied agricultural development economics at Wageningen University where he conducted research on agricultural credit in Togo and on community-based tsetse controls in Kenya. While working as desk officer for Zambia and Zimbabwe at the Netherlands Ministry of Foreign Affairs, he completed a PhD on the Zambian health sector from a political science/public administration perspective. He was then posted to the Netherlands Embassy in Kampala as the political officer and is currently working as First Secretary in Economic Development and Food Security at the Netherlands Embassy in Nairobi. E-mail: melle.leenstra@gmail.com.

André Leliveld

is a development economist at the African Studies Centre in Leiden. Since the late 1980s he has been conducting research on (informal) insurance and social-security support arrangements in Africa, particularly in Swaziland. His ongoing research is on community-based health insurance in Togo and the economic well-being and health status of its participants, and on social-security arrangements in rural Uganda. He had a leading role in the ASC's Tracking Development project that involved a comparative study of the development trajectories of four African and four South East Asian countries. He has recently been working on 'frugal innovations' in Africa and is also researching structural transformation processes in African economies and the effects on (youth) employment and labour protection. E-mail: leliveld@ascleiden.nl.

Olubunmi Abayomi Omotesho

is an agricultural economist. He joined the University of Ilorin as an assistant lecturer in March 1984 and was appointed professor in October 2000. He has served as head of department, a member of the Governing Council and Dean of Agriculture (2008–2012) and is the incumbent Dean of Student Affairs. He has over 100 scholarly publications to his name and has supervised numerous PhD students. His areas of specialization include agricultural resource economics using irrigation development, cooperatives and extension services to measure and promote the resource productivity of land, labour, water, credit and management. He is currently working on food security and poverty issues; the agricultural commodity trade and globalization; and measuring risk in agriculture and foreign direct investment and the agricultural sector, with particular reference to Kwara State, Nigeria. He is a consultant to many international organizations. E-mail: yomiomotesho@gmail.com.

Marcel Rutten

is a geographer at the African Studies Centre and Radboud University Nijmegen where he concentrates on research on natural resource management, notably land and water, in (semi-)arid Africa. In addition to his interests in (eco)tourism and Kenyan politics, he has been doing longitudinal research on land, water and food security in Africa since the early 1980s. His current research projects are on sustainable water management, which combines indigenous and modern technical and institutional knowledge to develop low-cost water sources (shallow wells); resource conflicts over land; and coping strategies related to famine and drought. His publications include *Out for the Count: The 1997 General Elections and Prospects for Democracy in Kenya* (with A. Mazrui & F. Grignon, 2001) and *Inside Poverty and Development in*

Africa; Critical Reflections on Pro-poor Policies (with A. Leliveld & D. Foeken, 2008). E-mail: rutten@ascleiden.nl.

Sebastiaan Soeters
has a PhD from Leiden University on urban institutional and socio-economic formation during the colonial period in Ghana, with a focus on Tamale, the regional capital of northern Ghana. His research interests relate to urban developments, with special emphasis on urban food security and the pressures and incentives faced by peri-urban communities due to urban sprawl in northern Ghana. At the ASC, he has been leading the Nairobi case study in an EU-funded project entitled Food Planning and Innovation for Sustainable Metropolitan Regions (FOODMETRES), which aims to model the environmental and livelihood implications of urban food-planning scenarios. He is currently a postdoc researcher at Utrecht University. He was the initiator of BlackStarBikes and is the Ghana country coordinator for the ASC Community. E-mail: srsoeters@gmail.com.

Kees van der Geest
is a human geographer who specializes in migration, climate change, the environment and development. He is a senior researcher in the Environmental Migration, Social Vulnerability and Adaptation Section at the United Nations University Institute for Environment and Human Security (UNU-EHS) in Bonn, Germany. He has been researching the impact of climate change, vulnerability, adaptation, rural livelihoods, the environment, development and human mobility for over fifteen years and has undertaken extensive fieldwork in Ghana and Burkina Faso. He is currently working as the science coordinator for nine case studies for a project entitled Loss and Damage from Climate Change in Vulnerable Countries. He has published two monographs and several articles in international journals. E-mail: kgeest@gmail.com.

Cornelia Francisca van Wesenbeeck
is a senior researcher at the VU University, Amsterdam. Her main research interests lie in general equilibrium theory and its applications, particularly the inclusion of institutions that facilitate cooperation between agents within or outside markets, and the modelling of price formation through negotiations and auction mechanisms. She is also interested in the biophysical characteristics of economic modelling, in spatially explicit models of production, transport and natural flows and has published widely on these topics. She has been involved in numerous research projects including the EU 7th Framework Project CLIMAFRICA, a collaborative project with the European

Joint Research Centre (JRC) and with the World Food Programme (WFP) on the design of a Food Atlas for Sub-Saharan Africa. She has sat on various commissions and panels dealing with food systems and food security. E-mail: c.f.a.van.wesenbeeck@vu.nl.

Koko Warner
is head of the Environmental Migration, Social Vulnerability and Adaptation Section at the United Nations University Institute for Environment and Human Security (UNU-EHS) and the Executive Director of its Munich Climate Insurance Initiative. Her research interests include the risk management strategies used by low-income communities to adapt to changing environmental and climatic conditions. She is a lead author for the Intergovernmental Panel on Climate Change's (IPCC) Fifth Assessment Report, Working Group 2 on Adaptation. She is also co-chair of the Working Group on Global Environmental Change and Migration, which is part of the World Bank's KNOMAD Network. She has published extensively on migration and new climate risk management strategies to highlight research findings for decision makers in (inter)national policy fora. E-mail: warner@ehs.unu.edu.

CHAPTER 1

Introduction

Akinyinka Akinyoade, Ton Dietz, Dick Foeken and Wijnand Klaver

Africa's Green Evolution

'Africa Needs a Green Revolution!' This is the message from the Alliance for a Green Revolution in Africa (AGRA), a high-level team that was headed by former UN Secretary General Kofi Annan until December 2013: http://agra-alliance.org. It started its activities in 2006 at a time when there was a global shift in thinking towards more pro-active engagement with global food production, and Africa's food sector in particular. The 2007/2008 food crisis, which saw steep increases in global food prices, was an additional wake-up call but, by then, the African Union had already formulated a more hands-on commitment to food production with the signing of the Maputo Declaration in 2003: www.nepad.org/nepad/knowledge/doc/1787/maputo-declaration. This formally encouraged member states to give more budgetary support to agricultural and rural development and to commit at least 10% of their annual budgets to this long-neglected policy domain.

South East Asian experiences from the 1970s and 1980s show, however, that this in itself is not enough and that a more explicit and substantial pro-poor policy has to go hand in hand with genuine commitment to rural transformation. States cannot do it alone. The 2013 African Dynamics volume entitled *Asian Tigers, African Lions: Comparing the Development Performance of South East Asia and Africa* (Berendsen *et al.* 2013) provides the major results of the African Studies Centre's Tracking Development research project that compared the development trajectories of Nigeria and Indonesia, Malaysia and Kenya, Tanzania and Vietnam, and Uganda and Cambodia between the 1950s and the 2000s. It highlighted the differences in development approaches between the 'developmental' states of South East Asia and the 'not-so developmental' states in Africa in the same fifty years. The main differences were the emphasis that the South East Asian states placed on a form of rural development and their support for agricultural transformation at the start of their economic breakthrough in the 1960s. This resulted in dramatic increases in agricultural productivity, better connections to markets, notable improvements in the living conditions of peasant families and significant poverty alleviation.

 | DOI 10.1163/9789004282698_002

One of the chapters in the 2013 African Dynamics volume discussed how Africa has in fact been showing agricultural growth since about 2000 and that this can be seen as a promising shift or breakthrough towards a Green Revolution 'African Style' (Dietz 2013). It could even be called 'Africa's Green Evolution'. And although progress is relatively slower than in South East Asia's cereal revolution, growth is part of a longer-term process and a wider array of 'breakthroughs'. Since the Tracking Development project ended, follow-up research has been done as part of the Initiating and Sustaining Developmental Regimes in Africa project. Breakthrough experiences or 'pockets of effectiveness' have been seen and ASC researchers are formulating hypotheses to explain the combination of increasing production figures and higher yield levels in a considerable number of agricultural products (www.institutions-africa.org; also Dietz & Leliveld 2014).

Three major 'drivers of change' have been highlighted: (i) the growth of Africa's metropolitan areas and, within these dynamic cities, an emerging middle class with its increasing demands for higher-quality food products; (ii) a growing entrepreneurial capability to better organize agricultural value chains; and (iii) more effective linkages between relevant agencies in agro-food clusters, often with innovation hubs in and around the metropoles. State-based agencies are playing a more benign role than before due to pressure from combinations of private-sector alliances and popular pressure. This is sometimes by effective 'threats to state and political stability' and sometimes as a result of pressure from elected political leaders who are mobilizing 'developmental pressure'. Non-state pressure tends to come from combined foreign agencies (multinationals, including South-African-owned supermarkets, but also international NGOs), and local agencies, entrepreneurs (that may become regional or multinational companies) and pressure groups. And, as in many aspects of economic and political life in Africa, there is a mix of formal and informal arrangements.

An overview of the long-term trends in Africa's farming sector in the chapter on agricultural performance in the 2013 African Dynamics volume (Berendsen *et al.*, op.cit.) showed changes in cropping areas, crop yields, livestock production and food energy values per capita based on population data and on Africa's production of cereals, pulses, roots and tubers. Using data from the Food and Agriculture Organization of the United Nations (FAO: http://faostat.fao.org), the overview spanned the period from 1961 to 2009 for Africa as a whole and gave details of the four African and four South East Asian countries that were compared in the Tracking Development project. This introductory chapter, at the start of the 2014 volume, presents updated and extended figures and compares 1961 and 2011/2012, highlighting crop details and outlining data for Africa in general as well as for the continent's five sub-regions, as defined by the FAO. Finally, in addition to production data, consumption data are also given. They were obtained by connecting farm-gate production estimates with import and export data, taking into account the fact that some production is used as the following year's seed supply, some for feed, some for industrial and other uses, and some is wasted. The data also give an idea of the long-term shifts in production-consumption linkages and show the importance of looking at agricultural dynamics from a food-security perspective as well as connecting this with nutrition dynamics.

Fifty Years of Agricultural and Food Dynamics in Africa[1]

Africa's population increased from 275 million inhabitants in 1961 to 994 million in 2012. In 51 years, this is an impressive 3.62-fold increase and is probably faster than any world region ever in history. Many observers have feared that Africa's rapid population growth – and its subsequent demand for food – would not be matched by food production increases and that this would create dramatic shortages of food and/or a massive drain on foreign exchange as food was imported, or there would be an ever greater dependence on food aid (Hilderink *et al.* 2012). The situation in the 1980s and 1990s did indeed look rather gloomy and Africa experienced a number of dramatic food crises and famines. However, looking at the last fifty years as a whole, most of Africa's food production has increased more than its 3.62-fold population growth (see Table B.1a in Appendix B), although its traditional export crops increased by less impressive percentages (see Tables B.1b and c in Appendix B). Africa's agricultural production volume has become more oriented to Africa itself, particularly so in the last decade. All the major basic food categories – cereals, pulses and particularly roots and tubers – have seen greater increases than population growth. In these fifty years, the group of roots and tubers grew most of all basic foods, followed by pulses and cereals. Appreciable differences can also be observed within these categories as they have all increased faster than population numbers (see Table B.1a in Appendix B). Crops that did not do so well in increased production volumes between 1961 and 2012 were fibre crops (cotton and sisal), oil crops and stimulants, (with the exception of tea), all of which were mainly grown for (inter)continental exports. Africa's livestock production volumes also increased considerably and often much faster than the population over the same period (see Tables B.1b and c in Appendix B).

1 Appendix A presents selected statistics on food and nutrition security in the major world regions and in four specific countries in Sub-Saharan Africa that serve as case studies in Chapters 3 and 4 in this volume.

There are differences in growth figures for agricultural produce within Africa. Figures 1.1a and 1.1b give the increases[2] in production volumes of cereals and roots & tubers between 1961 and 2012 by increase in population size for the five macro-regions set by the FAO. Results for cereals (Figure 1.1a) have been most impressive in North and West Africa where cereal growth rates are higher than population growth, and least so in Southern Africa. Cereal production in Middle and Southern Africa did not manage to keep pace with population growth. It is a different story for roots and tubers (Figure 1.1b). Here only Middle Africa failed to keep pace but all other regions had a higher or even much higher growth of roots and tubers than of its population, and the figures for North and West Africa are truly impressive.

As mentioned above, the production of traditional export crops, for example, cocoa, coffee, groundnuts, palm oil and cotton, did not grow as fast as food crops. The exceptions are tree nuts, fruits, spices and particularly milk and butter and tea. By comparing the volumes of exports between 1961 and 2011 for all categories of agricultural products and also the percentages of the crops exported to foreign markets (including neighbouring countries), it can be noted that Africa's markets for agricultural produce have always been mainly internal, and are increasingly so (see Table B.3 in Appendix B).

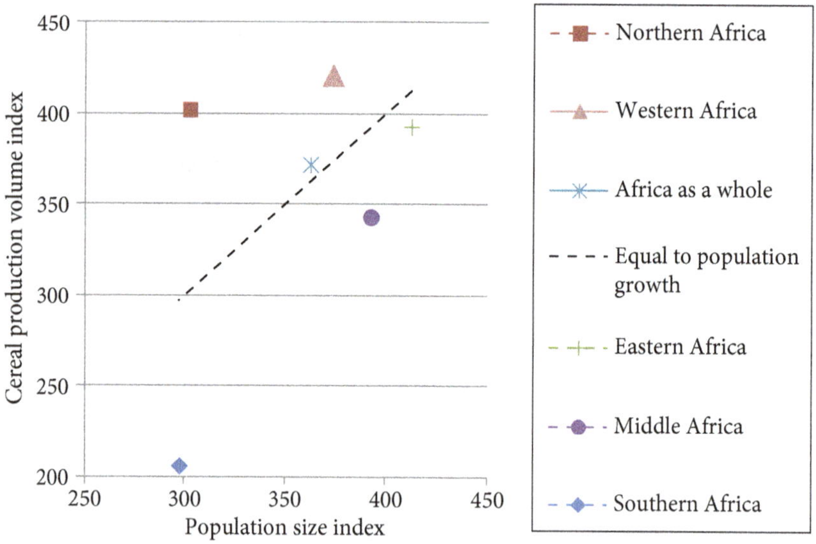

FIGURE 1.1A *Cereal production volume index, by population size index, 2012 (based on 1961).*
SOURCE: TABLE B.2 IN APPENDIX B

2 The increase is expressed as an index number, with the 1961 value set at 100.

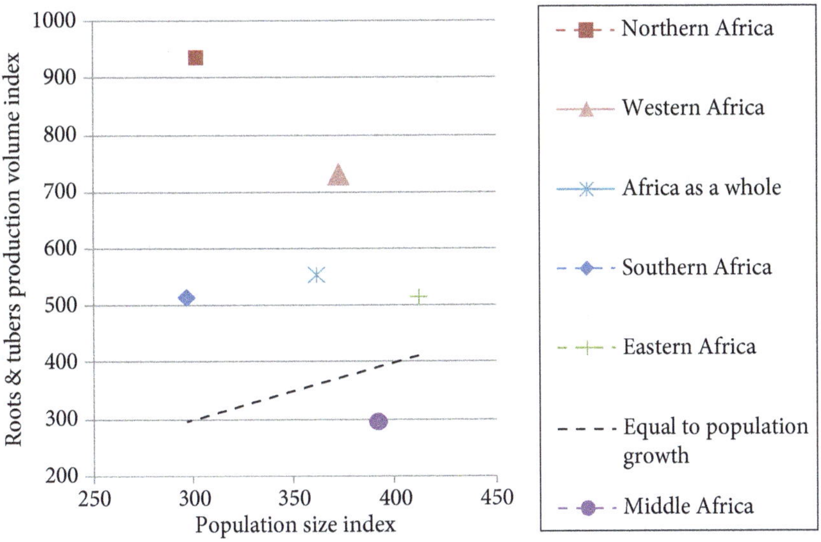

FIGURE 1.1B *Roots and tubers production volume index, by population size index, 2012 (based on 1961).*

Overall, it can be concluded that Africa's agricultural produce is overwhelmingly used for consumption in Africa itself, despite suggestions that some Asian countries have started using land and labour in Africa to produce food for Asian consumers. In absolute terms (volumes of total agricultural produce exported), exports have increased from 14.9 million metric tonnes to 33.6 million metric tonnes, more than twice the volumes seen in 1961. However, the total production exported has dwindled in relative terms: from 7.9% of all agricultural production in 1961 to only 4.1% in 2011. This is because total production volumes have increased much more than total export volumes: from 189 million metric tonnes in 1961 to 826 million metric tonnes in 2011, a more than fourfold increase.

But what about Africa's food imports? If food import figures for 1961 and 2011 are compared, a steep increase in imported agricultural products is noted. This implies additional (potential) food security but it also makes Africa more dependent on global food markets, which are expected to become ever more volatile. Africa is ultimately likely to lose out. In 1961, Africa imported 9.7 million metric tonnes of agricultural produce, half of which were cereals. That year, its export volumes were still higher than its imports: 14.9 million versus 9.7 million metric tonnes. Today's figures are very different. In 2011, Africa imported 122.9 million metric tonnes of agricultural produce, while it 'only' exported 33.6 million metric tonnes. So in these fifty years, import volumes have increased more than twelve times. The continent's agricultural imports are currently dominated by cereals (59%), mostly wheat but also large quantities of sugar crops, vegetable oils, milk and butter and fish and seafood. For most products, the relative importance of food imports as a percentage of available food has increased.

Animal production has increased significantly, so some of these increases in available food are likely to be as a result of feed produced in Africa itself. Part of Africa's primary food (locally produced as well as imported) is used as feed for animals. This feed is then not directly available for African consumers but only in the form of meat, offal, animal fat, milk, butter and eggs. If this use increases, it can be taken as a sign of shifts in consumption patterns, and shifts generally come with increasing wealth levels. This is exactly what has happened (see Table B.5 in Appendix B). Total volumes of feed have increased eightfold from 12.9 million metric tonnes to 103.3 million metric tonnes. In 1961, 7% of Africa's agricultural production and import volumes were used as feed for the continent's animals. By 2011, this had increased to more than 12%. One can safely assume that the increased feedstuffs mainly fed animals, whose products primarily fed the urban upper and middle classes. It remains to be seen how much this also improved the food security and food diversity of Africa's urban and rural poor.

Africa's total (potential) food availability can be expressed in terms of the food energy and proteins available for human consumption in Africa. The FAO's food balance sheets show production volumes, import volumes, export volumes, food used as feed and seed, food manufactured or used in other productive ways and, finally, food that is wasted between the farm gate and the retail level, i.e. ignoring waste on farms and in households. The food available for human consumption at retail level or at the level of farm households, if they consume what they produce, can be calculated in kilocalories and grams of protein. It is, therefore, possible to compare the food availability and diet composition for (average) African households if population numbers are taken into account (Table B.6 in Appendix B). Across the board, the food situation has improved from levels that were below WHO requirements for a healthy life to levels above these figures.

Figure 1.2a shows the increase in total food availability since 1961: energy per capita increased by 31% and protein per capita by 30%. Hunger and undernutrition in Africa are no longer a matter of limited total quantities available for human consumption but of inadequate distribution. It can thus be deduced that average wealth levels in Africa have improved if one not only considers the total calories and proteins consumed per capita (see Figure 1.2a) but also looks at dietary composition. The share of plant products decreased somewhat in favour of products of animal origin. The breakdown by food groups is given in Figures 1.2a, 1.2b and 1.2c.

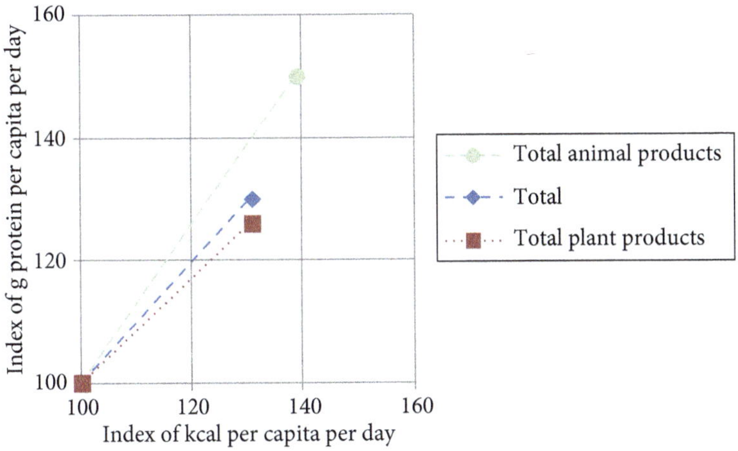

FIGURE 1.2A *Growth in Africa's food available per capita at retail level, 2011 (index compared to 1961 = 100).*

SOURCE: TABLE B.6 IN APPENDIX B

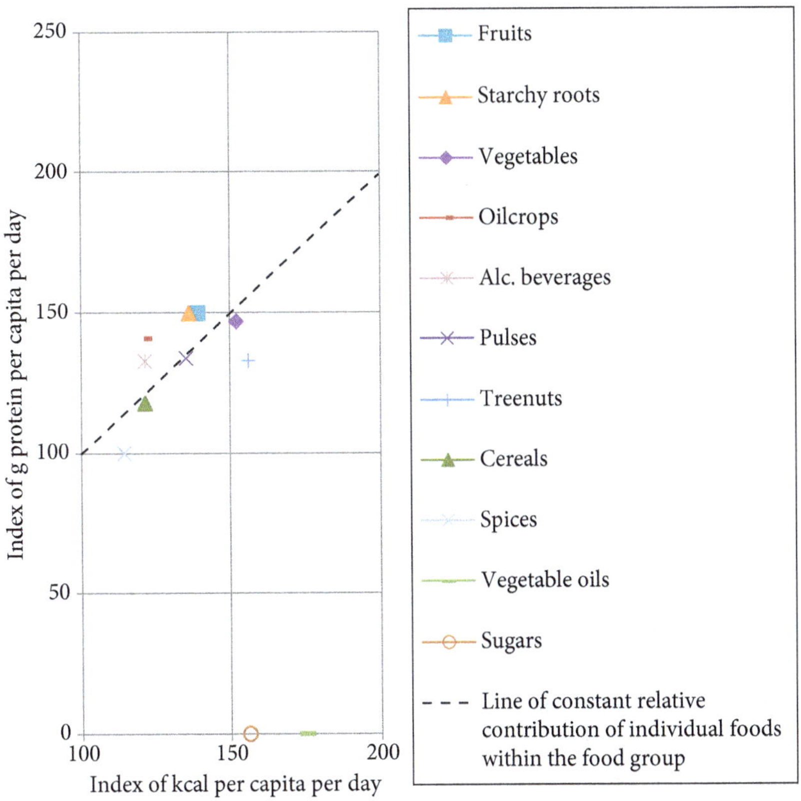

FIGURE 1.2B *Growth in Africa's plant food available per capita at retail level, 2011 (index compared to 1961 = 100).*

SOURCE: TABLE B.6 IN APPENDIX B

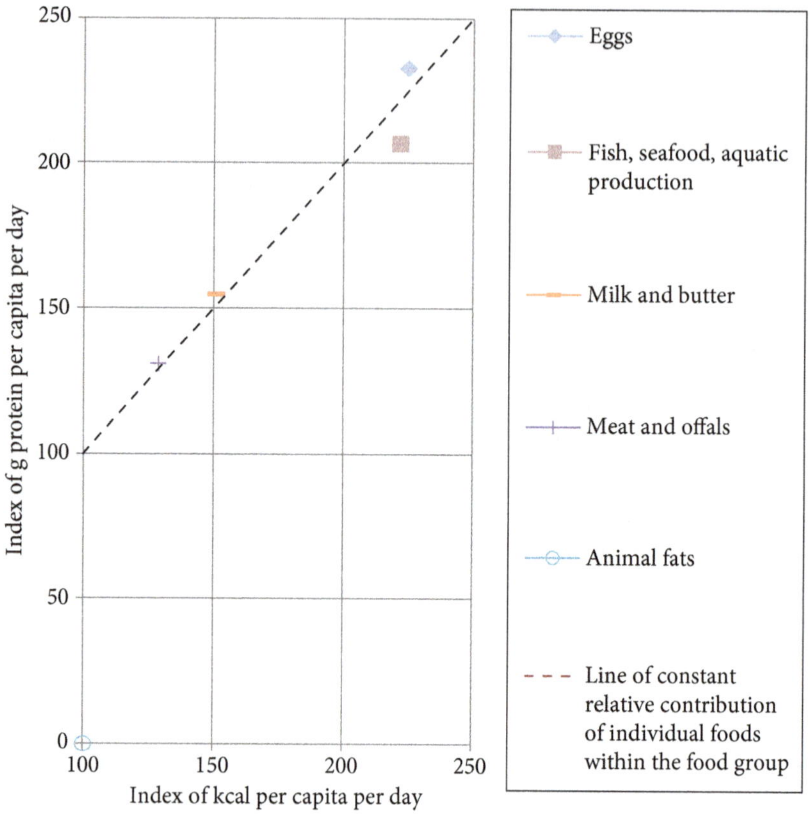

FIGURE 1.2C *Growth in Africa's animal food available per capita at retail level, 2011 (index compared to 1961 = 100).*

SOURCE: TABLE B.6 IN APPENDIX B

Note: For items above the dashed line, the food group has become richer in protein compared to energy due to changes in the relative contribution of different food items within the group. Fats, oils and sugars contain no protein so the protein index and the line of constancy are not applicable for these groups.

These are impressive improvements for Africa, a continent that has also experienced a tremendous population explosion. But behind the averages there is a lot of diversity and despite all the improvements in food availability, nutrition for the 'bottom half' of the population remains problematic. In the coming decades, all efforts should be directed at making growth and development more inclusive and more sustainable in both an environmental and a social-political sense.

Digging Deeper: Agricultural Dynamics and Economic Transformations in Africa

This volume attempts to dig deeper into what is currently happening in Africa's agricultural and rural sector and to convince policymakers and other stakeholders that it is important to look at current African rural dynamics in ways that connect metropolitan demands for food with value chain improvements and agro-food cluster innovations. It is essential to go beyond a 'development bureaucracy' and a state-based approach to rural transformation, such as the one that often dominates policy debate in African government circles and organizations like the African Union and the UN, and donor agencies.

The aim of this book is to link empirical accounts of agricultural dynamics to current policy debates on the need for economic transformation in Africa. It does this in ways that add to African attempts to understand and support 'transformation' and also engages in a critical debate about the emphasis in these circles on industrial development. One of the leading agencies in Africa in thinking about the much-needed structural transformation is the ACET, which is based in Accra, Ghana. In their African Transformation Report entitled 'Growth with Depth' (2014), only 20 of the 207 pages are devoted to agriculture in a section called 'Kickstarting Agroprocessing Value Chains'. This mainly focuses on export crops (coffee and fruits, as examples of old and new exports) and the only reference to food chains serving 1.1 billion Africans is to the soy-poultry chain. It is perhaps interesting to quote the start of the report's 'agricultural' chapter:

> Agriculture has the potential to contribute greatly to economic transformation, just as it did earlier in many developed countries. It can increase incomes in rural areas. It can increase exports and the foreign exchange needed to import machinery and other inputs for industry. It can supply the raw materials to support agricultural processing. It can release labor to manufacturing and other sectors of the economy. It can boost the supply of food to the growing urban areas and the growing industrial work force, thus moderating increases in the cost of living and thus wages. And it can expand markets for inputs and consumption goods and services for the non-agricultural sectors.
>
> ACET 2014: p. 10–11

All this is true but the framing is problematic. It starts with reference to transformation processes in 'many developed countries', assuming that Western Europe, the

US, Canada and Australia are the main countries meant. This is thought-provoking but it could, in fact, be more useful to look at recent transformations in East and South East Asia, in Latin America and in countries such as Turkey. The second observation is the order in which the points are mentioned, where overseas exports and 'earning foreign exchange to import… inputs for industry' are mentioned as almost the first point. Some of the contributors to this African Dynamics volume feel that agricultural exports will not necessarily be the main drivers of rural transformation but that this will be the provision of food to fulfil urban demands in Africa itself. The third observation is that South East Asian and East Asian evidence suggests that it is not just rural incomes that should be increased but that this has to happen in conjunction with massive poverty alleviation to help the rural poor. The rural population itself will thus also become a major consumer of much more, more diverse and higher-quality food that is produced in Africa. And finally, the issue of more inclusive development and connecting the 'bottom-of-the-pyramid' to economic transformation in ways that acknowledge their current multi-sector engagements to production-with-low-rewards and their often multi-spatial livelihoods and consumption and investment patterns should receive more emphasis.

The agricultural section in the ACET report continues by highlighting the 'challenges in modernizing agriculture'. The following are mentioned: (i) insufficient attention for research and technology; (ii) poor roads that cause many rural areas to be cut off from markets; (iii) a lack of reliable electricity, which makes agro-processing and cooling hazardous; (iv) lack of irrigation, which lowers productivity and makes agriculture risk-prone; (v) a lack of competition in the rural areas that encourages non-competitive behaviour among monopolistic or oligopolistic traders at the expense of farmers' profits; and (vi) forms of property rights that do not encourage the necessary private investments in farming (although the tensions that the social and cultural security 'customary systems' offer are acknowledged and described as an 'enigma'). These are all valid points, and the three case studies presented of coffee, fruit processing and soy-poultry value chains show the importance of holistic approaches to improving value chains and the importance of looking at all relevant stakeholders and their roles in the agro-food hub innovation clusters. And to do this properly, one has to be aware of the political economy aspects as well as the cultural economy aspects of food production-consumption chains.

Understanding the rural and agricultural sides of economic transformation demands going beyond agro-biological, technical and economic/business sciences and fully integrating social and cultural knowledge, not forgetting geographical and temporal scales and diversities. It is then evident that the list above of six major challenges has omitted the improvement of local capabilities for experimenting, learning and transferring knowledge, and the linkages between credit provision and the capability development of farmers, traders and other key players in agro-food innovation clusters. Here it is also important to understand indigenous knowledge, social capital formation and trajectories of linking and learning. And a keen eye is needed for social differentiation and the impact of rapid socio-economic change on the various brackets of income and wealth, where some are winners and others are losers. The Asian 'developmental states' ensured that massive rural poverty alleviation was part of the trajectory, although in some cases at major cost regarding human lives and human rights.

The Contributions

The chapters in this book have been divided into four main themes. The first section is on **mapping the evidence** and the first chapter, written by Lia van Wesenbeeck, is a mapping of the food economy of Sub-Saharan Africa. A model has been developed to support both national and international policymakers in their planning of interventions to enhance food security, also in emergency situations. Calibrated for the year 2005, this spatially explicit welfare model represents consumption, production and trade decisions as well as international food aid flows across the continent. The model is also potentially useful for assessing the impact of the 2007 food crisis, the budgetary effects of increased food and fuel prices on the provisioning of food aid and the possible impact of climate change on production and associated trade flows.

The subsequent chapter in this section is by Akinyinka Akinyoade, Ton Dietz and André Leliveld and offers a comparative inventory of agricultural 'pockets of effectiveness' in Kenya, Nigeria, Tanzania and Uganda since 2000. These four countries were the African part of the Tracking Development project at the African Studies Centre in Leiden. In its DRA-AFCA research,[3] rising agricultural production, improved food security and greater yields of many agricultural products have been seen in these countries since 2000. Comparison of the trends in food production has been in terms of the area harvested or stock size, yield and tonnes produced. Analysis indicates that it is important to go beyond studies of policy matters alone and to include metropole-hinterland dynamics and agro-food cluster institutions as drivers of change. The results have been compared to formulate a research agenda for the years to come.

3 The Developmental Regimes in Africa (DRA) project is a follow-up to the Tracking Development project and is a collaborative project involving the ODI in London and the ASC in Leiden. Like Tracking Development, funding has been generously provided by the Netherlands Ministry of Foreign Affairs. The DRA research at the ASC was undertaken as part of the Agro-Food Clusters in Africa (AFCA) collaborative research group and was partially funded by the ASC itself.

Wijnand Klaver then examines food production and consumption between 1961 and 2009 and the outcome in terms of food insecurity and undernutrition over the past fifteen years in Kenya, Nigeria, Tanzania and Uganda. This complements the previous chapter from a nutrition perspective. His first analysis is the consumption perspective, which uses the Food Balance Sheet (FBS) approach of the FAO. The FBS data (that were available up to 2009 at the time of this analysis) show the amount of food available for human consumption in a country in a year after corrections for imports and exports, stock changes and various other uses ('disappearances'). Per capita expression allows comparison over time which corrects for population size and growth. The FBS data also provide a breakdown by the types of foods consumed and thus give a picture of the whole diet. This chapter focuses on the FBS results in terms of dietary energy (kcal per capita). The second analysis from a nutrition angle focuses on two outcome indicators: (i) the result of the distribution of available food supplies among the population (Prevalence of Undernourishment: PoU), and (ii) the impact on child growth in terms of the prevalence of anthropometric failure, which is monitored by periodic Demographic and Health Surveys (DHS) among children aged under five.

The book's second section focuses on **agricultural production and effectiveness.** It starts with a chapter by Diederik de Boer and Jackson Langat on dairy clustering in Kenya that considers the dairy processing industry in the two main dairy regions in Kenya: the Northern Rift and the Central Rift regions. Gaps and opportunities with a focus on the dynamics and the linkages in the cluster are analyzed in an attempt to develop sustainable policies to increase productivity in dairy ventures. The results show a clear difference between the two areas: small-scale farmers dominate the Central Rift region while the Northern Rift region has more large-scale farmers. Dairy is the single largest agricultural sub-sector and accounted for about 3.5% of the country's GDP in 2008. A strong dairy sector in Kenya could thus be a sustainable source of economic growth for the decade to come. But the dairy industry's productivity figures are still low at 1500 litres per cow per year compared to South Africa where cows produce 5000 litres per year, and the Netherlands where the figure is 10,000 litres per cow per year. The low figures in Kenya are attributed to poor breeds and feeding regimes, both of which are a consequence of constraints that include poor access to information, the low and inconsistent quality of feeds, and lack of adherence to standards due to inadequate extension services and the high cost of inputs. These all have a negative impact on the quality of the milk produced but the constraints vary according to the region and strategies to overcome such constraints and increase productivity should therefore differ too.

Maru Shete and Marcel Rutten used secondary data to examine the status, challenges and contributions of biofuel feedstock production in Ethiopia. They noted that climate change and fuel price crises have spurred demand for land for the production of biofuel feedstock worldwide since the start of the 21st century. In Ethiopia, a considerable amount of land has been transferred to investors for biofuel feedstock production. The types of land conversion and feedstock produced are contributing both negatively and positively to the country's green economy and food-security efforts.

The second section concludes with a chapter by Joost Beuving on the mixed fortunes of young men working in the Nile perch sector at Lake Victoria. High-value food (fish, fruits, vegetables) exports to overseas destinations have emerged since the 1990s and appear to be replacing traditional cash crops as the mainstay of African economies. These exports are creating new opportunities but with mixed fortunes at the local level due to the contingencies associated with economic reforms and/or globalization. An actor-oriented view is taken with an ethnographic study of career histories in the Nile perch export sector at Lake Victoria, a location that is a melting pot for Africans from across East Africa. People are making ends meet there in a competitive environment where trading remains a challenge. The authors' findings raise critical questions about conventional interpretations of globalizing African food exports that emphasize institutions/structures over African agency.

The third section deals with the **drivers of food production**, with a focus on urban areas. Sebastiaan Soeters's chapter entitled 'Pressures and Incentives: Urban Growth and Food Production at Tamale's Rural–Urban Interface' concentrates on how the pressures and incentives arising from rapidly growing African cities are impacting on the ability of rural communities in the urban fringe to benefit from the growing demand for key food commodities. The conversion of rural land uses to urban land uses is one of the primary drivers of Africa's rapid urbanization. Land employed for farming by rural communities on the urban fringe is thus coming under increasing pressure from urban expansion. Based on data from interviews conducted in August 2011, this chapter presents the impact of urban pressure on agrarian productivity in Chanshegu, a rural settlement close to Tamale in Ghana's northern region. Analysis indicates that pressure on productive resources due to urban sprawl will be more likely to lead to economic and social collapse than to agricultural intensification.

Diana Lee-Smith's chapter in this section is on the dynamics of urban and peri-urban agriculture (UPA) and examines the complex dynamics of food production within urban areas. Urban food demand is the main driver of UPA and provides a good opportunity for income generation and a pathway out of poverty for the urban poor. But many people lack access to the land or open spaces that would make this possible, which is having a negative bearing on food security. UPA contributes to the nutritional needs of urban populations in general, and urban farmers in particular, by providing access to fresh vegetables and dairy products. As a result, children of urban livestock keepers are healthier because they have access to animal-sourced foods. African urban areas have a nutrient surplus while soil fertility is a major constraint in most of Africa's rural agriculture. Although African government policies in the region have mostly restricted UPA, the situation is evolving with new forms of institutions emerging in which farmers' engagement is required if they are to function well.

In the subsequent chapter, Melle Leenstra examines agriculture and diversified livelihoods among urban professionals, with particular reference to the evolution of suitcase farmers to telephone farmers. He argues that for Africa to unlock its agricultural potential, agriculture should move from being a semi-subsistence livelihood activity to a professional commercial venture. This chapter picks up the thread of the literature on rural livelihood diversification and deagrarianization. In the turbulent decades since decolonization, descendants of an erstwhile peasant class built livelihoods in which agriculture fitted into a range of income-generating activities. This has not only been the case for peasant farmers but also for those with apparently urban-professional identities. And agriculture now features in their portfolio of ventures. This phenomenon is called 'telephone farming' in Kenya and these farmers have better access to the resources needed in commercial agriculture and appear more interested in adopting non-traditional agricultural technologies compared to the average small-scale farmer. However, telephone farmers are facing serious constraints when it comes to creating profitable ventures as they lack the requisite social and human capital needed to make up for limited time. Additional research is thus needed to experiment with agricultural business models embedded in contemporary socio-economic realities.

The last section of this volume deals with **institutional issues**. The first chapter here is by Olubunmi Abayomi Omotesho and Abraham Falola and explores agricultural research systems in Africa, with examples drawn from Nigeria and the West African region. The authors review the activities of some notable agricultural research-based organizations *vis-à-vis* agricultural research endeavours and identify the challenges facing efficient agricultural research systems. It is argued that, despite the efforts made by relevant research entities, the agricultural research system in Africa is still deficient in certain aspects. Recommendations for improvement include strengthening research institutions; linking the agricultural research agenda to national development priorities; improving coordination, interaction, interlinkages, partnerships and networks among system agents; and securing sustainable sources of funding and resourcing mechanisms to support the system's various components.

The following chapter is by Sheu-Usman Akanbi and Akinyinka Akinyoade and discusses the contributions of small-and large-scale farms and foreign and local investments to agricultural growth. Using Nigeria as a case study, the analyses show that the major engine of rural growth and livelihood improvement in Nigeria is small-scale agriculture and that large-scale farming is still in its emergent stage. Despite the ubiquity of small-scale farming, the sector faces numerous constraints due to factors such as limited technical and financial support, indifference among the youth to farming, inadequate government policies and people's reliance on other livelihood sources. This leaves the country with few alternatives and leads to the conclusion that, to promote agricultural growth in Nigeria, sustained small-scale/local participation and foreign investment are needed to alleviate the fear of food insecurity if the country's oil-driven economy begins to stutter when oil reserves dwindle. The implications are onerous for other African countries with an agricultural sector structured in a similar way to Nigeria's.

The penultimate chapter is a contribution by Kees van der Geest and Koko Werner who have examined loss and damage from droughts and floods in rural Africa. 'Loss and Damage' refers to the adverse effects of climate variability and climate change that are occurring despite global mitigation and local adaptation efforts. Data analysis was based on fieldwork conducted in rural areas in Burkina Faso, Ethiopia, Kenya, Mozambique and The Gambia and included a questionnaire survey and participatory research tools. Their research goes beyond existing knowledge on adaptation and coping mechanisms by examining limits, constraints and the residual impacts of household measures to cope and adapt. In the survey, 28% of households managed to avoid loss and damage from droughts or floods but those who adapted through agricultural adaptations, such as planting drought-resistant crop varieties or by diversifying their livelihoods with non-farm activities, were significantly more successful than others. Out-migration was also associated with the failure to avoid loss and damage. Three 'loss and damage pathways' are examined in this chapter to illustrate the consequences of not being able to cope and adapt adequately. The findings presented indicate that climate-related losses and damages are already a reality in many rural African communities.

And finally, the chapter by Inge Brouwer describes the evidence on pathways that link agriculture to human nutrition. It starts by sketching the current situation concerning undernutrition and malnutrition among vulnerable population groups both in Africa and globally, and gives an update on current insights into how the various determinants interact and the merits of interventions in redressing the situation. Malnutrition remains a pervasive and damaging concern in many countries in Africa, with long-lasting and often irreversible consequences for health, cognitive ability, school performance and future earnings, and ultimately affects economic development. To reduce malnutrition, traditional nutrition-specific interventions should be aligned with so-called nutrition-sensitive activities in food security, agriculture, water and sanitation, and social protection/ safety nets. Scientifically, there is still a large gap in the evidence of efficacy of nutrition-sensitive agriculture, and this emphasizes the need for well-designed studies to test pathways and rigorous studies to measure impact using valid, and preferably simple, indicators of food and nutrient intake.

References

ACET (2014), Economic Transformation Report 2014: Growth with Depth. Accra: African Center for Economic Transformation.

Berendsen, B., T. Dietz, H. Schulte Nordholt & R. van der Veen (2013), *Asian Tigers, African Lions: Comparing the Development Performance of South East Asia and Africa*. African Dynamics; Issue: 12. Leiden: Brill.

Dietz, T. (2013), Comparing the Agricultural Performance of Africa and South East Asia over the Last Fifty Years. In: B. Berendsen, T. Dietz, H. Schulte Nordholt & R. van der Veen, *Asian Tigers, African Lions: Comparing the Development Performance of South East Asia and Africa*. African Dynamics; Issue: 12. Leiden: Brill, pp. 85–128.

Dietz, T. & A. Leliveld (2014), Agricultural 'Pockets of Effectiveness'. Kenya, Nigeria, Tanzania and Uganda since 2000. Developmental Regimes in Africa Policy Brief 06. London: DRA/Overseas Development Institute.

Hilderink, H., J. Brons, J. Ordonez, A. Akinyoade, A. Leliveld, P. Lucas & M. Kok (2012), *Food Security in Sub-Saharan Africa: An Explorative Study*. The Hague/Bilthoven: PBL Netherlands Environmental Assessment Agency.

SECTION 1

Mapping the Evidence

CHAPTER 2

Mapping the Food Economy in Sub-Saharan Africa

Lia van Wesenbeeck

This chapter describes a model for the food economy in Sub-Saharan Africa. It has been developed to support policy makers – national and international alike – in their planning of interventions to enhance food security, in particular in emergency situations. The model is a spatially explicit welfare model and represents consumption, production and trade decisions as well as international food-aid flows across the continent on a grid of 5 arc-minutes (approx. 100 km²). The model is calibrated for the year 2005 and generates as outputs, among others, estimates for undernutrition, which are an alternative for the indirect measures presented by the FAO (2012, 2013a), but also estimates for consumption and production quantities, and optimal flows of traded food and food aid. The chapter also presents possible applications for the model that range from an assessment of the impact of the food crisis in 2007 to the budgetary effects of increasing food and fuel prices on the provision of food aid and an assessment of the possible impact of climate change on production and associated trade flows.

Introduction[1]

Although many economies in Sub-Saharan Africa (SSA) are growing, there is still a great need for food aid. The increasing costs of providing food aid, caused by sharp increases in food and fuel prices, as well as critical reviews of the practice of food-aid provision (Barrett 2002a; IATP 2005; DISI 2006) have led to increasing pressure on the World Food Program (WFP) to show the efficiency of food-aid deliveries, especially in environments where the highest costs are incurred (Webb 2003) and to report on the effects of food-aid provision on local economies (Barrett 2002b; Abdulai *et al.* 2005). Although most studies conclude with general recommendations on how to improve the effectiveness and efficiency of food-aid deliveries, policy simulations that show the quantitative effects of policy changes are rare.

1 The author wishes to thank Wijnand Klaver, Diana Lee-Smith and the other participants at the 'Inside Africa's Agro-, Food and Nutrition Dynamics: Mapping Out the Evidence' workshop that was held at the African Studies Centre on 30–31 January 2014.

 | DOI 10.1163/9789004282698_003

A notable exception is the contribution focusing on the US food-aid policy for programmes and projects by Lentz & Barrett (2008), who formulate an integrated model of the food-aid distribution chain, from donor appropriations to household consumption choices. Three main actors are distinguished in the model: the donor government (the US), the operational agency (e.g. WFP) and a stylized household. The donor decides on the total budget for food aid and on the shares allocated to aid in kind and in cash. Given this budget, the operational agency then decides how and who the aid should be delivered to. Cash donations can be delivered as cash or used to purchase food aid for direct distribution, and donations in kind can be monetized or delivered directly. Errors in targeting (erroneous inclusion as well as erroneous exclusion) are explicitly allowed for. Finally, households maximize their welfare, defined over food and an aggregate of other goods, such as clothing, housing and fuel, under a given budget constraint. Analytically solving this maximization yields the household's optimal consumption of food, given prices, given cash and food aid transfers and given the locally available public goods that determine the household's non-aid income. Comparing this optimal level of consumption with some benchmark, each household can be classified as being 'food secure' or 'food insecure'. However the operational agency cannot assess the actual consumption of food with certainty, and targeting errors are thus made. The model is used to assess the effects of such targeting errors on market prices through the leakage of food aid to commercial markets (using assumed inverse elasticities of supply and demand) and on the nutritional status of households.

This chapter shares its focus on quantitative simulations with Lentz & Barrett (2008), but differs in important ways. First, it considers an explicit region, namely Sub-Saharan Africa, for which a detailed, spatially explicit[2] model is calibrated for the year 2005 at a resolution of 5 arcminutes (approx. 100 km^2). Secondly, although the model employed allows for simulations on the effects of 'aid in cash' versus 'aid in kind', this is not the focus of the chapter. As conflicts and widespread droughts are often the cause of hunger in Africa (Darfur was the most important destination for global food aid in 2005), the choice here for food aid in kind seems reasonable as markets are not

2 A model is spatially explicit when the units of analysis are very small, and location-specific characteristics – soil, climate, slope, height, etc. – are taken into account. Spatially explicit modelling is of particular importance in Sub-Saharan Africa because of the large variation in soil types and weather conditions and the high costs of transport from one place (village, market) to another.

functioning because of violence and lack of safety, or because areas are too remote to be reached profitably by commercial traders bringing in food from other regions. The focus here is thus on the provision of (emergency) food aid in kind.

Although the model is detailed in that roughly 255,000 different locations are distinguished, each with its own consumers, producers and possibilities for trade, the structure of the model remains simple and entirely within a theoretical welfare setting. This enables a clear interpretation in terms of the welfare effects of interventions and allows the application of the full range of economic theorems and instruments developed in this field. The main reasons for going beyond a standard, aggregated regional model are, firstly, that food aid is more often than not targeted at a relatively small area, reflecting the fact that many emergencies do not involve entire countries, let alone regions. This implies that a useful model cannot use averages for, say, consumption and production for large areas. Secondly, the specificity of the targeting implies that a model representing food-aid provision should be able to represent the actual physical process of procuring and delivering food as adequately as possible, and must therefore be calibrated at a level that is low enough to allow for such representativity. The possibility of representing model outcomes as maps addresses the possible objection against detailed modelling that the results are difficult to interpret and understand.

It is clear that, although the model is detailed by applied economic modelling standards, limitations remain. In the first place, the 100 km^2 grid is still too coarse to capture many underlying dynamics, including urbanization, as the area of many cities is less than 100 km^2 and urban sprawl cannot be represented. Secondly, detailed as it is, the model is only a first step in offering a platform for more detailed analyses of household characteristics, including the position of women within the household, the level of inter- and intra-household differences in income and nutritional status within the cell, and the options households have for coping with adverse conditions. The model framework should thus be seen as a first approximation of an entire continent, with the distinct possibility of adding additional knowledge in specific geographic areas of interest.

The structure of the chapter is as follows. The next section introduces the theoretical framework underlying the study, namely a spatially explicit welfare model, and the following section then focuses on the calibration procedures used, followed by a discussion of selected outcomes and some concluding remarks. The annex to this chapter provides details of the definition of aggregated regions and the data sources used.

A Food Atlas for Sub-Saharan Africa: Theory

The model introduced below builds on Keyzer (2004) and is an economic model in which the joint welfare of all the consumers is maximized under restrictions on production possibilities, trade and transport options, and the consumers' budgets. To limit the complexity of the model, it defines a number of agricultural goods and a large sector of another aggregated 'non-agricultural' good that represents the rest of the economy. In the model's optimal solution, all supply is consumed by someone, somewhere, at prices that allow for production to take place and transport costs to be covered. In the optimum, by definition, the consumers cannot make other choices that would increase their wellbeing given the income available to them. In technical terms, this implies that the model is a partial equilibrium welfare program in a spatially explicit setting that can accommodate all possible flows between a specific location on a grid of points and any immediate neighbours. The spatially explicit setting is needed to account for the large diversity in transport costs in Africa and particularly for the fact that, in many cases, the recipient populations are located in areas that are hard to reach (e.g. Darfur, Democratic Republic of Congo). Opting for an equilibrium formulation, where available income determines the demand of each individual and it is tacitly assumed that all markets function properly in bringing demand and supply together, may seem an odd choice to represent a food crisis in Africa, characterized as it is by drought, civil unrest and/or war. Yet efficiency in allocating food aid is one of the critical factors, especially in the current situation of limited budgets. As in Lentz & Barrett (2008), a welfare-maximizing approach is suitable to simulate the effects of transfers to households on their nutritional status as well as the effects on local markets. But more fundamentally, such an approach can also be used to simulate emergency situations, such as droughts, by considering the effects on the entitlements of consumers in that region.

The representation of the food-aid deliveries themselves is another aspect that deserves attention before introducing the formal model. As a demand category, food aid in kind differs from regular consumer demand in several ways. First, the channels along which it reaches the target groups are tightly controlled by the operational agency to prevent sales being made on the way. Secondly, the quantity provided is set by the operational agency and is not directly in response to consumer preferences. Finally, as described above, targeting errors may lead to leakages of food aid to commercial markets with associated effects on local food prices. A model that describes the process of food-aid provision for emergencies in its totality has to represent the three components that constitute the process: (i) a food-crisis component that imposes a shock on production jointly with its expression as an entitlement crisis; (ii) a needs assessment and aid-programming component that defines location-specific food insecurities in calorie terms; and (iii) an aid delivery component that allocates the food flows and assesses the adequacy of the envisaged aid programmes. This chapter concentrates on (iii), taking (i) and (ii) as givens for the base situation that represented the situation in SSA in 2005.

A geographical region consisting of a spatial grid with cells or sites of given shape is considered (in this application, cells are square), indexed s and $r = 1, \ldots, S$ and K commodities indexed k.[3] For scalar variables x_{ks}, w_{ks}, e_{ks}, ω_{ks} denoting consumption, production, demand for the commodity for use in production (input demand) and given stocks (endowments), respectively of commodity k at site s, and v_{ksr} the flow of commodity k from s to r, $\bar{y}_{ks}$ a non-negative bound on outflow, and with a physical loss rate ρ_{ks}, the spatially explicit equilibrium model may be formulated. Here, the bound on outflow may represent the maximum capacity of roads, trucks or other means of transport, while the physical loss rate refers to losses during transport because of spillage or the deterioration of quality such that part of the load must be considered unsuitable for consumption.

In accordance with standard theory for partial equilibrium models (see Ginsburgh & Keyzer 2011), the objective of the program is to maximize the sum of consumer surplus (welfare minus the cost of buying the commodities) and producer surplus (profit). In our application, this simplifies to maximization of the wellbeing (measured in monetary terms) minus total transport costs in Sub-Saharan Africa. However, three restrictions must be observed when maximizing total wellbeing: (i) at every site, total use of a commodity cannot exceed total availability, taking into account in- and outflows to other sites and losses in transport; (ii) total outflows from cell s cannot exceed the maximum capacity available for outflow (N.B. all outflows are positive; for technical reasons, any inflow is represented as the outflow of another site); (iii) at each site, available production technology bounds the maximum possible output that can be produced at a given level of inputs.

$$W = \max_{x_{ks}, e_{ks}, v_{ksr}, w_{ks} \geq 0} \textstyle\sum_s \beta_s u_s(x_{1s}, \ldots, x_{Ks}) - \sum_k \sum_s C_{ks}(v_{ks1}, \ldots, v_{ksS})$$

subject to

$$x_{ks} + e_{ks} + \textstyle\sum_r v_{ksr} \leq \sum_r \dfrac{1}{1+\rho_{kr}} v_{krs} + w_{ks} + \omega_{ks} \qquad (p_{ks}) \qquad (1)$$

$$\textstyle\sum_r v_{ksr} \leq \bar{y}_{ks} \qquad (\theta_{ks})$$

$$F_s((w_{1s}, \ldots, w_{Ks}), -(e_{1s}, \ldots, e_{Ks})) \leq 0$$

3 N.B.: The index 's' should not be interpreted as referring to sending and 'r' to receiving. The letters have been arbitrarily assigned and serve only to identify two different cells within the total range of S cells.

where β_s are given positive welfare weights that reflect the population densities in the different grid cells, u_s is the utility function of the consumer at site s, which relates his/her consumption of goods to his/her wellbeing (see below for mathematical requirements) and C_{ks} is the cost function of transport.

Program (1) represents a continent-wide economy, with representative consumers at every site (grid cell) maximizing given utility functions by choosing optimal consumption quantities of the available goods. In equilibrium, consumers pay the site-specific price for the commodities they consume, which is the cost of producing the good (or importing the good from outside the continent), plus the transport costs of bridging the distance between the production (or import) location and the consumption location.

Although a derivation of necessary assumptions on functions is beyond the scope of this chapter, for the sake of completeness, they are included here. See Keyzer & van Wesenbeeck (2011) for the formal arguments for the convergence of the model to an optimal solution. Utility, cost and transformation functions and endowments satisfy the following assumptions.

Assumption 1 (Utility Function) At every site s, the utility function $u_S : \mathbb{R}_+^K \to \mathbb{R}_+, u_S(x_S)$ is (i) strictly concave; (ii) increasing; (iii) differentiable; and (iv) has an upper bound on the derivative: there are positive constants $\bar{p}_k^x$ such that $\frac{\partial u_S(x_S)}{\partial x_{kS}} < \bar{p}_k^x$ for all x_S and all s.

This assumption implies that consuming more increases the utility of the consumer, but the higher the level of consumption, the less an additional unit adds to the utility: people become satiated by food. In addition, there is no good for which even a very small increase would lead to enormous increases in utility.

Assumption 2 (Transport Costs) The cost function of transport $C_{ks}(\nu_{ks1}, \ldots, \nu_{ksr}, \ldots, \nu_{ksS})$ of commodity k from site s to any other site r is (i) homogenous of degree one; (ii) continuously differentiable; (iii) strictly quasiconcave; and (iv) increasing with marginal costs exceeding some positive constants τ_k^o. Homogeneity implies that transporting twice as much does not increase or decrease the per unit cost of transport, while total costs increase if more is transported.

Assumption 3 (Transformation Function) At every site s, the transformation function $F_S : \mathbb{R}_+^K \times \mathbb{R}_-^K \to \mathbb{R}, F_S(w_S, -e_S)$ (i) is strictly quasiconvex non-decreasing; (ii) has the possibility to remain inactive $Fs(0, 0) \leq 0$; (iii) is bounded above: for all non-negative input prices p_s, $\sup_{v_S, e_S \geq 0} \{ \breve{p}_S{}' w_S - p_S{}' e_S \,|\, F_S(w_S, -e_S) \leq 0 \}$

attains its maximum, the profit $\Pi_s \bar{p}^x, p_s$; and (iv) marginal productivity of every input is bounded: $\partial\Pi_s(\bar{p}^x, p_s)/\partial p_{ks} = 0$ whenever $p_{ks} \geq \bar{p}_k^{\,e}$ for some constant $\bar{p}_k^{\,e}$.

This assumption ensures that producers cannot be forced to produce at a loss as there is always the option of stopping production. In addition, there are no 'magical inputs' that could increase production tremendously, and there is an optimal level of production that is finite.

Assumption 4 (Endowments) Endowments ω_{ks} are positive. Every consumer is assumed to own some assets (stocks of food) that can, in principle, be sold at a market to yield income.

Here, the focus is on the application of the theoretical welfare model to Sub-Saharan Africa, where 302,469 cells ('markets') are distinguished, 255,804 of which are 'active', i.e. have non-zero population or have roads. The latitude and longitude of each cell is known and given and any flow between non-adjacent cells takes place through intermediate cells, where the transport costs determine how the flows 'meander'.

Within this general modelling framework, food-aid interventions can be explicitly incorporated. Food aid has definite characteristics that need to be accommodated in this analysis. Essentially, food aid seeks to achieve a nutritional objective by providing adequate food to segments of the population that lack the purchasing power to buy this food with their own means and/or by supplying financial means to enable the population to buy their own food to satisfy their needs. The latter case – cash transfers – can be represented within a standard economic model, as it implies that the budget constraint of the individual is lifted to allow a person to buy more food. The former case – transfers in kind – requires a different representation, as the individual in a standard setting would have no instrument to 'pull' commodities towards him/herself. In the model, transfers in kind are therefore modelled as a response to demand exercised by a 'wealthy twin': an artificial second consumer that is located at the same spot as the target group for food aid, with the same preferences but without the budget constraints of the target group. This 'wealthy twin' ensures that food aid actually flows to the recipients. Since it is considered to be residual, covering the gap between available food and nutritional needs, it would be insufficient to focus on food aid only so a broadening of the scope to include the basics of the food economy is called for, which is why the model above is formulated broadly enough to allow for more than a single commodity.

The model distinguishes three types of commodities: local food, commercial food and food aid, all of which are measured in calories available for human consumption. Local food is produced and consumed within the same grid cell (i.e. within the 100 km² area). Commercial food is available for trading from surplus areas to deficit areas with consumers with purchasing power. Finally, food aid flows from ports or local procurement warehouses to the recipients.[4] Local and commercial food can be produced from land and other natural endowments (indicated as the additional commodity '0' (ω_{0s}) and commercial food can, in addition, be traded; food aid is 'produced' from commercial food through local and triangular purchases by, for example, the WFP and can also be imported through ports. Errors in targeting are represented by a factor $0 \leq \kappa \leq 1$ that represents leakage of food aid to the market for commercial food. Under perfect targeting, this factor will be equal to zero. Food aid uses transport means that are distinct from those used for commercial trade because they may be different physically and/or because the vehicles are not allowed to unload along the way for sales to local consumers or to buy at arbitrary locations. Hence, the commodity only has markets at designated sites $s \in S^+$, where food aid may be procured, and $s \in S^-$, where the food aid is to be delivered. Other sites are merely transport links. Welfare program (1) can now be written explicitly for the three commodities concerned, where Commodity 1 is food aid, Commodity 2 is local food and Commodity 3 is commercial food. To distinguish between the real consumer and the artificial consumer needed to 'pull' food aid towards the recipient, the former are denoted with a superscript *c* and the latter with a superscript *a*, and to simplify notation, total inflow in cell s is denoted q_s and total outflow y_s. As before, the symbols in brackets refer to the Lagrange multipliers of the restrictions that, because the restrictions are commodity balances, can be interpreted as equilibrium market prices.

$$\max_{e_{ks}, q_{ks}, v_{ksr}, w_{ks}, x_{ks}, y_{ks} \geq 0; k=1,2,3} \Sigma_s \left(\beta_s^c u_s^c \left(\kappa_{1s} x_{1s} + x_{2s} + x_{3s} \right) + \beta_s^a u_s^a \left(x_{1s} \right) \right) - \Sigma_s \left(C_{1s} \left(v_{1s1}, \ldots, v_{1sS} \right) + C_{3s} \left(v_{3s1}, \ldots, v_{3sS} \right) \right) \quad (2)$$

subject to

$$x_{1s} + y_{1s} = q_{1s} + w_{1s} + \omega_{1s} \quad (p_1 s) \quad \text{food-aid balance}$$

$$x_{2s} = w_{2s} + \omega_{2s} \quad (p_2 s) \quad \text{local food balance}$$

$$x_{3s} + e_{3s} + y_{3s} = q_{3s} + w_{3s} + \omega_{3s} \quad (p_3 s) \quad \text{commercial food balance}$$

$$F_{0s}\left(w_{s2}, w_{3s}, -\omega_{0s} \right) \leq 0 \quad \text{transformation restriction non-food aid}$$

$$F_{1s}\left(w_{1s}, -e_{3s} \right) \leq 0, \quad s \in S^+ \quad \text{transformation restriction food aid}$$

4 WFP reports were used to localize recipients and rations provided in 2005. See van Wesenbeeck & Merbis (2012) for country-specific details on data and information.

The spatial continuity requirement (the restriction of flows to adjacent cells only) restricts the number of routing options to be compared but, in any spatially explicit application with more than a few thousand sites, Program (2) still becomes much too large to be solved by regular convex optimization software and decomposition of the problem is essential. Although a complete description of the algorithm used to solve the model is clearly beyond the scope of this paper (see Keyzer & van Wesenbeeck 2011), a sketch is provided here to give a basic outline of the procedure, without going into a detailed description of the convergence of the procedure to a welfare optimum.

The algorithm consists of three parts. First, an optimal routing procedure is specified that, at given delivery prices (costs of production and transport), determines, in a finite number of iterations, the outflow shares to adjacent cells and the selling price of these outflows. Second, mild additional assumptions regarding production technologies and consumer preference functions are imposed to ensure that site- and commodity-specific optimizations can be solved to generate the optimal allocation of the available supply in the cell over consumption, input use and outflow to 'downstream' sites, i.e. sites with a higher price, as well as of marginal values of inflows. And third, delivery prices are adjusted iteratively towards these marginal values until they converge. Intuitively, goods flow from low to high prices just as water flows from higher to lower altitudes. The important difference is that, while the physical landscape is given – at least at a sufficiently short horizon, the price landscape adjusts in response to flows, i.e. large inflows lower prices.

A Food Atlas for Sub-Saharan Africa: Calibration

The model has been calibrated to represent production, consumption, food-aid flows and commercial deliveries for the entire sub-continent for 2005. The results were discussed with various units within the WFP and subsequently updated to reflect the actual situation for 2005 as accurately as possible. This was the latest year for which information for the entire continent was available, although some countries have been updated for 2010. The construction of the solution for 2005 proceeded in two phases. In the first stage, a database at national level was made for every country separately, where total supply had to balance total demand (including trade). Figure 2.1 summarizes the different elements included in this analysis.

First, total daily calorie intake per capita was estimated based on the recorded weights for women and children in the DHS surveys that cover 48 of the 50 countries in SSA. To ensure that sufficient observations were used for the

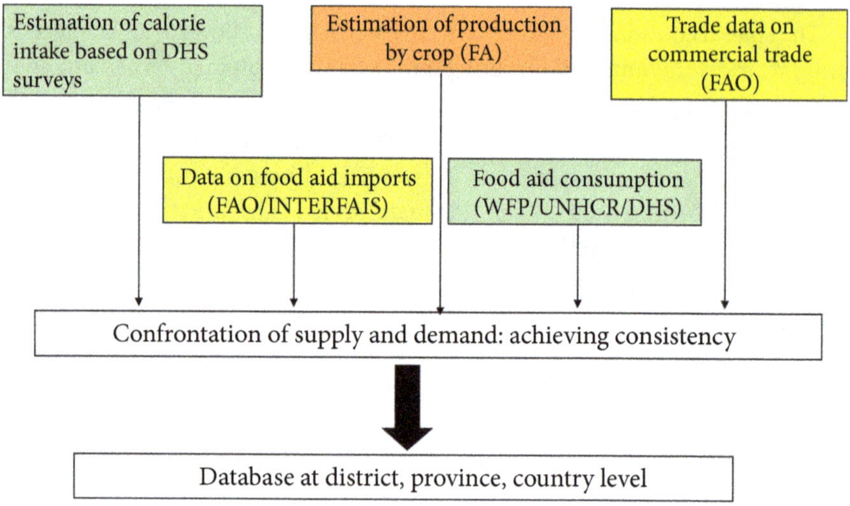

FIGURE 2.1 *Database at national level*

estimates, these were carried out at district level, with a distinction between urban and rural areas. In the estimate for 2005, 33 DHS surveys were used, complemented with WHO data for the remaining countries. The methodology for estimating undernutrition among adults closely followed the generally accepted method of estimating this number for children, where a direct link was made between (different types of) measures for being underweight and nutritional status. Van Wesenbeeck *et al.* (2009) describe the method used and the assumptions made in detail but here it suffices to state that the Body Mass Index (BMI) for women in the various age classes included in the survey is used as a basis for estimating undernutrition among adults of both sexes. These estimates are complemented with trade data and data on food-aid imports and confronted with FAO estimates of production by crop and WFP/UNHCR data on food-aid consumption. Consistency between supply and demand in this estimate was achieved by adjusting the estimates for production.[5] A final step was to decompose the total per capita consumption in calories per day with respect to the sources: subsistence production, commercial trade or food aid. WFP and UNHCR reports and data were used to locate the beneficiaries of food aid – refugees and internally displaced persons in camps as well as food-insecure populations in larger areas within countries in 2005.

5 Special assumptions were necessary for some countries to avoid very large adjustments in production. Details can be found in van Wesenbeeck & Merbis (2012) and are available from the authors upon request.

In a second stage, a spatially explicit database was constructed at a resolution of 5 arcminutes (approx. 100 km^2). The second stage is described in more detail in the sections below for consumption, production, transport possibilities, flows of food and food and food-aid prices by site in order to highlight assumptions made and enable the reader to assess the strengths and weaknesses of the analysis.

Calibration: Consumption

Starting from the estimated per capita consumption at district level (for rural and urban areas separately) that were obtained in the first stage, the second stage uses three population maps to 'distribute' these district averages over the population: (i) urban and peri-urban population; (ii) high-density rural population (i.e. population in rural towns); and (3) low-density rural population. Due to the relatively coarse grid used (100 km^2), rural and urban populations may coexist within one grid cell, with shares following from the aggregation of the original population maps from a finer resolution of 30 arcseconds (1 km^2). Multiplication of the urban (rural) population map by the estimate for urban (rural) per capita consumption, and summing the rural and urban maps led to the estimated consumption per grid cell, and division by the total population provided the average per capita consumption in each area of 100 km^2. It should be noted that the estimated per capita consumption at district level already accounted for the population composition in age and sex cohorts, and it was tacitly assumed that this composition was representative of every site in the country (van Wesenbeeck *et al.* 2009). It is clear that the effects of selective migration or disease (especially AIDS) may have caused substantial local deviations from the country average but because of a lack of reliable and continent-wide information on these effects, no local adjustments to population structure were made.

Calibration: Production

For the production of major crops, meat, fish, eggs and dairy products, a more intricate procedure was followed. The second stage took the estimates of the total calorie production obtained in the first stage, where production for 2005 acted as a balancing item, as a point of departure. Subsequently, shares of the major crops, livestock production and fisheries in total calorie production for 2005 were obtained from the FAOSTAT database (FAOSTAT 2013) and applied to the country totals. Next, the FAO's Agro-Ecological Zoning method (AEZ) (FAO 2013b) was applied on a crop-by-crop basis to distribute the crop totals over the grid cells within each country. Roughly speaking, the AEZ methodology relies on the calculation of potential crop-specific agro-climatic yields that are downscaled to reflect specific local constraints on productivity. The constraint-free crop yields calculated in the AEZ biomass model reflect yield potentials with regard to the temperature, radiation and moisture regimes prevailing in the respective grid cells. The reduction factors, which were successively applied to the potential yields, varied with crop type and the environment (in terms of climate, soil and terrain conditions) and depended on assumptions regarding the level of inputs/management (FAO 2013b). The resulting maps indicate where specific crops can be grown and provide an indication of the maximum yields that can be expected in each location. As an additional safeguard, the gridding of the total production by crop was limited in the calibration of production by the available rural population.[6] Finally, the procedure was iterative in the sense that, after each round of assignments, the remaining production possibilities at every location were updated to reflect that part of the available land that was already in use by other crops. The total production of all the different crops and other activities provided the total calorie production per cell. As a final check, these production quantities were converted into metric tons of cereal equivalents using the FAO's calorie conversion tables (FAO 1953) and divided by the population to assess whether the resulting production figures per capita were within reasonable limits.

Calibration: Transport

To assess the transport possibilities, maps of roads, railways, inland waterways and coastal waterways were compiled from a variety of sources where there was a particular challenge to ensure consistency of information across country borders. Transport costs on primary roads (well-maintained all-season roads that serve as corridors between countries and provide connections between main cities within countries) and secondary roads (all-weather roads connecting provincial centres with primary roads and with each other) were collected by the WFP's country offices through contacts with the Rome-based headquarters of logistics. This cooperation led to a unique database of transport costs that covers 30 of the 50 countries in Sub-Saharan Africa with assumptions applied to estimate the costs in the missing ones (see van Wesenbeeck & Merbis 2012 for details by country). The estimation of transport costs per tonne load per grid cell required the multiplication of the costs by road type with the length of each road within the cell.[7] Finally, the assumption was made that the cheapest available option per cell would be used for transport.

6 Using the rural population as the lower limit does not account for seasonal migration. However, the lower bound is applied only to prevent unrealistically high production in low-density areas, not to compute production.

7 Up to four classes of roads are distinguished, hence two additional classes in addition to primary and secondary roads, depending on the country. One cell may be crossed by more than one type of road, given the size of the cell.

Calibration: Flows and Prices

Consumption, production and transport possibilities and costs all concern grid-specific quantities. The final step in the calibration was to estimate the flows of goods that would be consistent with the distribution of consumption and production over the continent and to compute associated local prices that support these flows. Much as in aggregated models, net flows of international trade connect countries with production surpluses with countries where domestic demand exceeds supply. It is here that the model described above proved its strengths, as it was possible to compute optimal flows and associated prices, even for the large-scale application presented here. For calibration purposes, total consumption was subdivided into three categories (see above): food aid, including emergency aid, programme aid and project aid (based on data from the WFP on quantities and locations), local food (traded only within the cell), and commercial food (possibly traded between grid cells if prices are such that trade is profitable). For these given categories, supply sources were defined (1) for food aid (imports from outside the continent, at a given number of ports as well as local sourcing; (2) for local food production within the cell only, and (3) for commercial food imports as well as the production in each cell that was available after local consumption had been satisfied. Since for food aid the routes taken and the costs associated with provision were known as well as the quantities provided, this item was used to calibrate transport costs to reproduce the actual situation in 2005, in close consultation with the WFP's logistics department.

A Food Atlas for Sub-Saharan Africa: Selected Findings

This section presents some of the main findings of the study, starting with the direct estimates of undernutrition and moving through consumption, production and transport costs to flow and implied prices.

Undernutrition

As mentioned above, the prevalence of undernutrition was estimated using survey data on women and children in DHS surveys. Based on the district averages and applying the population composition by sex and age at country level, the number of people that are considered undernourished (measured for adults as having a Body Mass Index below 18.5 kg/m^2 and for children as having a weight more than 2 standard deviations below the standard weight for their age) is depicted in Figure 2.2. Well-known problem areas can be recognized, including Rwanda, Burundi, parts of Ethiopia and the northern areas of West Africa.

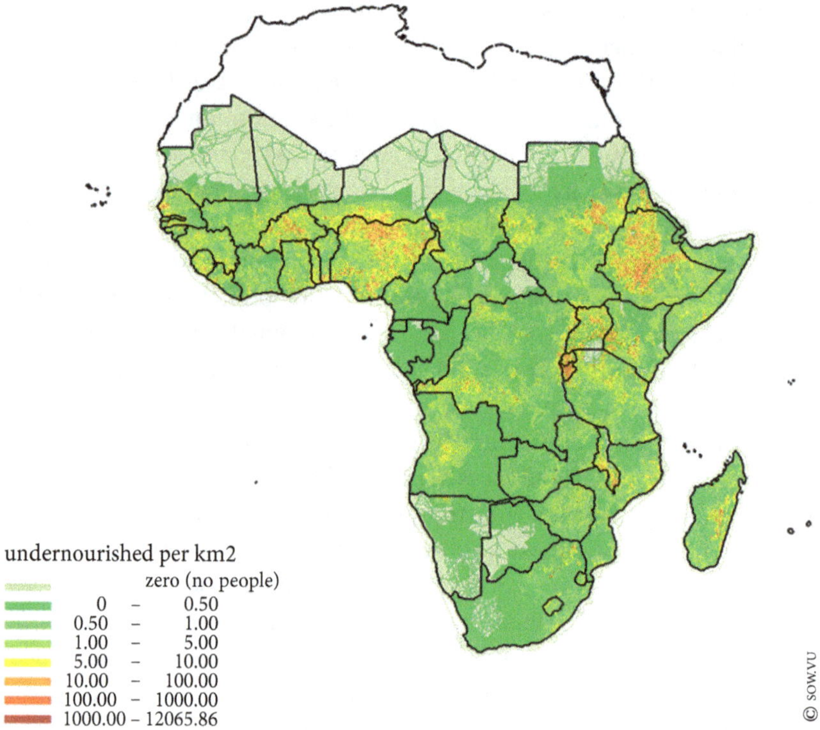

FIGURE 2.2 *Undernutrition in Sub-Saharan Africa (SSA) in 2005 (people per km²)*

For comparison, the same estimates were included as percentages of the population (Figure 2.3) to highlight the fact that the presentation of data may affect conclusions drawn as to where interventions should take place. The percentages indicate the severity of the problem but disregard whether it is a sparsely or densely populated area. For intervention purposes, one could factor in the issue of access, and preference could be given to high absolute numbers of needy people in an area, although this is not an *a priori* criterion.

New hotspots are appearing, such as parts of Chad, Niger and Eritrea, and are clearly marked red, with Burundi and Rwanda being among the average or even below-average countries in terms of prevalence of undernutrition. To enhance an understanding of the extent of the hunger, maps similar to the ones presented here can be included for severe undernutrition to expand the information base for policy makers planning interventions.

The results here differ substantially from those presented by the FAO, as can be seen in Table 2.1 that compares the outcomes for aggregated regions and the

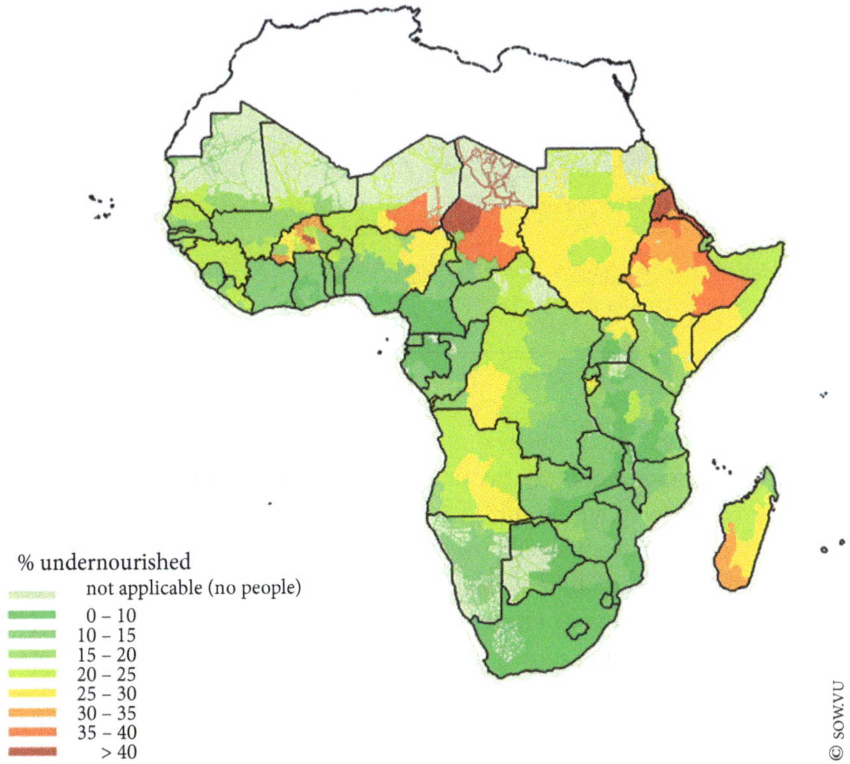

FIGURE 2.3 *Undernutrition in SSA in 2005 (% of population)*

whole of Sub-Saharan Africa. The DHS-based estimates are lower than the FAO estimates for each region but particularly so in East and Central Africa, where the prevalence of hunger is highest. This implies that, according to the DHS-based estimates, Africa is doing better than suggested in FAO reports (even after the 2012 revision in which the total for SSA was reduced from 221 m to 205 m). In turn, this means that production has kept up with population growth and urbanization to a much greater extent than is commonly assumed and that incomes – particularly in the cities – have increased enough to enable consumers to buy food in reasonable quantities. For a more detailed discussion of the differences, see van Wesenbeeck *et al.* (2009).

Two caveats should be mentioned. The analysis here does not include the quality of the diet in terms of the various nutrients, such as protein, lipids, dietary fibre, vitamins and minerals, as people's body weight does not convey information *per se* on a person's health. And child mortality in Sub-Saharan Africa is still very high, particularly when compared to South Asia, and this

TABLE 2.1 *Comparing outcomes: FAO vs DHS (millions of undernourished and % of population)*

	FAO, 2005	DHS-based	Difference
Central Africa	61 (46%)	20 (15%)	41
East Africa	114 (34%)	65 (20%)	49
Southern Africa	3 (5%)	1 (2%)	2
West Africa	43 (15%)	34 (12%)	9
Total	221 (28%)	120 (15%)	101

needs to be borne in mind when comparing rates of undernourishment among children across continents, as the ones who are most severely affected will, tragically, not show up in the statistics of undernourishment in SSA. However, the recorded difference between DHS-based and FAO estimates is too large to be explained by the combined effect of these two factors alone, so in order to substantiate why this team of researchers is confident that the anthropometric approach leads to a better understanding of the prevalence of undernutrition, a brief discussion of the FAO approach is included here.

The FAO measure of undernutrition (see also Chapter 4, this volume) is obtained by distributing food availability and measured as production plus net imports and stock holding, corrected for seed and feed use, processing and waste at national level, among household groups on the basis of information from household expenditure surveys. This yields food availability by household group, and comparison with a nutrition standard (i.e. a minimum calorie intake) leads to the gap, from which the headcount can be derived. The method has been criticized for being unnecessarily indirect and sensitive to assumptions (Klasen 2008; Nubé 2001; Svedberg 1999). In addition, its headcount and food gap are computed at national level only. Consequently, it is unable to pinpoint the 'hunger hotspots' where policy interventions, possibly including food aid, may be required. It is also unable to relate the problem to local conditions. Figure 2.4 shows the estimation process and the items used. It can be inferred that there are four items for which estimates are considered to be particularly problematic.

Production figures are derived from estimates of yields and cultivated area. For Sub-Saharan Africa, the FAO itself classifies data from 23 countries as weak, that of 25 countries as average and only 2 countries – South Africa and Mauritius – as good. Area estimates are based on country reports and brief missions, and while yields may vary widely from year to year, financial resources do not allow annual field surveys. Subsistence agriculture presents a special problem, as production by subsistence farmers is very hard to estimate but may represent a substantial

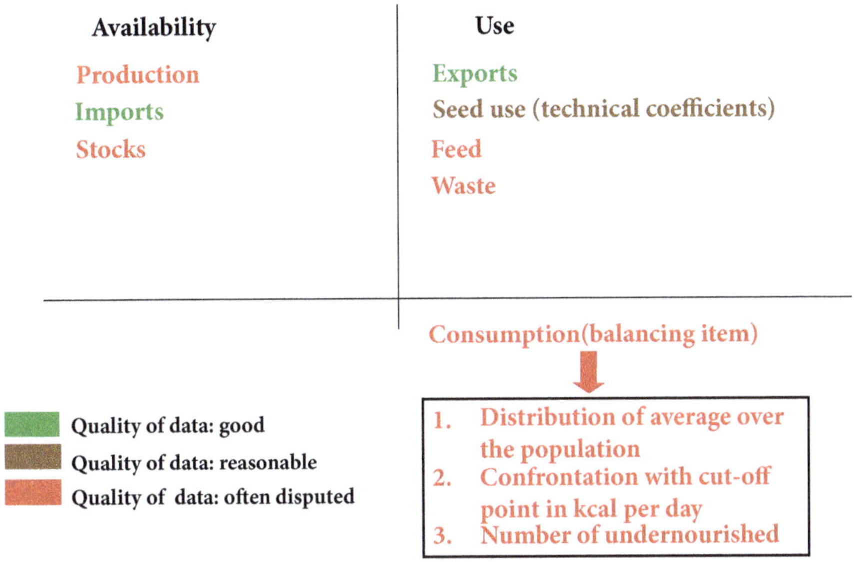

FIGURE 2.4 *Schematic representation of the FAO method of estimating undernutrition*
SOURCE: INTERPRETATION OF FAO INFORMATION (AUTHOR'S OWN VALUATION OF DATA QUALITY)

part of agricultural output in some countries or sub-national regions. By the same token, stock holding is not confined to publicly held reserves but also includes largely unknown quantities 'hidden' in distribution chains and private storage by firms and households. Feed use represents another item for which estimates are very hard to obtain, as livestock production systems vary hugely even within countries, and with it the use of feed per unit of meat, eggs or dairy produced. Estimates for post-harvest waste are scattered, and the FAO uses region-wide averages to approximate the implied losses. On the other hand, technical coefficients are used for seed use, based on knowledge of crop characteristics. Finally, official statistics on international trade are usually considered to be reliable.

In its 2013 publication on food insecurity (FAO 2013a), the organization implicitly acknowledges the weakness of its published data on the number of undernourished in the world, as it now presents a '*suite*' of indicators that jointly measure food security, although it is unclear how these indicators can be (or need to be) combined to yield insights that would be useful to policymakers. It thus seems that there is still a need to reconsider the way in which food security is measured and how progress in achieving MDG 1 is evaluated.

Total Consumption

According to the estimates presented here, average daily per capita calorie intake in SSA in 2005 is estimated at 2090 kcal/capita per day, while the average

TABLE 2.2 *Dietary energy intake by region in kcal/capita per day (based on DHS results) compared with minimum requirements*

	Average	Rural	Urban	MDER – PAL 1.75[8]
Central Africa	2056	2016	2132	1875
East Africa	2010	1989	2076	1895
Southern Africa	2434	2519	2325	1955
West Africa	2132	2127	2145	1865

based on the DHS surveys for the rural population is markedly lower than for the urban population (2072 vs. 2133 kcal/capita per day). There are also substantial differences in calorie-intake estimates between the four regions (see Table 2.2) but even the low estimates for East Africa would on average be (just) above minimal daily dietary requirements for a moderately active lifestyle (see right-hand column in Table 2.2). The distribution of calorie intake over the continent is uneven, with levels at 1730 -1795 kcal/capita per day in areas of the Democratic Republic of Congo, Burundi, Sudan and Ethiopia, while levels of around 2600 kcal/capita per day are derived from the DHS surveys in areas of Botswana, South Africa and Uganda for the same period.

Figure 2.5 depicts the outcome for specific areas for the continent as a whole, and for Ethiopia in particular, to highlight the possibilities of 'zooming in' and providing more detail in the within-country distribution.[9]

The possibility of focusing on individual countries and regions for more detailed analysis opens the way for special studies. In Ethiopia, for example, van Wesenbeeck *et al.* (2009) include an analysis of the effects of the 2007 food price crisis, using estimates on 2005 per capita consumption as benchmark values and simulating the effects of observed price changes in local markets on households' ability to maintain an adequate level of consumption. This study concluded that three types of areas can be distinguished. The first class comprises areas that are barely connected to the world market, where world price fluctuations have only a very limited effect on local markets; the second

8 Minimum Dietary Energy Requirement (MDER) – PAL 1.75, i.e. the amount of energy needed for the minimum acceptable weight for an attained height and moderate activity. Data from FAO (2013c) weighted by population size.

9 To determine rural and urban per capita consumption figures, district averages are used to prevent estimates from being based on too few observations.

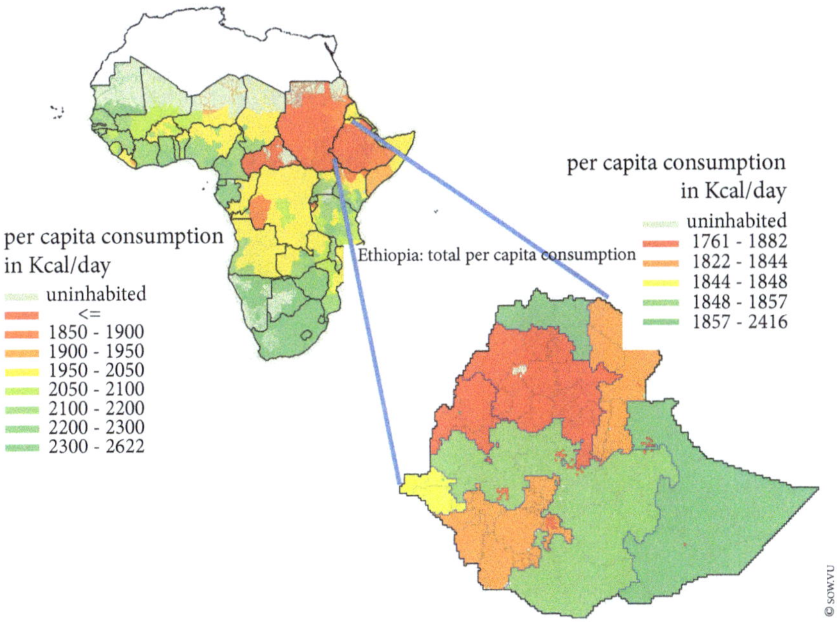

FIGURE 2.5 *Consumption in* SSA *and in Ethiopia in 2005 (in kcal per capita per day)*
SOURCE: AUTHOR'S OWN COMPUTATIONS

includes highly connected regions that are able to mobilize enough 'nuisance power' to pressurize governments into subsidizing consumption and cushioning the negative effects of rising world prices; and the third group represents regions that have been hit by price transmission but are too far from the centre of government power when it comes to receiving compensation. The Oromia region is typical of such an area: although it ranks much better (i.e. has lower prevalence) in terms of undernourishment than, say, the (remote) Afar or Somali regions, the effects of world price shocks are much more pronounced and there may be much more need for specific government interventions there than in the more remote areas where long-term poverty-reduction strategies aimed at improving the resilience of the mainly pastoralist population are more appropriate.

Production

The estimated average per capita production of food in cereal equivalents (see above) for SSA as a whole in 2005 is estimated at 420 kg per person per year, with district averages varying between almost zero and 3300 kg per person (see Figure 2.6 for SSA and for Nigeria in particular).

Surplus sites, i.e. sites where production exceeds local demand in calorie terms, are mainly found in West Africa but also in Northern Kenya, Uganda and

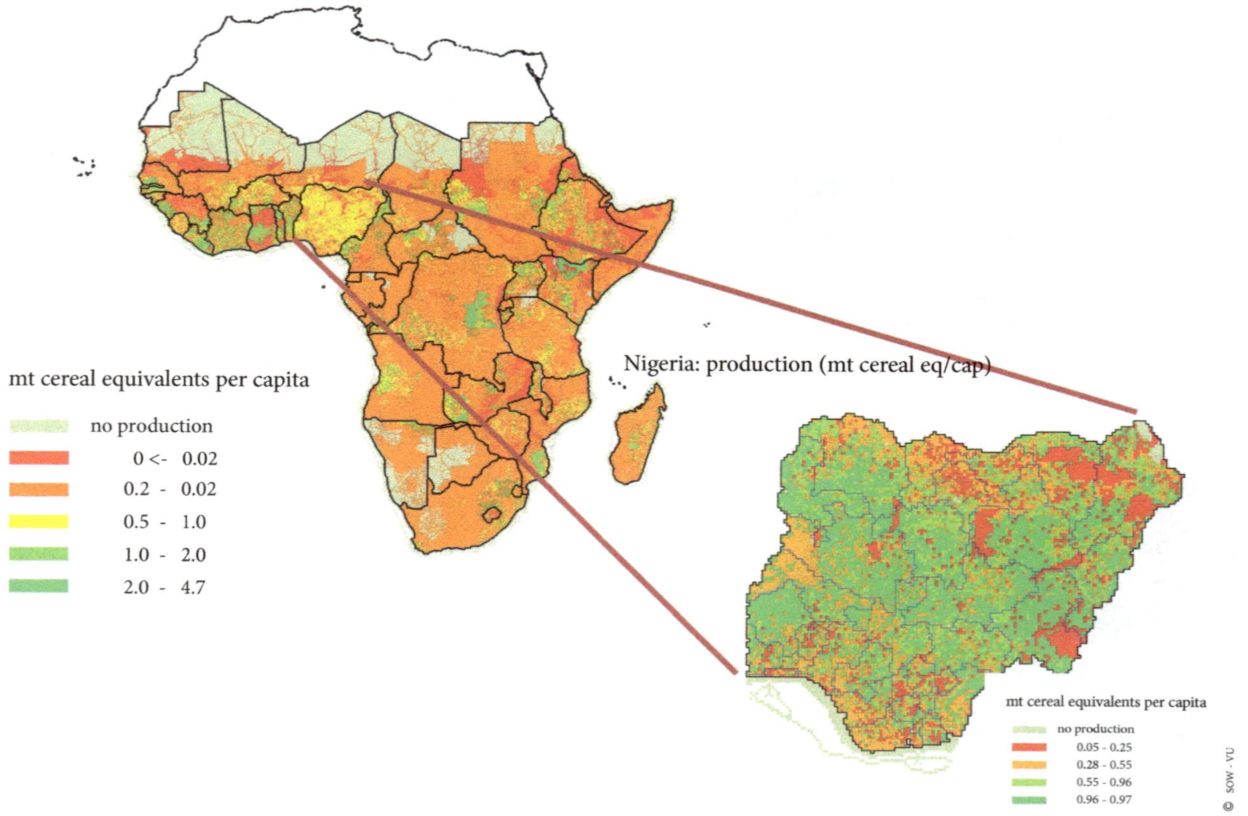

FIGURE 2.6 *Production of cereals in SSA and Nigeria, 2005 (in tonnes per capita)*
SOURCE: AUTHOR'S OWN COMPUTATIONS

even parts of the Democratic Republic of Congo (Figure 2.7a). Another measure of surplus is depicted in Figure 2.7b where the difference is measured between actual production and a normative measure of consumption that is based on standard weights among children and a BMI of 18.5 kg/m^2 for adults with a moderately active lifestyle. This map differs from Figure 2.7a in Sudan, Ethiopia and the Central African Republic where major parts of the surplus areas have disappeared. In fact, these surpluses are 'artificial' as they represent areas where the population is not able to secure available production for consumption due to a lack of income (entitlement crisis) or because they are physically unable to reach markets where food can be bought (or both).[10]

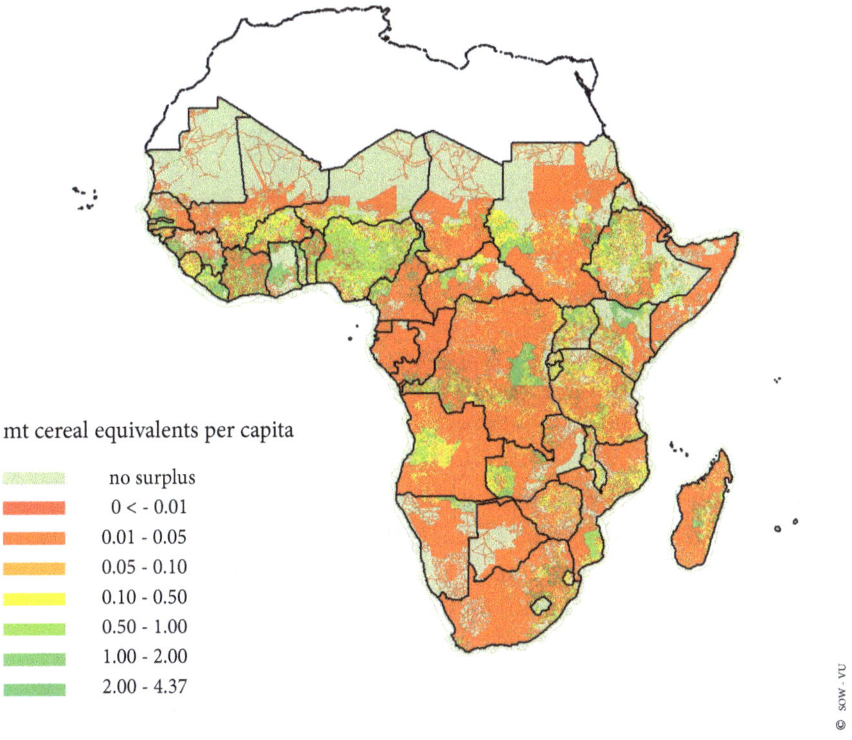

FIGURE 2.7A *Production surpluses in SSA in 2005 (in tonnes of cereal equivalents per capita relative to observed consumption)*
SOURCE: AUTHOR'S OWN COMPUTATIONS

10 It is admitted that because of the coarseness of the grid cells, transport costs between cells may hide these micro-level obstacles, and the cell average of consumption merely reflects underlying dynamics that cannot be modelled within the current framework. In this sense, model outcomes may serve to signal specific problem areas where more detailed research is needed to clarify the nature of the food insecurities observed.

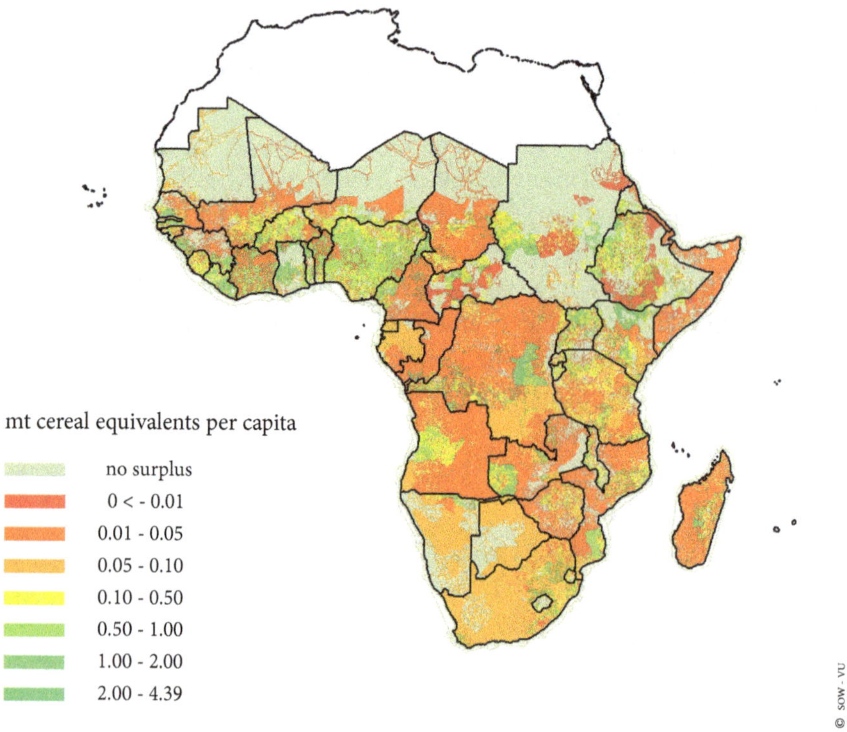

FIGURE 2.7B *Production surpluses in SSA in 2005 (in tonnes of cereal equivalents per capita relative to norm consumption)*
SOURCE: AUTHOR'S OWN COMPUTATIONS

Deficit sites (where consumption exceeded local production in 2005) are mainly located in the Horn of Africa, Malawi and in northern parts of West Africa (Figure 2.8). In the future, it is important to combine information on deficit areas with predictions about the impact of climate change, as in Liu *et al.* (2008), to guide policymakers in targeting preventive and adaptive policies.

The design of such policies naturally requires more than simply assessing production possibilities, as the resilience of the households in deficit areas also depends on their sources of income and existing coping mechanisms, while the indirect effects of climate change, such as migration away from affected areas, can also harm regions that seem 'safe' based on the analysis presented in this chapter. Combining production indicators with agro-ecological indicators and household characteristics extracted from geo-referenced surveys provides a promising avenue to enrich any analysis of the likely impact of climate change on the African continent.

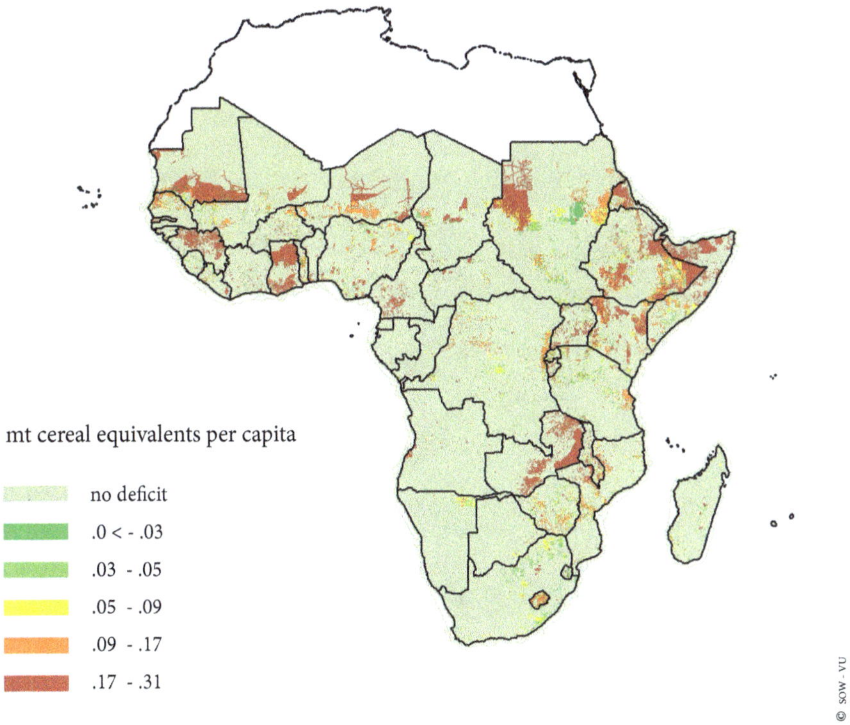

FIGURE 2.8 *Production deficits in SSA in 2005 (in tonnes of cereal equivalents per capita)*
SOURCE: AUTHOR'S OWN COMPUTATIONS

Transport

The availability of high-quality roads is a major problem in Sub-Saharan Africa. As Figure 2.9a shows, there are only a limited number of primary roads and these are mainly concentrated in West Africa and South Africa, leaving the interior of the content largely inaccessible.[11] The availability of secondary roads follows the same pattern (Figure 2.9b), with high densities in West Africa and South Africa, and fewer roads in the rest of the continent.

Transport cost estimates for food aid reflect the availability of roads to a large extent (Figure 2.10), although security in the country and climatic conditions (seasonal variations) play an important role in explaining the estimated pattern: data on costs (an inclusive term that included fuel costs, costs incurred by driving

11 The number of railroads and inland waterways in Sub-Saharan Africa is very limited, although efforts to restore and upgrade old railways have recently been made.

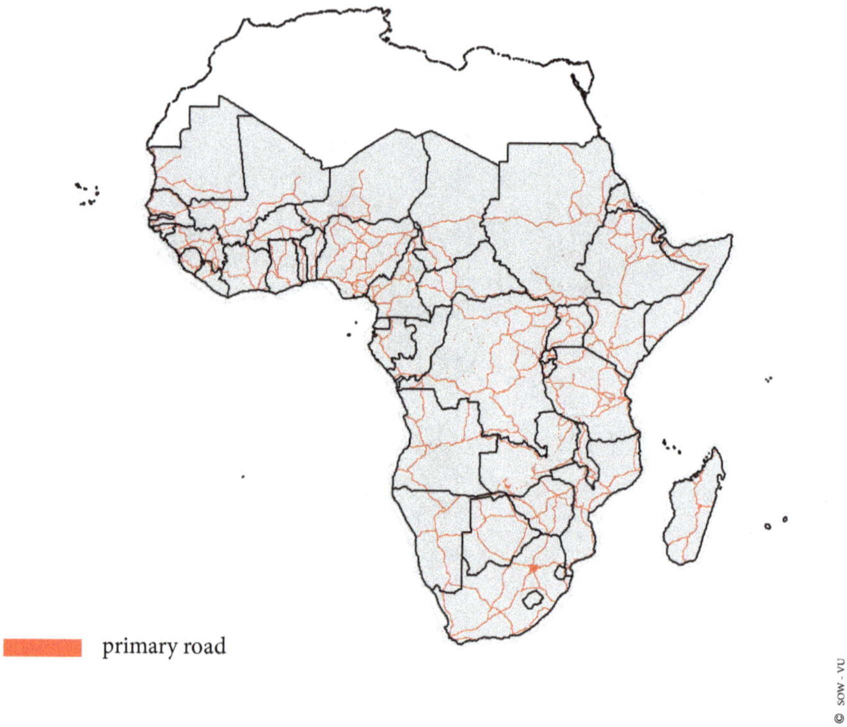

FIGURE 2.9A *Sub-Saharan Africa's primary roads, 2005*
SOURCE: AUTHOR'S OWN COMPUTATIONS

in convoy, [un]official tolls and other costs) by road type sometimes differ substantially between countries and even regions within countries. For commercially traded food, no reliable continent-wide estimates of transport costs exist and the estimates for food aid are thus also applied to commercial food.

Again, the possibility of selecting a particular country (e.g. Ghana as in Figure 2.10) for further analysis opens the way to specific studies. In West Africa, for example, the cultivation and marketing of vegetables, particularly those that can be cultivated counter-seasonally such as tomatoes in Burkina Faso and northern Ghana, is attracting a great deal of attention from policy-makers. This is seen as a potentially promising way of promoting rural development and alleviating poverty since small-scale production can be profitable. However, a major bottleneck is the bad quality of many roads and the presence of (un)official roadblocks that cause spoilage of fresh produce. Using the database on transport costs as the basis for the analysis, additional survey information on roadblock costs and delays can be added to assess the profitability of the tomato marketing chain and present possible options for improvement (ongoing research).

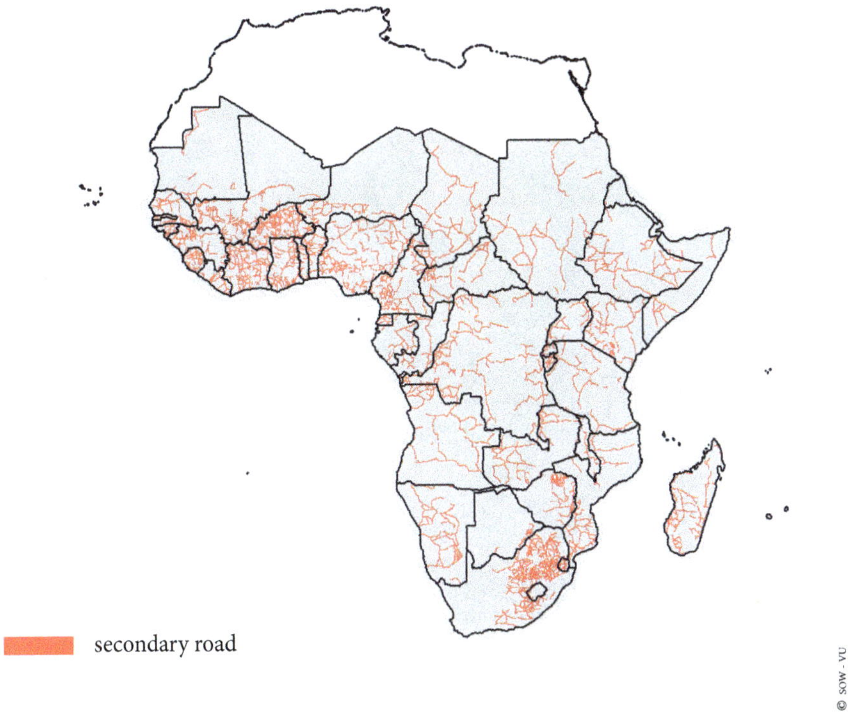

FIGURE 2.9B *Sub-Saharan Africa's secondary roads, 2005*
SOURCE: AUTHOR'S OWN COMPUTATIONS

Trade and Local Consumption

As explained above, the consumption needs of each person in the model are met by three sources: locally produced foods and local stocks of foods; foods produced outside the cell that are made available through trade; and food aid. Given the high transport costs in large areas of the continent, by far the largest share of needs are satisfied by locally produced foods within the grid cell of 100 km^2. In coastal areas and areas with a high degree of urbanization in surroundings with limited production capacity, the share of consumption of traded food is highest. On the other hand, the set of sites with high commercial consumption also includes areas where food production is low but cash crops or other activities, such as mining or other industrial activities, provide the population with enough income to buy food produced elsewhere. On the whole, areas with a substantial share of trade with other cells are relatively rare (Figure 2.11). West Africa (and particularly Nigeria) stands out as the region with the highest share of commercial consumption: it is supported by relatively good transport possibilities and ongoing efforts at regional integration and trade have developed

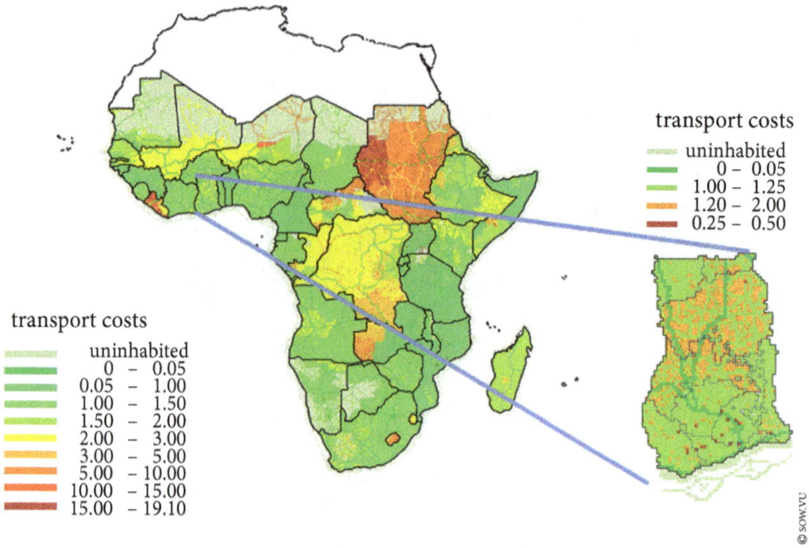

FIGURE 2.10 *Transport costs in Sub-Saharan Africa and Ghana 2005 (in* US$ *per tonne-km)*
SOURCE: AUTHOR'S OWN COMPUTATIONS

mainly along north–south lines where differences in climate provide opportunities for northern producers to supply the relatively affluent urban coastal centres in the south.

Food Aid

Sub-Saharan Africa was the largest recipient of emergency food aid in 2005 (Figure 2.12). Drought and other emergencies led to shipments of 0.9 m tonnes of food aid (in cereal equivalents) to Ethiopia, while the WFP responded to the crisis in Darfur and also to food security concerns in the areas of Kassala and Blue Nile, with the shipment of 0.67 m tonnes of food. Refugees and internally displaced persons present a major issue for Sudan too. In 2005, there were three camps for refugees and IDPs (Ardamata Camp with 35,000 registered refugees/IDPs; Mornei Camp with 61,000 and Wad Sherife with 33,000 inhabitants), while there were 933,000 registered refugees and IDPs living outside camps in both rural and urban areas. The large majority of these people were partly or completely dependent on food aid and, given the bad quality of infrastructure and general insecurity in the areas concerned, the costs of providing food aid to those in need were high.

The concentration here on Sudan and Ethiopia is reflected in the spatial distribution of food aid across the continent in 2005, as depicted in Figure 2.13. However Kenya, parts of the Democratic Republic of Congo and Malawi have

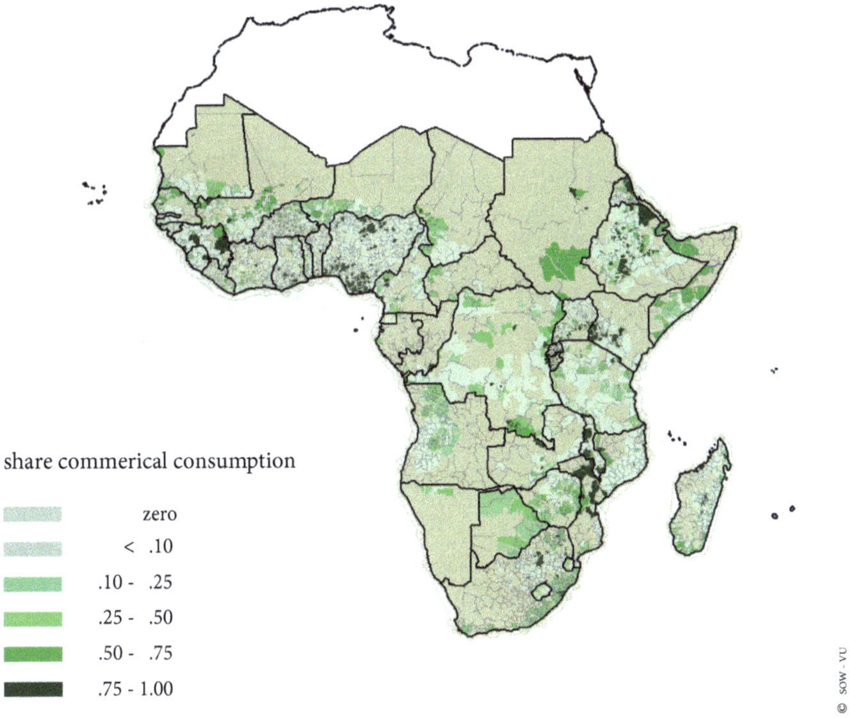

FIGURE 2.11 *Share of commercial consumption in SSA, 2005 (in total %)*
SOURCE: AUTHOR'S OWN COMPUTATIONS

also received substantial support from the WFP. As Figure 2.13 shows, per capita food-aid provision varies substantially, from being only a minor addition to the daily intake to a full daily ration. As mentioned above, the data for the provision of food aid reflect the actual operations undertaken by the WFP in 2005 and do not take into account any judgement of the need, effectiveness or timeliness of these operations.

Figure 2.14 presents the optimal flows of food aid, with the red flows representing large flows and the green smaller ones, showing the 'fanning out' of aid from primary roads to the actual locations of the recipients. Many ports in Africa act as points of entry, notably those in South Africa that serve a large part of Southern Africa, but Dar es Salaam in Tanzania and Mombasa in Kenya are major sources for further distribution too. As can be seen in the Africa-wide picture, the Darfur region is supplied mainly through the port of Douala in Cameroon because of the high costs associated with transport through Sudan itself. Zooming in on Tanzania (Figure 2.14) highlights how the country is used to transport food aid to Burundi, Rwanda and the DRC, but also to bring food aid to the

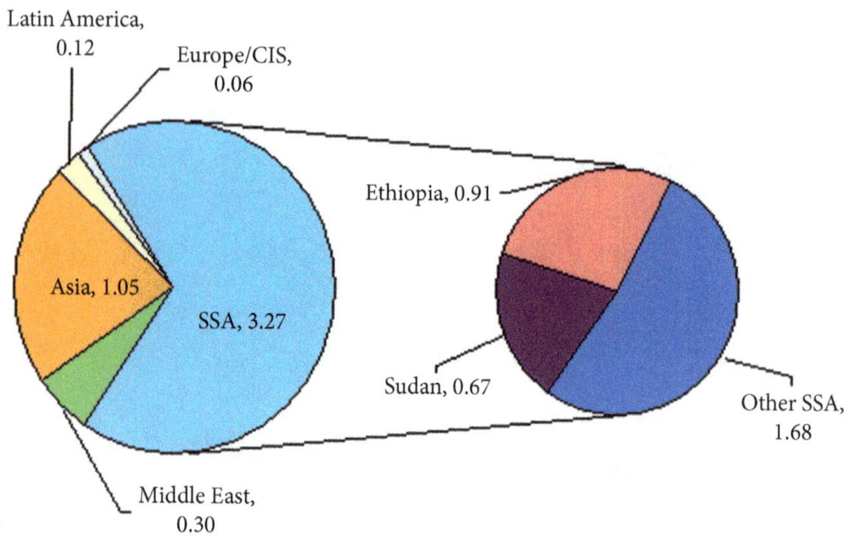

FIGURE 2.12 *Emergency shipments of food aid in the World, 2005 (in million tonnes of cereal equivalents)*
SOURCE: WFP (2007)

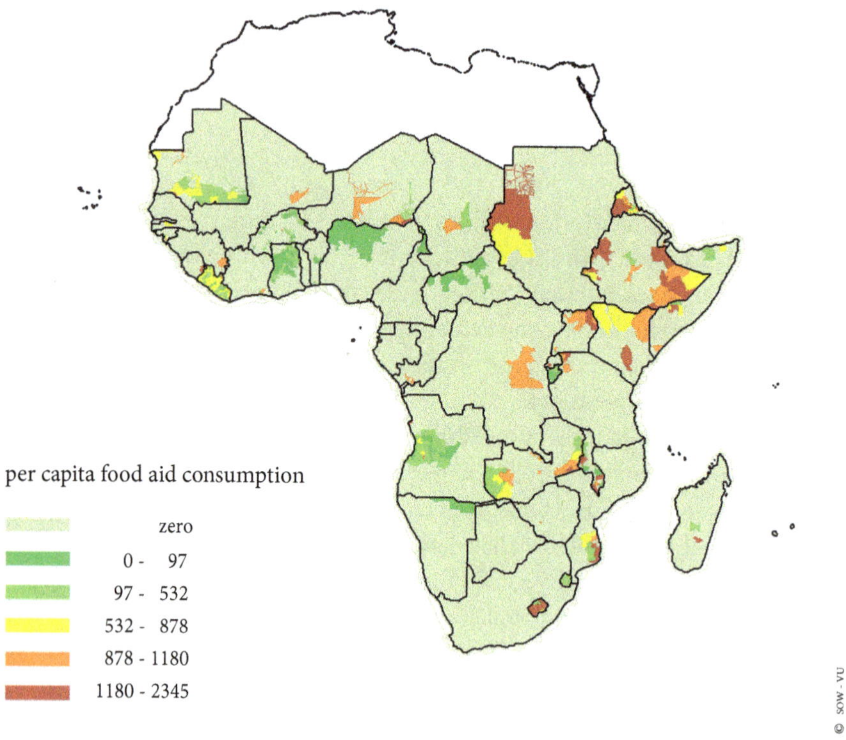

FIGURE 2.13 *Sub-Saharan Africa's per capita food aid consumption, 2005 (in kcal per day)*
SOURCE: AUTHOR'S OWN COMPUTATIONS

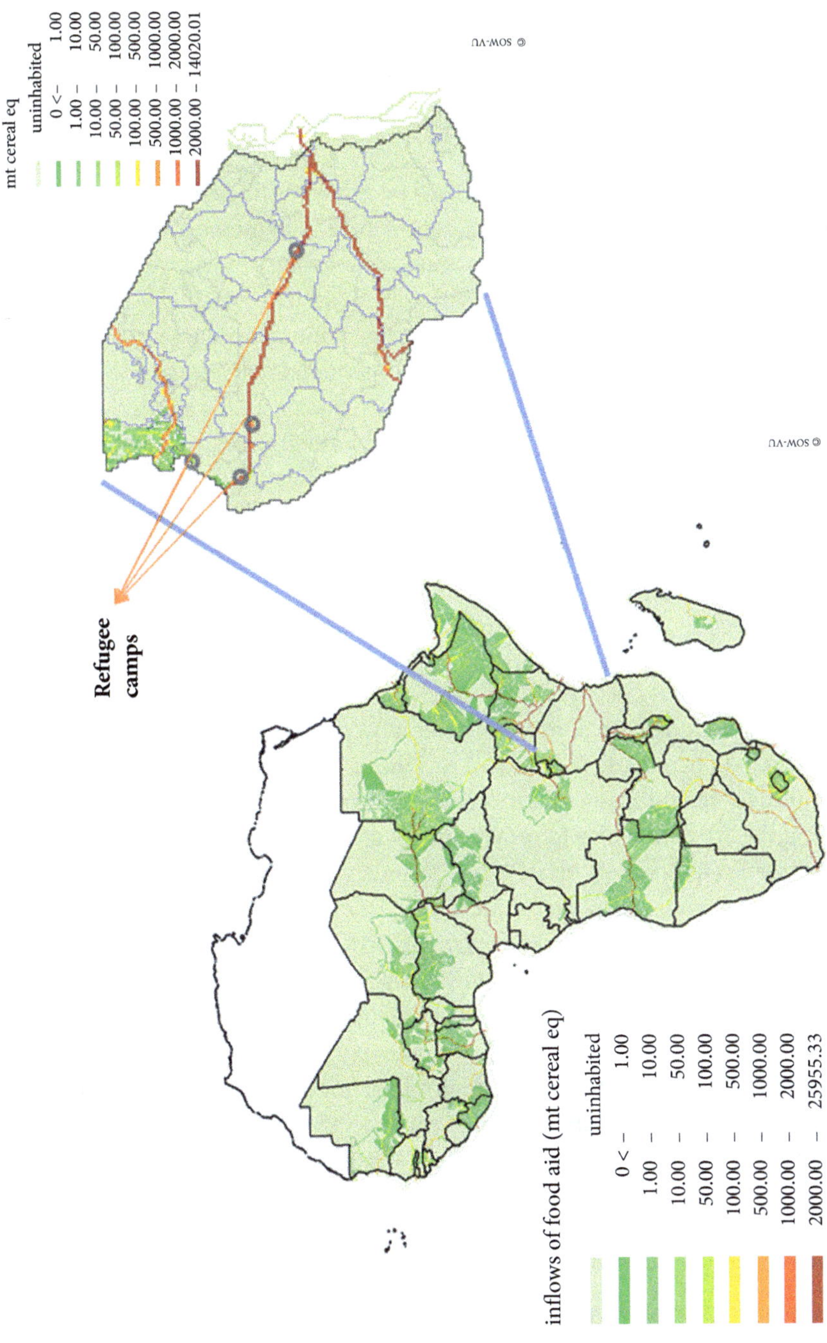

FIGURE 2.14 *Optimal procurement and routing of food aid to recipients,* SSA *and Tanzania, 2005*
SOURCE: AUTHOR'S OWN COMPUTATIONS

nine refugee camps in four different grid cells that jointly house almost 350,000 people and to the 201,000 registered refugees outside camps who are also provided by food-aid transports through Kenya.

It should be noted that the outcomes presented are part of the optimal solution for the welfare program described above. The choice of ports and the routing from the source to the recipients is made endogenously[12] to minimize total transport costs. Alternative scenarios could be run with the model, for example simulating an increase in prices of cereals or fuel, increased options for local sourcing of food aid or the optimal location of emergency stocks. An attractive feature of the model is that it can be rerun with alternative specifications of, say, per capita consumption in specific areas within 10 minutes on a standard PC.

Finally, Figure 2.15 summarizes the costs of food-aid deliveries, i.e. the total of procurement costs and the costs of transport. Sudan stands out as the most expensive area but areas located far from ports are also costly to reach. The distribution channels for Ethiopia are clearly recognizable: the road corridors are dark green and costs increase as transports move further along secondary and tertiary roads to reach all the intended recipients.[13]

Concluding Remarks

As poverty and hunger are basic yardsticks of underdevelopment and destitution, the need for reliable statistics in this domain is self-evident, not only to feed international indicators such as those selected for monitoring progress in reaching the MDGs but also to provide information for policymakers that will enable them to plan appropriate interventions. This chapter has shown how direct anthropometric data on women's weights and BMI in DHS surveys – the same surveys that are used for the anthropometric assessment of the growth of children – can be used to provide estimates for undernutrition and calorie intake among the total population in Sub-Saharan Africa. It is recommended that more and better use be made of anthropometric data in estimates of undernutrition worldwide as a refined alternative to the aggregate food-production-based indicators that probably overstate the hunger problem. In addition, the chapter has highlighted the potential of a spatially explicit

12 It is not derived from any 2005 data external to the model but is the result of the welfare maximization function.

13 In principle, similar maps could be shown for optimal flows and prices of commercial food but, given the uncertainty with respect to transport costs, more research would be required to present those outcomes with any degree of confidence.

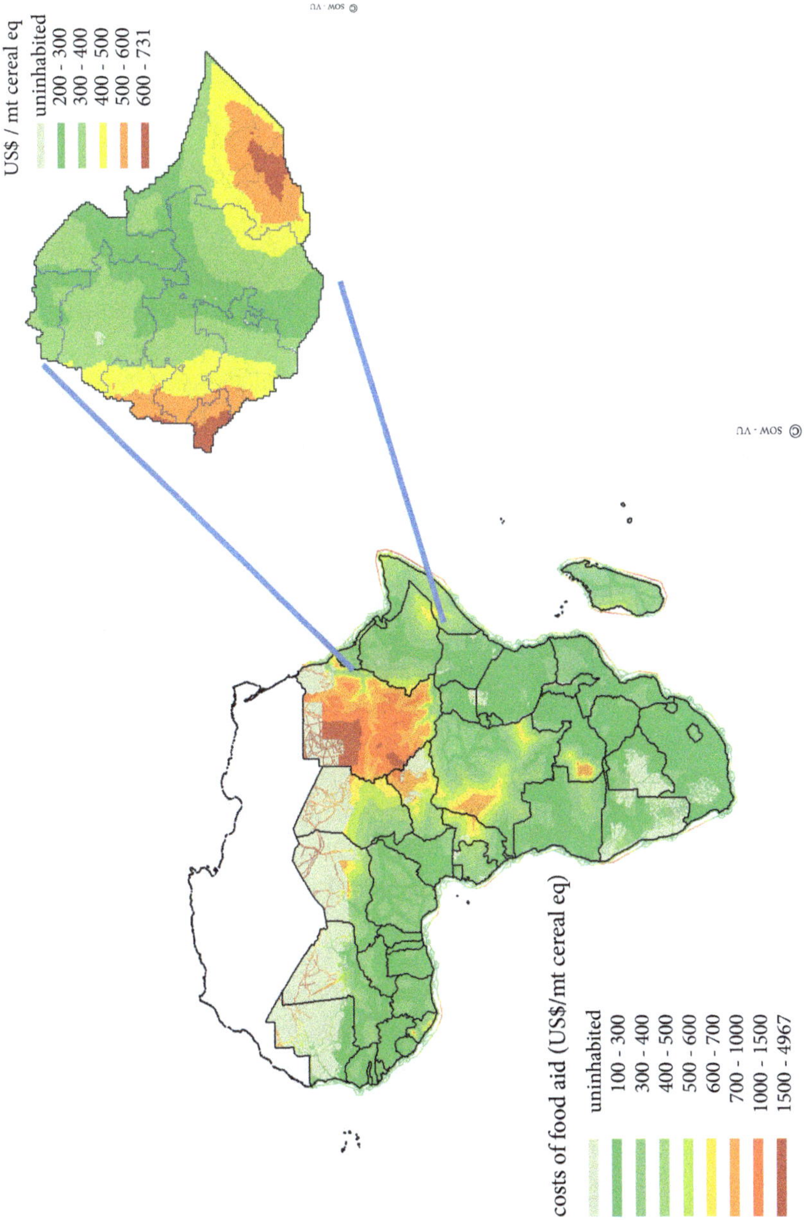

FIGURE 2.15 *Cost of procuring and delivering food aid to recipients in SSA and Ethiopia, 2005*

SOURCE: AUTHOR'S OWN COMPUTATIONS

welfare model to assess possibilities for trade and costs of food aid by simulating different types of shocks to the system and assessing the opportunities and bottlenecks for the further development of food production systems in Sub-Saharan Africa. In this sense, the chapter also serves to show the relevance of fundamental research in representing the many different aspects of food security in its spatial and social context.

References

Abdulai, A., C.B. Barrett & J. Hoddinott (2005), Does Food Aid *Really* Have Disincentive Effects? New Evidence from Sub-Saharan Africa. *World Development* 33(1): 1689–1704.

Barrett, C.B. (2002a), Food Aid and Commercial International Food Trade. Paper prepared for the Trade and Markets Division, OECD. Cornell University Working paper WP 2002–39. http://dyson.cornell.edu/faculty_sites/cbb2/files/papers/BarrettOECDReportMar2002.pdf. Accessed 21 August 2014.

Barrett, C.B. (2002b), Food Aid Effectiveness: It's the Targeting, Stupid!. Report to the Policy Service, Strategy and Policy Division. Rome: WFP. http://dyson.cornell.edu/faculty_sites/cbb2/files/papers/WFPPaperDec2002.pdf. Accessed 21 August 2014.

DHS (2013), Various Demographic and Health Surveys: http://measuredhs.com/.

DISI (2006), Comprehensive Food Security and Vulnerability Analysis: An Internal Review. Prepared by Development Information Services International (DISI) for WFP, Rome.

FAO (1953), Food Composition Tables for International Use. Rome: FAO.

FAO (2012), The State of Food Insecurity in the World, 2012. Rome: FAO.

FAO (2013a), The State of Food Insecurity in the World, 2013. Rome: FAO.

FAO (2013b), Global Agricultural Zones. http://gaez.fao.org/Main.html#. Accessed 21 August 2014.

FAO (2013c), Food Security Indicators: A Core Set of Food Security Indicators. Rome: FAO. http://www.fao.org/economic/ess/ess-fs/en/ Accessed 6 December 2013.

FAOSTAT (2013), "New FAOSTAT" data available at: http://faostat3.fao.org/. Accessed 21 August 2014.

Ginsburgh, V. & M.A. Keyzer (2011), *The Structure of General Equilibrium Models*. Cambridge, MA: MIT Press.

IATP (2005), *US Food Aid: Time to Get it Right*. Minneapolis: Institute for Agriculture and Trade Policy.

Keyzer, M.A. (2004) Food crisis management: a spatially explicit equilibrium approach. Unpublished manuscript, SOW-VU.

Keyzer, M.A. & C.F.A. van Wesenbeeck (2011), A Gradient-Related Algorithm for Solving Large Scale, Spatially Explicit, Welfare Models with Transportation. Numerical Analysis and Applied Mathematics ICNAAM 2011. American Institute of Physics Conference Proceedings, Volume 1389, pp. 1743–1745. http://dx.doi.org/10.1063/1.3636946. Accessed 18 December 2013.

Klasen, S. (2008), Poverty, Undernutrition and Child Mortality: Some Inter-regional Puzzles and Their Implications for Research and Policy. *Journal of Economic Inequality* 6(1): 89–115.

Lentz, E.C. & C.B. Barrett (2008), Improving Food Aid: What Reforms Would Yield the Highest Pay-off? *World Development* 36(7): 1152–1172.

Liu, J., S. Fritz, C.F.A. van Wesenbeeck, M. Fuchs, L. You, M. Obersteiner & H. Yang (2008), A Spatially Explicit Assessment of Current and Future Hotspots of Hunger in Sub-Saharan Africa in the Context of Global Change. *Global and Planetary Change* 64(3): 222–235.

Nubé, M. (2001), Confronting Dietary Energy Supply with Anthropometry in the Assessment of Undernutrition Prevalence at the Level of Countries. *World Development* 29(7): 1275–1289.

Svedberg, P. (1999), 841 Million Undernourished?. *World Development* 27: 2081–2098.

van Wesenbeeck, C.F.A., M.A. Keyzer & M. Nubé (2009), Estimation of Undernutrition and Mean Calorie Intake in Africa: Methodology, Findings and Implications. *International Journal of Health Geographics* 8(1). http://www.ij-healthgeographics.com/content/8/1/37. Accessed 21 August 2014.

van Wesenbeeck, C.F.A. & M.D. Merbis (2012), *Data Repository of the Food Economy of Sub-Saharan Africa*. Amsterdam: Centre for World Food Studies.

Webb, P. (2003), Food as Aid: Trends, Needs and Challenges in the 21st Century. WFP Occasional Papers No. 14. Rome: WFP.

WFP (2007) Food Aid Monitor. Rome: WFP. http://www.wfp.org/fais/. Accessed 21 August 2014.

Annex

Technical Notes

1 Countries Included

Angola	Benin	Botswana
Burkina Faso	Burundi	Cameroon
Cape Verde	Central African Republic	Chad
Comoros	Congo Republic	Côte d'Ivoire
Democratic Republic of Congo	Djibouti	Equatorial Guinea
Eritrea	Ethiopia	Gabon
Gambia	Ghana	Guinea
Guinea-Bissau	Kenya	Lesotho
Liberia	Madagascar	Malawi
Mali	Mauritania	Mauritius
Mozambique	Namibia	Niger
Nigeria	Réunion	Rwanda
Sao Tome and Principe	Senegal	Seychelles
Sierra Leone	Somalia	South Africa
St Helena	Sudan	Swaziland
Tanzania	Togo	Uganda
Zambia	Zimbabwe	

Regional Aggregates

West Africa

Benin	Burkina Faso	Cape Verde	Côte d'Ivoire
Gambia	Ghana	Guinea-Bissau	Guinea
Liberia	Mali	Mauritania	Niger
Nigeria	St Helena	Senegal	Sierra Leone
Togo			

East Africa

Burundi	Comoros	Djibouti	Eritrea
Ethiopia	Kenya	Madagascar	Malawi
Mauritius	Mozambique	Reunion	Rwanda
Seychelles	Somalia	Sudan	Tanzania
Uganda	Zambia	Zimbabwe	

Southern Africa

Botswana	Lesotho	Namibia	South Africa
Swaziland			

Central Africa

Angola	Cameroon	Central African Republic	Chad	Congo Republic
Democratic Republic of Congo	Equatorial Guinea	Gabon	Sao Tome and Principe	

2 Population Estimates

The basis for the population estimates is the 2005 Landscan database (http://web.ornl.gov/sci/landscan/), corrected for the presence of refugee camps (UNHCR Statistical Yearbook 2006: http://www.unhcr.org/478cda572.html) and outflow of refugees.

3 Consumption Estimates

For consumption estimation, the major sources used were the Demographic and Health Surveys (DHS: http://www.measuredhs.com). If the DHS survey was not available or was too old, other sources were used.

4 Production Estimates

The main source for the estimates of production was the FAO Agromaps data set: http://www.fao.org/landandwater/agll/agromaps/interactive/page.jspx. The conversion from metric tons of crop produce to calories was done using the calorie conversion table:

http://www.fao.org/docrep/x5557e/x5557e00.htm#Contents. These conversion factors (in kcal per mt edible wet produce) are summarized below. For the conversion from calories in to metric tons cereal equivalents, a calorie content of 3000 kcal per kg was used.

Wheat:	3340	Rice:	2800	Barley:	3320
Maize:	3560	Popcorn:	3560	Rye:	3190
Oats:	3850	Millet:	3400	Sorghum:	3430
Buckwheat:	3300	Quinoa:	3420	Fonio:	3380
Triticale:	3270	Canary seed:	3880	Mixed grains:	3400
Cereals NES:	3400	Potatoes:	670	Sweet potato:	920
Cassava:	1090	Yautia:	1090	Taro:	860
Yam:	1010	Roots and tubers NES:	910	Sugarcane:	300
Dry beans:	3410	Dry broad beans:	3430	Dry peas:	3460
Chickpeas:	3580	Dry cowpeas:	3420	Pigeon peas:	3430
Lentils:	2420	Bambara beans:	3650	Vetches:	3150
Lupins:	4000	Pulses NES:	3400	Soybeans:	3350
Groundnuts in shell:	4140	Oil-palm fruit:	1580	Sunflower seeds:	3080
Sesame seed:	5730	Cotton seed:	2530	Green beans:	500
Green peas:	310	Bananas:	700	Plantains:	750
Beef & veal:	2250	Palm kernels:	5140	Olives:	1750
Tomatoes:	170	Onions:	240	Other vegetables:	220
Lemons & limes:	150	Grapefruit:	160	Other citrus fruit:	260
Apples:	480	Pineapples:	260	Dates:	1560
Grapes:	530	Other fruit:	360	Rapeseed:	4940
Other oil crops:	3870	Oranges & mandarins:	340		

5 Transport Costs

The following types of infrastructure are included in the database:

- primary roads
- secondary roads
- tertiary roads
- railways
- inland waterways
- coastal waterways

The sources for the raw data are: (i) roads: Nelson, A. (2004) Updated Digital Chart of the World (DCW), supplied by the United States Geological Survey: http://www.usgs.gov/ and (ii) railways, inland waterways and coastal waterways: compiled from different sources by R.L. Voortman (SOW-VU).

For country-specific sources of information, see van Wesenbeeck & Merbis (2012).

CHAPTER 3

Agricultural Pockets of Effectiveness in Africa

A Comparative Inventory of Nigeria, Kenya, Tanzania and Uganda since 2000

Akinyinka Akinyoade, Ton Dietz and André Leliveld[1]

Increasing pockets of agricultural effectiveness are emerging in Nigeria, Uganda, Tanzania and Kenya, the four countries that made up the African part of the Tracking Development project. Research at the African Studies Centre in Leiden in its DRA-AFCA project[2] *has found rising agricultural production, improved food security and greater yields for many agricultural products since 2000. The study suggests that it is important to go beyond researching policy matters alone and to include metropole-hinterland dynamics and agro-food cluster institutions as drivers of change. This chapter offers an inventory of the evidence presented in the four research papers produced to date. The results have been compared to formulate a research agenda for the years to come.*

Introduction

The Tracking Development research project (2007–2012) compared the long-term[3] developments in South East Asia and Sub-Saharan Africa to determine how certain countries, all with low levels of socio-economic performance in the 1950s, could differ so much in their economic performance over the following decades. A team of researchers from Leiden's Royal Netherlands Institute of South East Asian and Caribbean Studies (KITLV) and the African Studies Centre (ASC) as well as researchers and PhD students in four South East Asian and four African countries compared various 'twinned' countries: Nigeria and Indonesia; Uganda and Cambodia; Kenya and Malaysia; and Tanzania and Vietnam (Berendsen *et al.* 2013).

1 The authors were assisted in the preparation of this chapter by members of the African Studies Centre's Agro-Food Clusters in Africa Collaborative Research Group (AFCA): Dick Foeken, Blandina Kilama, Wijnand Klaver, Heleen Smits, Sebastiaan Soeters and Merel van 't Wout.

2 The Developmental Regimes in Africa (DRA) project is a follow-up to the Tracking Development project and is being carried out by the ODI in London and the ASC in Leiden. Like Tracking Development, funding has been generously provided by the Netherlands Ministry of Foreign Affairs. The DRA research at the ASC was undertaken by the AFCA collaborative research group and was partially funded by the ASC.

3 'Long-term' in this case meant a focus on the second half of the 20th century.

 | DOI 10.1163/9789004282698_004

One of the main conclusions drawn by project leaders David Henley (KITLV) and Jan Kees van Donge (ASC) was that the economic breakthrough in South East Asia can only be understood if one looks at the massive state-led rural development campaigns that were imposed from the 1960s onwards and that resulted in a major agricultural revolution and the generally successful alleviation of large-scale rural poverty. This was much less the case in Africa where political leaders in many post-colonial governments made different choices, neglected rural peasants and implemented an elite-based industrialization strategy that ultimately had disappointing results (Henley & van Donge 2012; Vlasblom 2013). Similar conclusions were drawn by the DfID-funded Africa Power and Politics Project (APPP) that focused on Africa's ruling elites who exploited or ignored the rural masses and can be considered to have been responsible for the subsequent economic stagnation, widespread poverty and hunger. An important scientific and policy question should be whether Africa would have achieved a 'growth miracle' by combining an agriculture-based rapid-growth strategy with a successful poverty-alleviation strategy, particularly in the rural areas, if it had put more emphasis on its agricultural sector, as South East Asia did from the 1960s onwards.

Although the main conclusions were agreed by most of the participants in the Tracking Development team, there has equally been discussion about the causal factors and more recent trends (Dietz 2013). Based on statistical evidence from FAO sources (FAOSTAT), four new research reports have focused on Africa's agricultural dynamics (Akinyoade *et al.* 2013 for Nigeria; Leliveld *et al.* 2013a for Uganda; Leliveld *et al.* 2013b for Tanzania; Dietz *et al.* 2014 for Kenya). Recent trends show that Africa's agricultural performance was not entirely gloomy between 1960 and 2000, and that major breakthroughs can be seen that suggest that the continent's agricultural sector has been improving since 2000 and, in some cases, that Africa has recently been experiencing an agricultural revolution, although one rather different from South East Asia's 'Green Revolution'.

The research reports discuss five major datasets: (i) FAO-based data on (staple) food production between 1961 and 2011; (ii) data linking food production and food consumption in the same period based on so-called food balance sheets; (iii) data on breakthrough crops and livestock products between 2000 and 2010; (iv) data on child under-nutrition; and (v) maps presenting a detailed geographical picture of many of these variables. This chapter focuses on a comparison of the data available for the period between 2000 and 2010. One notable caveat concerns the reliability of the FAO data Extra note with "One notable caveat concerns the reliability of the FAO data[4] as doubts have been expressed as doubts have been expressed about their statistics, for example the

4 See also the discussion about Figure 2.4 in Chapter 2, this volume.

figures relating to maize in Uganda or potatoes in Kenya, that are too worrying to ignore. This chapter attempts, however, to show what emerges when the FAO data are used to assess the current situation.

Nigeria

The most successful crops in Nigeria were defined as those with: (i) an area of more than 150,000 ha in 2010; (ii) a growth in production between 2000 (=average 1999–2001) and 2010 (=average 2009–2011) that was above the population growth for the same period (28%); and (iii) a yield increase of 20% or more over the decade. Successful livestock were considered as those species whose numbers grew faster than the population in the same period. For Nigeria, the most successful crops were maize, cassava, rice, melon seed, potatoes and pineapples, and the most successful livestock were chickens, pigs, sheep and goats. Milk and egg production and fish also more than kept pace with population growth. Nigeria did, however, see some of its crops, namely millet, oil palm, cocoyam and karité nuts, decrease in absolute terms regarding both yield and production (Tables 3.1, 3.2 & 3.3).

TABLE 3.1 *Performance of Nigeria's major crops, 2000–2010 (population growth 2000–2010: 28%)*

Crop > 150,000 ha in 2010 (highest acreage first) Green: Most successful crop Red: Problem crop	Ha in 2010 (x 1000)	Production [2010 index number compared to 2000] Green >128 Turquoise: 120–128 Red <100	Yield [2010 index number compared to 2000] Green >128 Turquoise: 120–128 Red < 100	Area [2010 index number compared to 2000] Green >128 Turquoise: 120–128 Red < 100
Sorghum	4863	87	118	73
Maize	4503	171	130	137
Millet	3680	65	86	70
Cassava	3449	136	132	103
Oil palm	3200	99	96	104
Yams	2845	128	119	108
Cowpeas	2790	118	155	78
Groundnuts	2592	115	82	139

(cont.)

Crop > 150,000 ha in 2010 (highest acreage first) Green: Most successful crop Red: Problem crop	Ha in 2010 (x 1000)	Production [2010 index number compared to 2000] Green >128 Turquoise: 120–128 Red <100	Yield [2010 index number compared to 2000] Green >128 Turquoise: 120–128 Red < 100	Area [2010 index number compared to 2000] Green >128 Turquoise: 120–128 Red < 100
Rice	2283	135	129	105
Cocoa	1299	129	89	146
Sweet potatoes	1100	134	100	134
Citrus	788	116	110	106
Vegetables (fresh)[1]	661	140	126	110
Taro (cocoyam)	486	80	97	83
Melon seed	476	138	166	83
Plantain	450	138	120	114
Soybeans	391	100	140	76
Okra	388	145	108	135
Cashew	362	180	127	142
Rubber	345	134	124	108
Cotton	344	106	161	65
Karité (shea) nuts	343	89	60	149
Sesame	319	197	92	214
Tomatoes	269	141	91	148
Kola nuts	257	186	65	284
Potatoes	257	181	148	122
Fruit (other)	181	75	149	50
Green maize	172	74	166	45
Pineapples	162	143	130	109
Onions (dry)	159	215	63	346
Fonio	150	108	98	110

[1] Not elsewhere specified.

SOURCE: FAOSTAT | © FAO STATISTICS DIVISION 2013. UPDATED 8 AUGUST 2013. HTTP://FAOSTAT.FAO.ORG/SITE/339/DEFAULT.ASPX – CROPS. ACCESSED 17 SEPTEMBER 2013. [AS OF SEPTEMBER 2014: HTTP://FAOSTAT3.FAO.ORG/FAOSTAT-GATEWAY/GO/TO/DOWNLOAD/Q/QC/E].

TABLE 3.2 *Performance of Nigeria's main livestock, 2000–2010 (population growth 2000–2010: 28%)*

Product/ Type of animal	No. of animals producing or slaugh-tered in 2010 (x 1000)	Production [2010 index number compared to 2000]	Weight of milk/meat/ eggs per animal [2010 index number compared to 2000]	Offtake (% of animals producing or slaughtered out of total stock) [2010 index number compared to 2000]	Headcount [2010 index number compared to 2000]
Green: Most successful species and products Red: Problem species		Green >128 Turquoise: 120–128 Red <100	Green >128 Turquoise 120–128 Red < 100	Green >128 Turquoise 120–128 Red <100	Green >128 Turquoise 120–128 Red < 100
Cow's milk	2,097	132	100	117	113
Hen's eggs	156,000	147	99	93	159
Chickens	280,000	163	100	103	159
Goats	22,678	130	100	98	132
Sheep	14,876	141	100	101	140
Pigs	5,183	147	100	100	148
Cattle	2,630	115	98	104	113

N.B. The index number of total production is the index for headcount multiplied by the two indices for 'yield' (offtake and weight per animal) and divided by 10,000.

SOURCE: FAOSTAT | © FAO STATISTICS DIVISION 2013. UPDATED 8 AUGUST 2013. HTTP://FAOSTAT.FAO.ORG/SITE/339/DEFAULT.ASPX - LIVE ANIMALS, LIVESTOCK PRIMARY AND LIVESTOCK PROCESSED. ACCESSED 19 SEPTEMBER 2013. [AS OF SEPTEMBER 2014: HTTP://FAOSTAT3.FAO.ORG/FAOSTAT-GATEWAY/GO/TO/DOWNLOAD/Q/*/E].

Maize is an example of a very successful crop that has become important as a staple crop (Akinyoade *et al.* 2013). Maize production stagnated in the 1960s and was slightly below population growth. The 1970s saw yields start to improve but the area harvested declined dramatically, resulting in decreasing overall maize production per capita. From about 1980 onwards, the areas harvested started to increase and, with stabilized yield levels, total production grew much faster than Nigeria's population. The mid-1970s and early 1980s were the years of 'Operation Feed the Nation' and the start of the Green Revolution

TABLE 3.3 *Nigeria's most, less and least successful crops and livestock species, 2000–2010* *

Yield in 2010 as a % of yield in 2000	Production in 2010 as a % of production in 2000			
	<100%	100–120%	>128%	
>128%	**Fruit (other)** **Green maize**	**Cowpeas** **Soybeans** **Cotton**	**Maize** **Cassava** **Rice** **Melon seed** **Potatoes** **Pineapples**	
120–128%			**Vegetables (fresh)** **Cashew**	
100–120%	Sorghum	Citrus	Yams Sweet potatoes Plantain Okra Rubber	Milk Chicken Goats Sheep Pigs Fish[1]
<100%	Millet Oil palm Taro (cocoyam) Karité (shea) nuts	Groundnuts Cattle Fonio	Cocoa Sesame Tomatoes Kola nuts Onions	Eggs

* In **bold**: most successful crops and livestock species.

1 No information is available for fish that could serve to calculate an index of yield.

SOURCE: TABLES 3.1 AND 3.2.

programmes respectively. However, the areas harvested decreased considerably in the 1990s, although yield levels increased slowly. As a result, per capita maize availability dropped before a major improvement was seen. Since 2000, maize has clearly been a very successful crop with cropping areas easily keeping pace with population growth and steady improvements in yield levels resulting in a considerable growth in maize availability per capita. The data for cassava, rice, melon seed, potatoes and pineapples speak for themselves (Akinyoade *et al.* 2013).

TABLE 3.4 *Performance of Uganda's major crops, 2000–2010 (population growth 2000–2010: 38%)*

Crop > 35,000 ha in 2010 (highest acreage first) Green: Most successful crop Red: Problem crop	Ha in 2010 (x 1000)	Production [2010 index number compared to 2000] Green >138 Turquoise 120–138 Red <100	Yield [2010 index number compared to 2000] Green >138 Turquoise 120–138 Red < 100	Area [2010 index number compared to 2000] Green >138 Turquoise 120–138 Red < 100
Plantains	1,699	102	96	106
Maize	1,012	219	136	161
Beans (dry)	953	102	75	136
Sweet potatoes	587	112	106	106
Cassava	417	101	94	107
Groundnuts (with shell)	391	204	105	194
Sorghum	353	100	79	127
Coffee, (green)	303	93	85	108
Sesame seed	301	198	128	155
Sunflower seed	208	319	118	272
Vegetables (primary)	194	169	81	209
Millet	177	47	102	46
Soybeans	150	154	108	142
Bananas	142	96	92	105
Vegetables (fresh)[1]	129	180	79	226
Potatoes	105	151	98	154
Seed cotton	103	139	335	41
Pigeon peas	97	118	96	124
Rice paddy	88	207	170	122
Cowpeas (dry)	71	145	131	111
Onions (dry)	61	134	75	178
Oilseed[1]	48	157	136	115
Cocoa beans	46	401	119	338
Sugarcane	42	182	85	213

[1] Not specified elsewhere.

SOURCE: FAOSTAT | © FAO STATISTICS DIVISION 2013. UPDATED 8 AUGUST 2013. HTTP://FAOSTAT.FAO.ORG/SITE/339/DEFAULT.ASPX – CROPS. ACCESSED 17 SEPTEMBER 2013. [AS OF SEPTEMBER 2014: HTTP://FAOSTAT3.FAO.ORG/FAOSTAT-GATEWAY/GO/TO/DOWNLOAD/Q/QC/E].

TABLE 3.5 *Performance of Uganda's main livestock, 2000–2010 (population growth 2000–2010: 38%)*

Product/ Type of animal	No. of animals producing or slaughtered in 2010 (x 1000)	Production [2010 index number compared to 2000]	Weight of milk/meat/ eggs per animal [2010 index number compared to 2000]	Offtake (% of animals producing or slaughtered out of total stock) [2010 index number compared to 2000]	Headcount [2010 index number compared to 2000]
Green: Most successful species and products Red: Problem species		Green >138 Turquoise 120–138 Red <100	Green >138 Turquoise 120–138 Red < 100	Green >138 Turquoise 120–138 Red <100	Green >138 Turquoise 120–138 Red < 100
Cow's milk	3,367	229	100	176	130
Hen's eggs	4,733	114	104	86	127
Chickens	45,167	135	100	106	127
Goats	2,750	131	98	96	140
Sheep	648	168	100	100	168
Pigs	1,895	146	100	100	146
Cattle	863	132	100	101	130

N.B. The index number of total production is the index for headcount multiplied by the two indices for 'yield' (offtake and weight per animal) and divided by 10,000.

SOURCE: FAOSTAT | © FAO STATISTICS DIVISION 2013. UPDATED 8 AUGUST 2013. HTTP://FAOSTAT.FAO.ORG/SITE/339/DEFAULT.ASPX - LIVE ANIMALS, LIVESTOCK PRIMARY AND LIVESTOCK PROCESSED. ACCESSED 19 SEPTEMBER 2013. [AS OF SEPTEMBER 2014: HTTP://FAOSTAT3.FAO.ORG/FAOSTAT-GATEWAY/GO/TO/DOWNLOAD/Q/*/E].

In the livestock sector, poultry was the most successful agricultural product with its related production of chicken meat and eggs.

Uganda

Major differences can be seen in Uganda's agricultural successes if crop production between 2000 (a three-year average for 1999–2001) and 2010 (a three-year average for 2009–2011) are considered. Tables 3.4 and 3.5 show crops and livestock

TABLE 3.6 *Uganda's most, less and least successful crops and livestock species, 2000–2010**

Yield in 2010 as a % of yield in 2000	Production in 2010 as a % of production in 2000		
	<100%	100–138%	>138%
>138%			**Cotton** **Rice** **Milk**
120–138%			**Maize** **Sesame** **Cowpeas** **Oilseed**
100–120%	Millet	Sweet potatoes Chicken (meat) Cows (meat)	Groundnuts Sunflower Soybeans Cocoa Pigs Sheep Fish[1]
<100%	Coffee Bananas	Plantains Dry beans Cassava Sorghum Pigeon peas Onions Eggs Goats	Vegetables (total) Vegetables[2] Irish potatoes Sugarcane

* In **bold**: most successful crops and livestock species.
[1] No information is available for fish that could serve to calculate an index of yield.
[2] Not elsewhere specified.
SOURCE: TABLES 3.4 AND 3.5.

data and Table 3.6 shows a matrix with growth in production and yield improvements for crops and information on livestock dynamics. Crops are presented in the order of magnitude of the area harvested in 2010. The top right-hand cell shows the most successful crops that saw total faster production growth than population growth over the decade (38%) and yield increases of more than 20%

in the same period. Maize, sesame, seed cotton, rice and cowpeas are crops that are relatively widely grown and are significant for Uganda, although oilseed is of less importance. Uganda's agricultural success is not particularly evident in the figures from other sources and the yield figures for maize need to be seriously questioned as there are large discrepancies between FAO and 'local' figures.

The most successful crops together represented 21% of Uganda's harvested crop area in 2010, with maize accounting for about 63%. There is, however, a reason why these crops should not be presented as 'miracle' crops. The successful crops, namely cotton, rice, maize, cowpeas and oilseed, are mostly those that experienced a 'peace dividend'. Maize, cotton and cowpeas, and to a lesser extent rice and oilseed, are cultivated and/or dominant in crop and food patterns in northern Uganda. After the 2006 ceasefire and the withdrawal of the Lord's Resistance Army (LRA) from the region, internally displaced people returned *en masse* to their land and this caused a major increase in the national production figures of these crops. This also applies to goats, which are the most affordable form of livestock (and a kind of savings account), especially in comparison with cattle, which are much more expensive.

Some livestock species show remarkable growth figures over the last decade: sheep and particularly pigs made up 16% of the total tropical livestock units in Uganda in 2010. Cattle are increasingly being kept for their milk and have shown a concomitant production increase. Freshwater fish production has more than doubled in the past decade as well. Uganda's 'problem crops', i.e. bananas and coffee with their negative production and yield trends between 2000 and 2010, accounted for only 2% of the area under harvest in 2010. There were no 'problem' livestock species. For details of the long-term trends of high-growth crops and livestock species, see Leliveld *et al.* (2013a).

Tanzania

Many crops in Tanzania have seen remarkable increases in production in the last decade. Three crops with more than 500,000 ha in 2010 demonstrated particularly high growth figures between 2000 and 2010: sweet potatoes, groundnuts and sunflowers. Table 3.7 classifies all the major crops (with more than 45,000 ha in 2010, in decreasing order) on two indexes: those for production and yield since 2000. If the production increase was higher than population growth (32% for the decade) *and* yield increases were more than 20% (and preferably higher than population growth), these crops were regarded as being very successful and worthy of further analysis. There are seven highly successful crops: sweet potatoes, groundnuts, bananas, coconut, cowpeas, pigeon peas and sesame. Sunflower and also 'other pulses' and tobacco have seen considerable increases in yields too, but they have

TABLE 3.7 *Performance of Tanzania's major crops, 2000–2010*[1] *(population growth 2000–2010:*[2] *32%)*

Crops > 45,000 ha in 2010 (highest acreage first)	Harvested area in 2010 (x 1000)	Production [2010 index number compared to 2000]	Yield [2010 index number compared to 2000]	Area [2010 index number compared to2000]
Green: Most successful crop Red: Problem crop		Green >132 Turquoise 120–132 Red <100	Green >132 Turquoise 120–132 Red < 100	Green >132 Turquoise 120–132 Red < 100
Maize	3100	176	53	330
Rice paddy	1020	262	101	255
Beans (dry)	938	143	105	140
Cassava	898	100	79	127
Sorghum	768	125	116	110
Coconut	671	155	114	135
Sweet potatoes	642	430	263	161
Groundnuts	529	440	154	309
Sunflower	525	343	131	250
Bananas	486	422	246	172
Seed cotton	389	148	110	141
Millet	357	157	95	165
Sesame	289	634	164	368
Vegetables[3]	282	164	83	197
Plantains	274	107	95	109
Cashew	245	78	41	270
Cowpeas	202	167	143	134
Pigeon peas	196	222	148	152
Coffee	195	107	77	158
Potatoes	191	216	87	250
Peas	165	251	105	239
Pulses (other)	114	224	130	168
Wheat	104	126	76	172
Tobacco	101	281	123	242
Sisal	57	107	82	131
Chickpeas	48	168	213	78

1 2000 = average of 1999–2001; 2010 = average of 2009–2011.

2 Estimates for the years 2000 and 2010 were used for population size.

3 Fresh vegetables other than tomatoes, onions, cabbages, chillies, peppers, garlic, green maize, green legumes and watermelons.

SOURCE: FAOSTAT | © FAO STATISTICS DIVISION 2013. UPDATED 8 AUGUST 2013. HTTP://FAOSTAT.FAO.ORG/SITE/339/DEFAULT.ASPX – CROPS. ACCESSED 17 SEPTEMBER 2013. [AS OF SEPTEMBER 2014: HTTP://FAOSTAT3.FAO.ORG/FAOSTAT-GATEWAY/GO/TO/DOWNLOAD/Q/QC/E].

TABLE 3.8 *Performance of Tanzania's main livestock, 2000–2010 (population growth 2000–2010: 32%)*

Product/Animal Green: Most successful species and products Red: Problem species	Number of animals producing or slaughtered in 2010 (x 1000)	Production* [2010 index number compared to 2000] Green >132 Turquoise 120–132 Red <100	Weight of milk/meat/eggs per animal [2010 index number compared to 2000] Green >132 Turquoise 120–132 Red < 100	Offtake (% of animals producing or slaughtered in total stock) [2010 index number compared to 2000] Green >132 Turquoise 120–132 Red <100	Headcount [2010 index number compared to 2000] Green >132 Turquoise 120–132 Red < 100
Cow's milk	6,867	226	170	117	114
Goat's milk	2,684	112	94	119	100
Hen's eggs	12,333	95	80	119	100
Duck's eggs	500	99	93	106	100
Chickens	60,033	131	110	119	100
Ducks	1,320	106	99	106	100
Cattle	2,743	107	102	117	90
Goats	2,770	113	95	119	100
Sheep	1,009	117	90	129	100
Pigs	350	111	101	110	100

* The index number of total production is the index for headcount multiplied by the two indices for 'yield' (offtake and weight per animal) and divided by 10,000.

SOURCE: FAOSTAT | © FAO STATISTICS DIVISION 2013. UPDATED 8 AUGUST 2013. HTTP://FAOSTAT.FAO.ORG/SITE/339/DEFAULT.ASPX - LIVE ANIMALS, LIVESTOCK PRIMARY AND LIVESTOCK PROCESSED. ACCESSED 19 SEPTEMBER 2013. [AS OF SEPTEMBER 2014: HTTP://FAOSTAT3.FAO.ORG/FAOSTAT-GATEWAY/GO/TO/DOWNLOAD/Q/*/E.]

TABLE 3.9 *Tanzania's most, less and least successful crops and livestock species, 2000–2010* *

Yield in 2010 as a % of yield in 2000	Production in 2010 as a % of production in 2000		
	<100%	100–138%	>138%
>138%			**Sweet potatoes** **Groundnuts** **Bananas** **Sesame** **Cowpeas** **Pigeon peas** **Chickpeas** **Cow's milk**
120–138%			**Sunflower** **Pulses**[1] **Tobacco**
100–120%		Sorghum Chicken meat Cattle meat Pork	Rice Beans Coconut Cotton Peas Fish[2]
<100%	Cashew Hen's eggs Duck's eggs	Wheat Cassava Plantains Coffee Sisal Goat's milk Duck meat Goat's meat Sheep meat	Maize Millet Vegetables[3] Potatoes

* In **bold**: most successful crops and livestock species.

1 Not elsewhere specified.

2 No information is available for fish that could serve to calculate an index of yield.

3 Fresh vegetables other than tomatoes, onions, cabbages, chillies, peppers, garlic, green maize, green legumes and watermelons.

SOURCE: TABLES 3.7 AND 3.8.

TABLE 3.10 *Performance of Kenya's major crops 2000–2010*[1] *(>45,000 ha in 2010) (population growth 2000–2010:*[2] *29.6%)*

Crop > 45,000 ha in 2010 (highest acreage first) Green: Most successful crop Red: Problem crop	Ha in 2010 (x 1000)	Production [2010 index number compared to 2000] Green >130 Turquoise 120–130 Red <100	Yield [2010 index number compared to 2000] Green >130 Turquoise 120–130 Red < 100	Area [2010 index number compared to 2000] Green >130 Turquoise 120–130 Red < 100
Maize	2008	128	100	128
Beans (dry)	896	136	125	109
Sorghum	218	137	84	163
Tea	173	140	98	143
Cowpeas (dry)	164	151	109	141
Coffee, (green)	160	60	64	94
Wheat	141	149	133	109
Pigeon peas	139	112	126	86
Potatoes	122	261	248	106
Millet	105	122	110	110
Pulses[3]	93	88	65	137
Oilseed[4]	89	112	109	103
Sweet potatoes	74	150	120	125

Bananas	72	139	146	95
Vegetables (fresh)[5]	67	102	114	90
Sugarcane	66	140	110	127
Cassava	64	107	113	94
Coconut	49	112	39	284
Mangoes, mangosteens, guavas	48	387	132	302

1 2000 = average for 1999–2001; 2010 = average for 2009–2011.

2 For population size, estimates for 2000 and 2010 were used.

3 Pulses other than beans (and not shown elsewhere).

4 Oilseed other than sesame, sunflower, seed cotton, groundnuts, linseed and soybeans (and not shown elsewhere).

5 Fresh vegetables other than tomatoes, onions, cabbages, chillies, peppers, garlic, green maize, green legumes and watermelons.

SOURCE: FAOSTAT | © FAO STATISTICS DIVISION 2013. UPDATED 8 AUGUST 2013. HTTP://FAOSTAT.FAO.ORG/SITE/339/DEFAULT.ASPX – CROPS. ACCESSED 17 SEPTEMBER 2013. [AS OF SEPTEMBER 2014: HTTP://FAOSTAT3.FAO.ORG/FAOSTAT-GATEWAY/GO/TO/DOWNLOAD/Q/QC/E].

TABLE 3.11 *Performance of Kenya's main livestock, 2000–2010 (population growth 2000–2010: 29.6%)*

Product/Animal Green: Most successful species and products Red: Problem species	Number of animals producing or slaughtered in 2010 (x 1000)	Production [2010 index number compared to 2000] Green >130 Turquoise 120–130 Red <100	Weight of milk/meat/eggs per animal [2010 index number compared to 2000] Green >130 Turquoise 120–130 Red < 100	Offtake (% of animals producing or slaughtered from total stock) [2010 index number compared to 2000] Green >130 Turquoise 120–130 Red <100	Headcount [2010 index number compared to 2000] Green >130 Turquoise 120–130 Red < 100
Cow's milk	5,545	170	135	85	148
Goat's milk	3,400	175	97	81	223
Camel's milk	50	124	100	57	219
Sheep's milk	1,538	102	100	47	219
Hen's eggs	18,300	153	112	101	134
Chickens	17,500	156	102	113	134
Camels	83	110	100	50	219
Goats	4,300	118	100	53	223
Sheep	2,900	141	123	62	184
Pigs	195	123	98	131	96
Cattle	3,050	172	110	103	148

N.B. The index number of the total production is the index for headcount multiplied by the two indices for 'yield' (offtake and weight per animal) and divided by 10,000.

SOURCE:FAOSTAT | © FAO STATISTICS DIVISION 2013. UPDATED 8 AUGUST 2013. HTTP://FAOSTAT.FAO.ORG/SITE/339/DEFAULT.ASPX - LIVE ANIMALS, LIVESTOCK

TABLE 3.12 *Kenya's most and least successful crops and livestock species, 2000–2010**

Yield in 2010 as a % of yield in 2000	Production in 2010 as a % of production in 2000		
	<100%	100–130%	>130%
>130%		Pork	**Wheat** **Potatoes** **Bananas** **Mangoes**
120–130%		Pigeon peas	**Beans** **Sweet potatoes**
100–120%		Maize Millet Vegetables Oilseed[2] Cassava Marine fish[3]	Cowpeas Sugarcane Chicken Hen's eggs Cattle Cow's milk Freshwater fish[3]
<100%	Coffee Pulses[1]	Coconut Camel milk CamelsGoats Sheep's milk	Sorghum Tea Goat's milk Sheep

* In **bold**: most successful crops and livestock species.
1 Fresh vegetables other than tomatoes, onions, cabbages, chillies, peppers, garlic, green maize, green legumes and watermelons.
2 Not elsewhere specified.
3 No information is available for fish that could serve to calculate an index of yield.
SOURCE: TABLES 3.10 AND 3.11.

been lower than the population growth figures. Table 3.8 shows the same details for Tanzanian livestock: none of the livestock species experienced a growth in numbers higher than the population growth over the decade. Table 3.9 summarizes the findings presented in Tables 3.7 and 3.8. Details of the long-term trends of high-growth crops can be found in Leliveld *et al.* (2013b).

Kenya

If Kenya's crop production for 2000 (the average for the years 1999–2001) and 2010 (the average for the years 2009–2011) is considered, major differences in agricultural success can be seen. Tables 3.10 and 3.11 show the data for crops and livestock and Table 3.12 presents a matrix that classifies them according to growth in production by yield improvement for crops or similar information for livestock. The cells in Table 3.12 give the crops in order of magnitude of area harvested in 2010, with the top right-hand cell showing the most promising crops. For these crops, total growth in production was faster than population growth over the decade (30%) and yield increased by 20% or more. The following are the most successful crops for Kenya: beans, wheat, potatoes, sweet potatoes, bananas and mangoes and they covered a relatively high number of hectares (ca. 23% of Kenya's total harvested area in 2010). As for livestock and livestock products, the promising ones that saw increases in production of over 30% over the decade have only modest or even negative yield increases, as in the other three research countries as well. For details of these crops and livestock species, see Dietz *et al.* (2013). Some of these figures can, however, be contested. The very high yield increases for potatoes, for instance, are due to a sudden increase in the mid-2000s that other sources do not accept. A recent study by Janssens *et al.* (2013: 53) apparently does not believe FAOSTAT's sudden more than doubling of calculated figures of potato yield from 8 to 20 ton/ha in 2005 and assumed instead a gradual more than halving of potato yield since 2005 down to 2.9 ton/ha in 2010, only one seventh of the FAO value.

TABLE 3.13 *Comparison of agricultural high-growth products in all four countries studied, 2000–2010*

Agricultural product	Nigeria PG: 28%			Uganda PG: 38%			Tanzania PG 32%			Kenya PG: 30%		
	T	P	Y	T	P	Y	T	P	Y	T	P	Y
Grains												
Maize	++	171	130	+*	219	136	+-	176	53	*o	128	100
Rice	++	135	129	++	207	170	+o	262	101	<		
Wheat	<			<			*-	126	76	++	387	132
Pulses												
Beans (dry)	<			o-	102	75	+o	143	105	+*	136	125
Cowpeas	o+	118	155	+*	145	131	++	167	143	+o	151	109

Agricultural product	Nigeria PG: 28%			Uganda PG: 38%			Tanzania PG 32%			Kenya PG: 30%		
Pigeon peas	<			o-	118	96	++	222	148	o*	112	126
Other pulses	<			<			+*	224	130	<		
Roots and tubers												
Cassava	++	136	132	o-	101	94	o-	100	79	oo	107	113
Potatoes	++	181	148	+-	151	98	+-	216	87	++	261	248
Sweet potatoes	+o	134	100	oo	112	106	++	430	263	+*	150	120
Other food crops												
Sesame seed	+-	197	92	+*	198	128	++	634	164	<		
Oil-palm seeds	<			+*	157	136	++	430	263	oo	112	109
Groundnuts	o-	115	82	+o	204	105	++	440	154	<		
Sunflower seeds	<			+o	319	118	++	343	131	<		
Melon seed	++	138	166	<			<			<		
Pineapples	++	143	130	<			<			<		
Bananas	<			--	96	92	++	422	246	++	139	146
Mangoes	<			<			*o	126	115	++	387	132
Non-food crops												
Seed cotton	o+	106	161	++	139	335	+o	148	110	<		
Livestock products												
Cattle		115		*	132			107		+	172	
Cow's milk	+	132		+	229		+	226		+	170	
Goats	+	130		*	131			112			118	
Sheep	+	141		+	168			117		+	141	
Pigs	+	147		+	146			111		*	123	
Chickens	+	163		*	135		*	131		+	156	
Hen's eggs	+	147			114		-	95		+	153	

Explanation of symbols and colours: P=production; Y=yield; T=typology of P and Y together
+ production or yield growth faster than population growth (PG)
* production or yield growth between 20% and PG
o production or yield growth between 0% and 20%
- production or yield growth below 0
< minor crop in this country

Comparison

Table 3.13 highlights all the crops and livestock products that can be defined as 'agricultural high-growth' (or 'most successful') products in at least one of the four research countries. The growth index figures are shown for total production and for average crop yields between 1999–2001 and 2009–2011.

If these four African countries are compared, it is important to first look at population dynamics. These differ considerably. Uganda has seen explosive population growth (38%) over the last decade, while the other three countries grew considerably less rapidly but still had high growth: Tanzania 32%, Kenya 30% and Nigeria 28%.

Another contextual comparison that is relevant is each country's relative potential food security. The assessment presented in Table 3.14 assumes that staple foods would provide (on average) 1430 kcal/cap/day, i.e. around 65% of all food requirements if average daily food requirements were 2200 kcal/cap/day. The 'potential' in food security means that only national production figures are considered and food harvests used as seed and feed, food exports, imports, waste and stock movements are ignored. The same goes for the differences between areas and between wealth and other categories within the four countries. However, these differences are considerable and can explain the hunger and under-five undernutrition situation in some countries and in periods when the total figures suggest there has been food sufficiency. For a more detailed explanation of earlier years, see Akinyoade *et al.* (2013) for Nigeria; Leliveld *et al.* (2013a) for Uganda; Leliveld *et al.* (2013b) for Tanzania; and Dietz *et al.* (2014) for Kenya. Potential food sufficiency based on national food production can be regarded as having been sufficient in three of the four countries in both 2000 and 2010. Kenya's figures show a more problematic situation, although improvements can be noted there between 2000 and 2010.

TABLE 3.14 *Potential food security in Nigeria, Uganda, Tanzania and Kenya, 2000–2010*

	Food coverage in kcal/cap/day		**Potential food sufficiency (%)**	
Country	2000	2010	2000	2010
Nigeria	2846	2720	199	190
Uganda	2782	2328	195	163
Tanzania	1580	2251	110	157
Kenya	974	1255	68	88

The list of most successful agricultural products includes all the products for which the particular crop or livestock product was found to be a high-growth product (according to the definitions used here) in at least one of the four countries. But this does not necessarily mean that it has been a high-growth product in all four of the countries under study. The yield developments of the crops on the list between 1999–2001 and 2009–2011 in the four countries are compared below and the findings interpreted for the various crop groups (Table 3.15).

TABLE 3.15 *Crop yields (kg/ha) in 2000 and 2010 for most successful crops in at least one of the four countries studied*

Crop	Nigeria		Uganda		Tanzania		Kenya	
	2000	2010	2000	2010	2000	2010	2000	2010
Maize	1433	1858	1758	2400	2532	1332	1541	1534
Rice	1431	1847	1470	2496	1980	1999	3722	3380
Wheat	1466	1389	1766	1701	1200	912	1731	2302
Beans (dry)	-	-	633	477	803	842	428	536
Cowpeas	600	928	969	1247	731	1044	407	442
Pigeon peas	-	-	1000	958	650	965	439	551
Pulses (other)	567	407	-	-	677	882	378	220
Cassava	9634	12,669	12,961	12,165	7124	5654	8308	9375
Potatoes	2815	3554	7001	6877	7881	6857	8159	20,239
Sweet potatoes	3000	2889	4362	4641	1457	3831	9473	11,401
Sesame	480	442	501	640	405	685	388	415
Oil-palm seeds	-	-	686	931	657	998	391	411
Groundnuts	1517	1251	700	733	593	914	1745	736
Sunflower	-	-	959	1244	651	851	1023	1026
Melon seed	600	996	-	-	-	-	-	-
Pineapples	6142	6990	8562	8207	7630	3805	46,345	34,767
Bananas	-	-	4531	4168	2624	6460	14,130	20,678
Mangoes	5594	6747	-	-	8719	10,058	9576	12,641
Seed cotton	742	1191	269	901	565	627	530	514

The Most Successful Cereal Crops and Pulses

Maize has become the most important food crop in each of the four countries studied. In Nigeria, and to a lesser degree in Uganda, yield increases have been high over the last decade but they have dropped in Tanzania from a relatively high level, and yields have barely stabilized in Kenya. Yield levels are currently highest in Uganda, followed by Nigeria. Average yield levels in Tanzania have dropped to alarmingly low levels and are significantly lower than previously recorded. With world averages much higher than the highest recorded yield figures in any of the four countries (close to 2500 kg/ha in Tanzania in 2000 and currently at this level in Uganda), further improvements are possible. For rice, Nigeria and Uganda are the leaders in yield increases, but here Kenya still leads on average yield levels, although these are lower than before. Most of Kenya's rice comes from its well-managed irrigated schemes and this may explain the relatively high levels seen there. The recent performance of these irrigation schemes should be studied to determine why yield levels have been deteriorating. It is also interesting to study the very diverse performance of wheat, where only Kenya seems to be doing well. Yield levels in the other countries have all declined. Among the pulses, the most interesting crop to study would be cowpeas as yield levels have improved everywhere, particularly in Uganda and Tanzania. For pigeon peas, dry beans and other pulses, the situation is more diverse and generally less encouraging.

The Most Successful Root Crops and Tubers

Potatoes could be a particularly interesting crop for further study. Although experts seriously question current FAO figures, their yield levels were good in Kenya up to 2004 and yields have increased in Nigeria (even though they are much lower there) but they have declined in Uganda and Tanzania. Cassava yields did well in Nigeria where it is a very important crop both as a staple food and as animal feed. Sweet potatoes did well in Tanzania and Kenya where very high yield levels were recorded, but did less well in Nigeria.

The Most Successful Other Crops

Mangoes, oil-palm seeds and sunflower seeds have shown steady increases in yield in all the countries and would be interesting crops for further comparative study. Bananas are displaying another trend, with ever-higher yields in Tanzania and especially in Kenya, but declining (and even low) yields in Uganda. Yields of pineapples, which are mostly a plantation crop, are increasing in Nigeria, but the situation is not as good elsewhere.

A Final Note: The Way Forward with Policy-Oriented Research

Future research could involve investigating the factors behind the relatively successful agricultural production rates in the four countries studied in this chapter to see why they have been so successful over the last decade. Was this due to market expansion, institutional arrangements, such as value-chain and agro-support institutions including business development, and/or state support? It would be good to compare the circumstances surrounding agricultural production as it has become blatantly clear that the successful agricultural products are very country-specific. They cannot be regarded as an overall success story in all four countries.

Market expansion has mainly involved expansion of the internal markets in the four countries. According to FAOSTAT data, very little food is exported, even though crops and livestock are regularly traded (largely unrecorded) across borders. And countries like Kenya have become famous for their successful horticultural exports although this is a very limited percentage of the country's total food production. As everywhere in Africa, the urban populations in all four countries are rapidly increasing even though the current level of urbanization is still relatively low. The last ten years have seen the economies – and particularly the urban economies – in all four research countries improve considerably in spite of some ups-and-downs. For example, Kenya had major problems between 2007 and 2009 as a result of its post-election violence. Urban consumers are gradually placing increased demands on their urban hinterlands today and cities are becoming markets that demand growth in agricultural production and innovation (see Maps 3.1 and 3.2). Food insufficiency in nearby countries, like South Sudan and Somalia (for Kenya and Uganda), and Rwanda and Burundi (for Tanzania), can be expected to result in increased demands for the agricultural produce of the four countries studied here (Map 3.1).

A study of relevant institutional arrangements for agriculture in general (input support, quality assurance, education/training/extension support including the role of the main extension delivery service, credit, marketing support, logistical improvements, etc.) and for the most successful agricultural products in particular would be beneficial. Government-based institutions are important in all four of the countries studied but the private sector, which tends to be locally owned but may have some foreign influence, has become important as well.

The following questions have been formulated to guide systematic follow-up research:

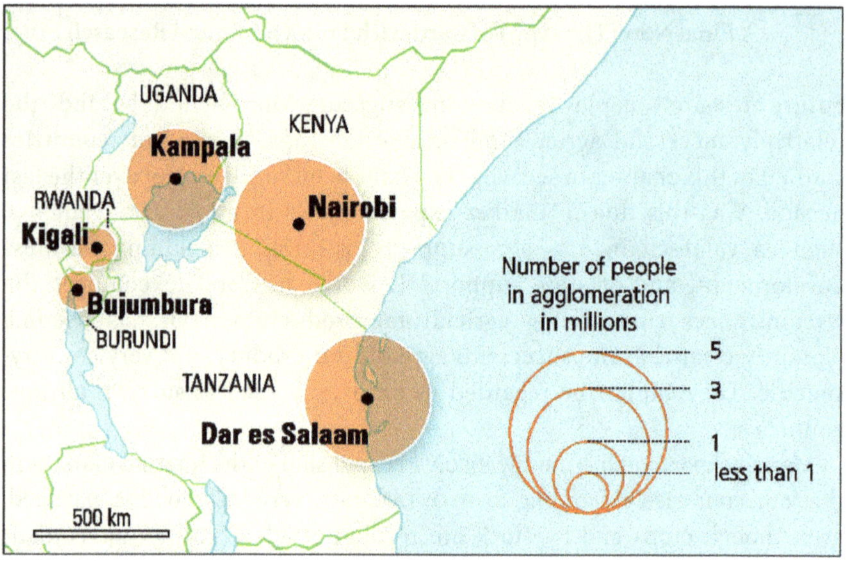

MAP 3.1 *Urbanization in Kenya, Uganda, Tanzania and neighbouring countries*
SOURCE: 'AFRICA: FROM A CONTINENT OF STATES TO A CONTINENT OF CITIES'. ASC THEMATIC MAP. HTTPS://OPENACCESS.LEIDENUNIV.NL/HANDLE/1887/20017. ACCESSED 23 AUGUST 2014.

- What are the value chains for successful agricultural products like? Which are the main production and consumption areas and how are they linked? Who are the major stakeholders in the chain?
- Which are the main supporting agencies and institutions (government, business and/or others) and how do they assess the performance of successful agricultural products?
- What are the local, national and international elements in the chain of innovation and how are they related?
- What have the major incentives and disincentives been in recent production and yield increases according to farmers and stakeholders in the production-consumption chains?

To understand the link between potential food sufficiency, average food security and the nutritional impact of food expansion, access to these most successful commodities by the poorest quintile of food consumers in the country could be investigated. Future research might include analysis of explicit government poverty-alleviation policies and other relevant policy agencies in the four countries in general terms and also of the most successful commodities. It is important to see how and where agricultural products produced in these four countries are being exported to as well. Is Africa's food industry part of a new scramble for the continent's resources? And what is the recent history of food imports? Where do they come from and what role do policy and policy

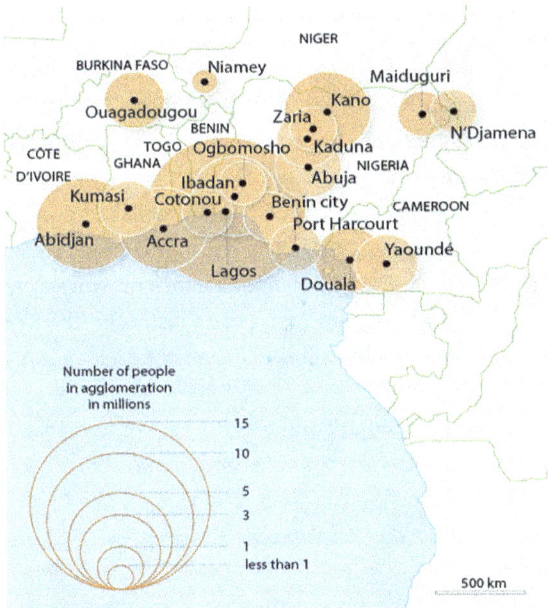

MAP 3.2 *Cities and hinterlands in the wider West African region*
SOURCE: 'AFRICA: FROM A CONTINENT OF STATES TO A CONTINENT OF CITIES'. ASC THEMATIC MAP. HTTPS://OPENACCESS.LEIDENUNIV.NL/HANDLE/1887/20017. ACCESSED 23 AUGUST 2014.

formulation and implementation play regarding imports and exports and also investment incentives and disincentives?

Recognition of the link between urbanization, food provisioning and agricultural innovation is growing. There is an increasing body of knowledge relating to urban agriculture in Africa, most of it starting in the 1990s (e.g. Duchhart & Grootenhuis 1993; Obudho & Foeken 1999; De Zeeuw *et al.* 2000). Nairobi and Nakuru are prominent as case-study cities in Kenya (e.g. Freeman 1991; Lee-Smith & Memon 1994; Mwangi & Foeken 1996; Foeken & Mwangi 2000; Foeken & Owuor 2006) but attention is also being paid to cities in Nigeria (e.g. Gbadegesin 1991; Gefu 1992; Ajaegbu, Grossman & van den Berg 2000), in Uganda (e.g. Maxwell 1995; Atakunda & Maxwell 1996) and in Tanzania (e.g. Jacobi, Amend & Kiango 2000; Foeken, Sofer & Mlozi 2004). Another type of study is when researchers consider how urban households access a proportion of their food through their linkages with rural households or by (part-time) farming themselves in the rural areas they originally come from (e.g. Foeken & Owuor 2001). But for the purposes of our proposed study design, it is important to include all the relevant (rural and urban) food production and (urban) food consumption linkages and to do so by using a political-economy approach to agricultural innovation clusters of the most relevant agro-food value chains. Studies like Satterthwaite, McGranahan and Tacoli (2010) can lead the way in this respect.

References

Ajaegbu, H.I., D. Grossman & L.M. van den Berg (2000), *Market Gardening, Urban Growth and Sustainable Income Generation on the Jos Plateau, Nigeria*. Amsterdam: Royal Tropical Institute.

Akinyoade, A., T. Dietz, D. Foeken & W. Klaver in collaboration with A. Leliveld, H. Smits, S. Soeters & M. van 't Wout (2013), Agricultural Dynamics and Food Security Trends in Nigeria. Developmental Regimes in Africa (DRA) Project. ASC-AFCA Collaborative Research Group Agro-Food Clusters in Africa (AFCA). Research Report 2013-ASC-1. London/Leiden. https://openaccess.leidenuniv.nl/handle/1887/22942. Accessed 23 August 2014.

Atakunda, G. & D. Maxwell (1996), Farming in the City of Kampala: Issues in Urban Management. *African Urban Quarterly* 11(2–3): 261–272.

Berendsen, B., T. Dietz, H. Schulte Nordholt & R. van der Veen (2013), *Asian Tigers, African Lions: Comparing the Development Performance of South East Asia and Africa*. Leiden: Brill.

De Zeeuw, H., S. Gruendel & H. Waibel (2000), The Integration of Agriculture in Urban Policies. In: N. Bakker *et al.* (eds), *Growing Cities, Growing Food. Urban Agriculture on the Policy Agenda*. Feldafing: Deutsche Stiftung für internationale Entwicklung (DSE), pp. 161–180.

Dietz, T. (2013), Comparing the Agricultural Performance of Africa and South East Asia over the Last Fifty Years. In: B. Berendsen, T. Dietz, H. Schulte Nordholt & R. van der Veen (eds), *Asian Tigers, African Lions: Comparing the Development Performance of South East Asia and Africa*. Leiden: Brill, pp. 85–130.

Dietz, T., D. Foeken, S. Soeters & W. Klaver in collaboration with A. Akinyoade, A. Leliveld, H. Smits & M. van't Wout (2014), Agricultural Dynamics and Food Security Trends in Kenya. Developmental Regimes in Africa (DRA) Project. ASC-AFCA Collaborative Research Group Agro-Food Clusters in Africa (AFCA). Research Report 2013-ASC-4. London/Leiden. https://openaccess.leidenuniv.nl/handle/1887/23291. Accessed 23 August 2014.

Duchhart, I. & F. Grootenhuis (1993), Urban Agriculture, Incentive Planning and Green Towns. *Urban Perspectives* 4(1): 9–10.

Foeken, D. & A.M. Mwangi (2000), Increasing Food Security through Urban Farming in Nairobi. In: N. Bakker *et al.* (eds), *Growing Cities, Growing Food. Urban Agriculture on the Policy Agenda*. Feldafing: Deutsche Stiftung für internationale Entwicklung (DSE), pp. 303–328. https://openaccess.leidenuniv.nl/handle/1887/4670. Accessed 23 August 2014.

Foeken, D. & S.O. Owuor (2001), Multi-spatial Livelihoods in Sub-Saharan Africa: Rural Farming by Urban Households: The Case of Nakuru Town, Kenya. In: M. de Bruijn, R. van Dijk & D. Foeken (eds), *Mobile Africa: Changing Patterns of Movement in Africa and Beyond*. Leiden: Brill, pp. 125–140. https://openaccess.leidenuniv.nl/handle/1887/4676. Accessed 23 August 2014.

Foeken, D. & S.O. Owuor (2006), *'To Subsidise My Income': Urban Farming in an East-African Town*. Leiden: Brill. https://openaccess.leidenuniv.nl/handle/1887/14742. Accessed 23 August 2014.

Foeken, D., M. Sofer & M. Mlozi (2004), *Urban Agriculture in Tanzania: Issues of Sustainability*. Leiden: African Studies Centre. https://openaccess.leidenuniv.nl/handle/1887/4678. Accessed 23 August 2014.

Freeman, D.B. (1991), *A City of Farmers: Informal Urban Agriculture in the Open Spaces of Nairobi, Kenya*. Montreal: McGill University Press.

Gbadegesin, A.S. (1991), Farming in the Urban Environment of a Developing Nation. A Case Study from Ibadan Metropolis in Nigeria. *The Environmentalist* 11: 105–111.

Gefu, J.O. (1992), Part-time Farming as an Urban Survival Strategy: A Nigerian Case Study. In: J. Baker & P.O. Pedersen (eds), *The Rural-Urban Interface in Africa: Expansion and Adaptation*. Uppsala: Scandinavian Institute of African Studies, pp. 295–302.

Henley, D. & J.K. van Donge (2012), *Policy for Development in Africa: Learning from South East Asia*. London: Developmental Regimes in Africa Policy Brief 01. http://www.institutions-africa.org/filestream/20120318-dra-policy-brief-01. Accessed 23 August 2014.

Jacobi, P., J. Amend & S. Kiango (2000), Urban Agriculture in Dar es Salaam: Providing an Indispensable Part of the Diet. In: N. Bakker *et al.* (eds), *Growing Cities, Growing Food. Urban Agriculture on the Policy Agenda*. Feldafing: Deutsche Stiftung für internationale Entwicklung (DSE), pp. 257–283.

Janssens, S.R.M., S.G. Wiersema, H. Goos & W. Wiersma (2013), The Value Chain for Seed and Ware Potatoes in Kenya. Opportunities for Development. LEI Memorandum 13-080, August. The Hague: WUR-LEI. http://edepot.wur.nl/279729. Accessed 23 August 2014.

Lee-Smith, D. & P.A. Memon (1994), Urban Agriculture in Kenya. In: A.G. Egziabher *et al.*, *Cities Feeding People: An Examination of Urban Agriculture in East Africa*. Nairobi/Ottawa: IDRC, pp. 67–84. http://www.idrc.ca/EN/Resources/Publications/openebooks/706-x/index.html#page_63. Accessed 23 August 2014.

Leliveld, A., T. Dietz, D. Foeken & W. Klaver in collaboration with A. Akinyoade, H. Smits, S. Soeters & M. van't Wout (2013a), Agricultural Dynamics and Food Security Trends in Uganda. Developmental Regimes in Africa (DRA) Project. ASC-AFCA Collaborative Research Group Agro-Food Clusters in Africa (AFCA). Research Report 2013-ASC-2. London/Leiden. https://openaccess.leidenuniv.nl/handle/1887/22943. Accessed 23 August 2014.

Leliveld, A., T. Dietz, W. Klaver, B. Kilama & D. Foeken in collaboration with A. Akinyoade, H. Smits, S. Soeters & M. van't Wout (2013b), Agricultural Dynamics and Food Security Trends in Tanzania. Developmental Regimes in Africa (DRA) Project. ASC-AFCA Collaborative Research Group Agro-Food Clusters in Africa

(AFCA). Research Report 2013-ASC-3. London/Leiden. https://openaccess.leidenuniv.nl/handle/1887/22944. Accessed 23 August 2014.

Maxwell, D. (1995), Alternative Food Security Strategy: A Household Analysis of Urban Agriculture in Kampala. *World Development* 23(10): 1669–1681.

Mwangi, A.M. & D. Foeken (1996), Urban Agriculture, Food Security and Nutrition in Low-Income Areas in Nairobi, Kenya. *African Urban Quarterly* 11(2–3): 170–179. https://openaccess.leidenuniv.nl/handle/1887/4667. Accessed 23 August 2014.

Obudho, R.A. & D. Foeken (1999), *Urban Agriculture in Africa. A Bibliographical Survey*. Leiden/Nairobi: African Studies Centre/Centre for Urban Research. https://openaccess.leidenuniv.nl/handle/1887/387. Accessed 23 August 2014.

Satterthwaite, D., G. McGranahan & C. Tacoli (2010), Urbanization and its Implications for Food and Farming. *Philosophical Transactions of the Royal Society B: Biological Sciences* 365(1554): 2809–2820. http://rstb.royalsocietypublishing.org/content/365/1554/2809.full. Accessed 23 August 2014.

Vlasblom, D. (2013), *The Richer Harvest. Economic Development in Africa and South East Asia Compared*. Leiden: African Studies Centre. https://openaccess.leidenuniv.nl/handle/1887/20579. Accessed 23 August 2014.

CHAPTER 4

Food Production and Consumption in Relation to Food Insecurity and Undernutrition in Kenya, Nigeria, Tanzania and Uganda

Wijnand Klaver

This chapter complements the results presented in the previous chapter of the comparison of trends in food production in Kenya, Nigeria, Tanzania and Uganda over the past 10–50 years by analyzing the subject from a nutritional angle. It looks at the consumption perspective, using the FAO's Food Balance Sheet (FBS) approach. Food availability for human consumption is a vital determinant of food security and FBS data show how much food is available for human consumption in a country after corrections have been made for imports and exports, changes in the volume of food stocks and other uses. The FAO expresses the final results of the FBS in terms of dietary energy (kcal), g protein and g fat per capita. Per capita statistics allow comparisons over time which corrects for population growth. Dietary energy is used in this chapter as a common unit for the different components of the FBS. Two nutritional outcome indicators are considered: the Prevalence of Undernourishment (PoU), which is the result of the distribution of available food supplies among the population; and the Prevalence of Anthropometric Failure (inadequate child growth) among the under-fives, based on the results of national Demographic and Health Surveys (DHS).

Introduction

This chapter focuses on a comparison of trends over the past 10–50 years in the availability of food for human consumption, the prevalence of undernourishment (derived from the total amount of food available) and the nutritional status of young children in Kenya, Nigeria, Tanzania and Uganda (Dietz *et al.* 2013; Akinyoade *et al.* 2013; Leliveld *et al.* 2013a; Leliveld *et al.* 2013b). It thus complements the analyses of food production in the previous chapter in this volume.

The UN's Food and Agriculture Organization (FAO) and the World Health Organization (WHO) are hosting the second International Conference on Nutrition 21 years later (ICN + 21) entitled 'Better Nutrition, Better Lives' in

 | DOI 10.1163/9789004282698_005

Rome on 19–21 November 2014. FAO/WHO member countries will be represented by top-level officials from the health and food and agriculture sectors and will consider the progress made in terms of political commitment and operational capacity since the 1992 International Conference on Nutrition (ICN) that unanimously adopted a World Declaration and Plan of Action for Nutrition. The 2014 conference will propose a flexible policy framework to address the major nutrition challenges of the coming decades. Country papers (FAO/WHO 2012) will be an essential input at the conference that will also identify priorities for cooperation on nutrition in the near and medium term.

In the run-up to ICN + 21, the FAO has provided an international set ('*suite*') of core indicators of food insecurity to assess aspects of food insecurity across regions and over time (FAO 2013). These indicators are in line with four pillars of food security – availability, access, utilization and stability. To facilitate their interpretation, they are also classified as indicators of determinants or of outcomes of food insecurity. Determinants, which can be either static or dynamic, refer to structural conditions that worsen food insecurity in the absence of adequate policy interventions, including emergency assistance. Outcome indicators capture results in terms of inadequate food consumption or anthropometric failure. Appendix A (this volume) provides time series information on a number of major indicators for different world regions and also for Kenya, Nigeria, Tanzania and Uganda.

Two of the basic determinants of availability are Dietary Energy Supply Adequacy and Share of Dietary Energy Supply Derived from Cereals, Roots and Tubers. The first represents the quantitative dimension of food availability, while the latter is a way of presenting the monotony of the diet because when people's incomes rise, they tend to consume a more varied diet. This will include more non-staple foods like vegetables, nuts, fruit and animal products, while the percentage of staple foods they consume decreases.

One of the outcomes of access is the Prevalence of Undernourishment (PoU), the food-based estimated of the percentage of the population that is chronically hungry, while the percentage of children under five with anthropometric failure (i.e. who are stunted, wasted and/or underweight, respectively) is an outcome indicator for the utilization of food. This chapter presents current and trend data on these indicators for four case-study countries: Kenya, Nigeria, Tanzania and Uganda (see also Chapter 3, this volume). It starts by looking at food availability (as the net result of its various determinants) and then considers the outcomes in terms of the PoU and the percentage of under-fives with anthropometric failure. Particular interest is devoted to the extent to which anthropometric failure reflects the trend in food-based indicators.

National Food Availability

Food production and food consumption are connected and the two perspectives are also connected in the FAO's Food Balance Sheet (FBS) approach. FBS data show the amount of food in a country in a year that is available for human consumption after correcting for a number of 'disappearances': imports and exports, changes in the volume of food stocks and other destinations (seed, animal feed, processing and waste).[1] Results are expressed in terms of dietary energy (kcal), g protein and g fat per capita. By expressing the results in per capita terms, it is possible to make comparisons over time wich are corrected for population growth. This chapter uses kcal per capita as the common unit to express all the different components of the FBS. As a result, this gives an indirect estimate of the annual food balance, namely the amount of food available for consumption at retail level. Figures 4.1a–4.1d illustrate the food supply and utilization accounts for Kenya, Nigeria, Tanzania and Uganda in 1000 kcal per capita per year. To reduce errors due to problems in recording annual food stock changes, three-year averages have been used. The annual food balance is depicted in the first bar below the zero line. In general, the food balance follows domestic agricultural production but there are also periods with, for example, higher food imports or with less waste or disappearance that tend to cushion the effects of fluctuations in domestic food production on the country's food supply.

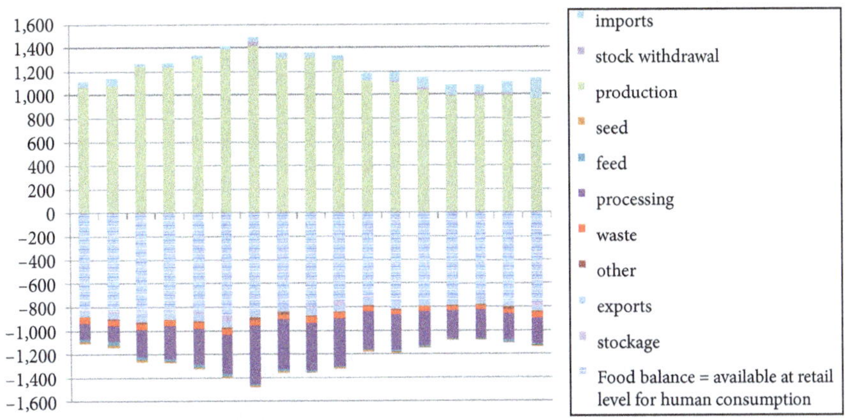

FIGURE 4.1A *Total food supply and utilization (all foods of plant and animal origin), 1961–2009, three-year averages (x 1000 kcal per capita per year) for Kenya*
SOURCE: FAOSTAT FOOD BALANCE SHEETS | © FAO STATISTICS DIVISION 2013. UPDATED: 29 JUNE 2012. HTTP://FAOSTAT.FAO.ORG/SITE/368/DESKTOPDEFAULT.ASPX?PAGEID=368. ACCESSED 24 JUNE 2013. [AS OF SEPTEMBER 2014: HTTP://FAOSTAT3.FAO.ORG/FAOSTAT-GATEWAY/GO/TO/DOWNLOAD/FB/FBS/E].
Legend: see Figure 4.1d.

1 Animal feed and processing can mean indirect food availability but this can also be (partly) exported.

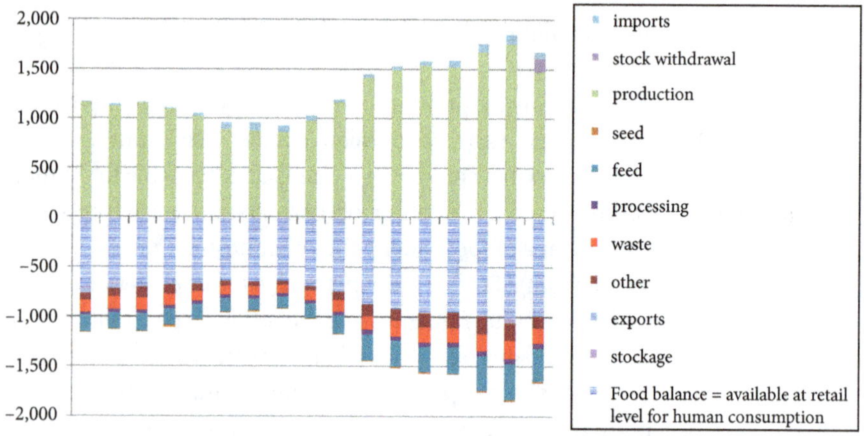

FIGURE 4.1B *Total food supply and utilization (all foods of plant and animal origin), 1961–2009, three-year averages (x 1000 kcal per capita per year) for Nigeria*
SOURCE: FAOSTAT FOOD BALANCE SHEETS | © FAO STATISTICS DIVISION 2013. UPDATED: 29 JUNE 2012. HTTP://FAOSTAT.FAO.ORG/SITE/368/DESKTOPDEFAULT.ASPX?PAGEID=368. ACCESSED 23 APRIL 2013. [AS OF SEPTEMBER 2014: HTTP://FAOSTAT3.FAO.ORG/FAOSTAT-GATEWAY/GO/TO/DOWNLOAD/FB/FBS/E]
Legend: see Figure 4.1d.

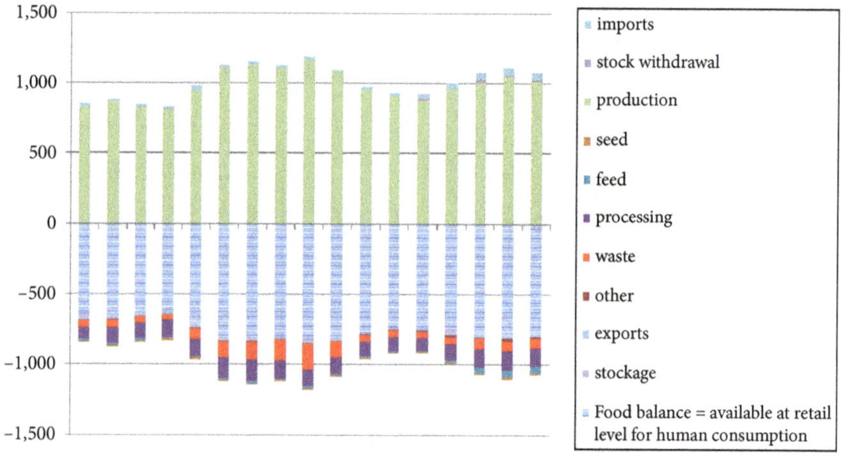

FIGURE 4.1C *Total food supply and utilization (all foods of plant and animal origin), 1961–2009, three-year averages (x 1000 kcal per capita per year) for Tanzania*
SOURCE: FAOSTAT FOOD BALANCE SHEETS | © FAO STATISTICS DIVISION 2013. UPDATED: 29 JUNE 2012. HTTP://FAOSTAT.FAO.ORG/SITE/368/DESKTOPDEFAULT.ASPX?PAGEID=368. ACCESSED 1 JULY 2013. [AS OF SEPTEMBER 2014: HTTP://FAOSTAT3.FAO.ORG/FAOSTAT-GATEWAY/GO/TO/DOWNLOAD/FB/FBS/E].
Legend: see Figure 4.1d.

Figure 4.2 gives the food balance data (first bar below the zero line in Figures 4.1a–4.1d), expressed in kcal per capita per day. It shows a serious long-term food deficit in Tanzania in the 1960s and in Nigeria up to the 1980s, sensitive

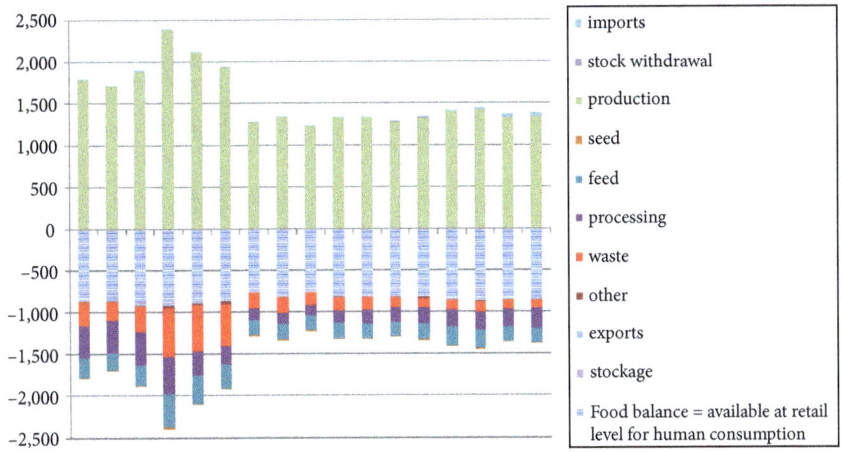

FIGURE 4.1D *Total food supply and utilization (all foods of plant and animal origin), 1961–2009, three-year averages (x 1000 kcal per capita per year) for Uganda.*

Legend to Figures 4.1a–d

- Positive scale: three sources of food supply: production + withdrawal from stocks + import
- Negative scale: seven 'disappearances' into uses other than human consumption: putting into stocks + export + other + waste (down to retail level) + processing + feed + seed
- Amount remaining (shaded part: the 'food balance') = *indirect* estimate of food available at retail level for human consumption.
- The bars are for the 3-year averages, from left to right: 1961/3, 1964/6, 1967/9, 1970/72, 1973/75, 1976/78, 1979/81, 1982/4, 1985/7, 1988/90, 1991/3, 1994/6, 1997/9, 2000/2, 2003/5, 2006/8 and the 1-year result of 2009.

SOURCE: FAOSTAT FOOD BALANCE SHEETS | © FAO STATISTICS DIVISION 2013. UPDATED: 29 JUNE 2012. HTTP://FAOSTAT.FAO.ORG/SITE/368/DESKTOPDEFAULT.ASPX?PAGEID=368. ACCESSED 3 APRIL 2013. [AS OF SEPTEMBER 2014: HTTP://FAOSTAT3.FAO.ORG/FAOSTAT-GATEWAY/GO/TO/DOWNLOAD/FB/FBS/E].

Note 1: 2009 is the most recent year for which FAOSTAT Food Balance Sheets are available.

Note 2: Disappearance data are positive amounts but are represented on the negative scale here, adding up to the same as the food supply.

Note 3: The legend shows uses in reverse order due to technical constraints in constructing this 'mirror image'.

vagaries in Uganda in the 1980s, a strong upward trend in Nigeria since the mid-1980s and a worsening situation in Kenya compared to requirements since the mid-1990s. WHO requirements correspond to roughly 800,000 kcal per capita per year, or 2,250 kcal per capita per day (averaged over all age groups)[2].

2 Table A.4b in Appendix A gives more specific average dietary energy requirements (as used by the FAO) for each of the four countries studied. They are more conservative

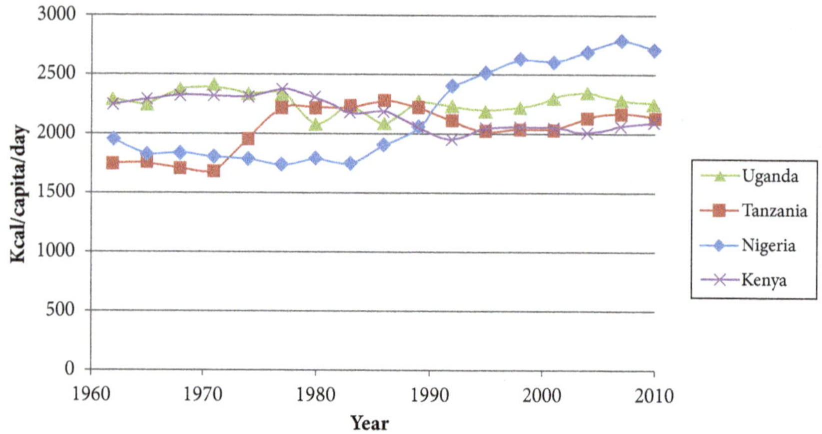

FIGURE 4.2 *Food balance of Kenya, Nigeria, Tanzania and Uganda, 1961–2009 (three-year averages)*
SOURCE: FAOSTAT FOOD BALANCE SHEETS | © FAO STATISTICS DIVISION 2013. UPDATED: 29 JUNE 2012. HTTP://FAOSTAT.FAO.ORG/SITE/368/DESKTOPDEFAULT.ASPX?PAGEID=368. [AS OF SEPTEMBER 2014: HTTP://FAOSTAT3.FAO.ORG/FAOSTAT-GATEWAY/GO/TO/DOWNLOAD/FB/FBS/E].

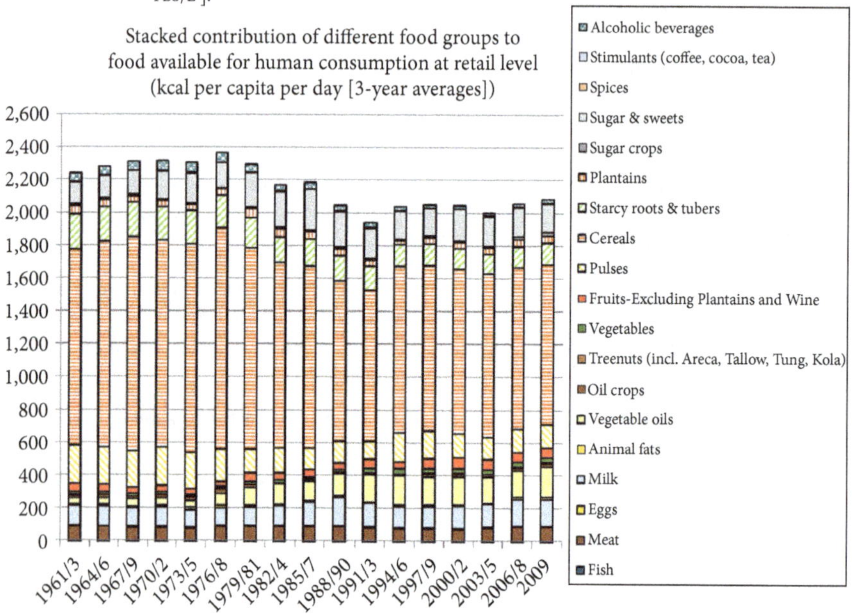

FIGURE 4.3 *Kenya: Food available for human consumption: Composition of the food basket at retail level, 1961–2009*
SOURCE: FAOSTAT FOOD BALANCE SHEETS | © FAO STATISTICS DIVISION 2013. UPDATED: 29 JUNE 2012. HTTP://FAOSTAT.FAO.ORG/SITE/368/DESKTOPDEFAULT.ASPX?PAGEID=368. [AS OF SEPTEMBER 2014: HTTP://FAOSTAT3.FAO.ORG/FAOSTAT-GATEWAY/GO/TO/DOWNLOAD/FB/FBS/E]

(2,100–2,150 kcal per capita per day) as they accept some degree of stunting in the population. The analyses in this volume use the higher figure as a general rule of thumb.

Figure 4.3 shows Kenya's food balance broken down into the food groups used in FAOSTAT. It can be noted that there has been a considerable food gap of about 10% (when compared with the daily requirements of 2,250 kcal per capita) since the late 1980s. The food gap was most serious in 1991–1993 but has not widened since then so food availability has kept pace at this restrained level with population growth in Kenya since the mid-1990s. The share of cereals has somewhat decreased over the years, while the share of nutritious non-staple foods (especially fats and oils, milk and vegetables & fruit) has increased slowly.

Figure 4.4 gives the breakdown of the total food balance by staple foods (cereals, roots & tubers, plantains and pulses), nutritious (from fish to fruit) and less nutritious (sugar to alcohol) non-staple foods as a percentage of the total

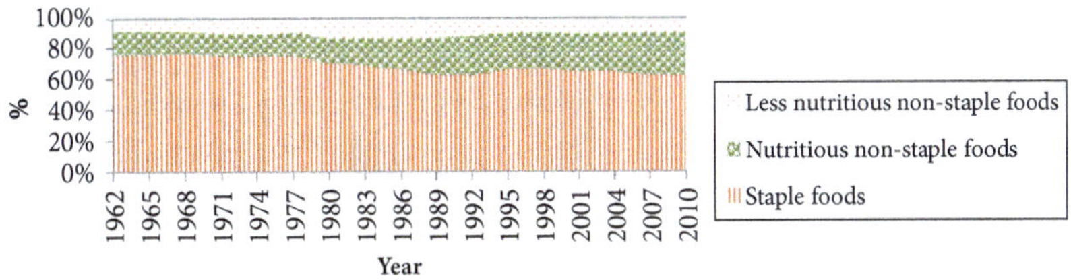

FIGURE 4.4A *Composition of food balance for Kenya by staple foods (cereals, roots & tubers, plantains and pulses), nutritious and less-nutritious non-staple foods (% of total)*

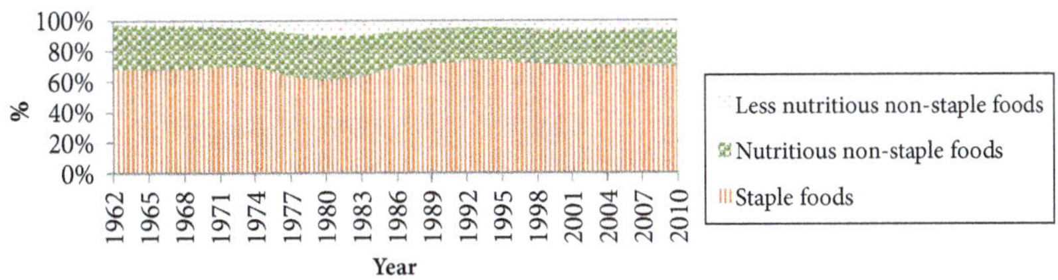

FIGURE 4.4B *Composition of food balance for Nigeria by staple foods (cereals, roots & tubers, plantains and pulses), nutritious and less-nutritious non-staple foods (% of total)*

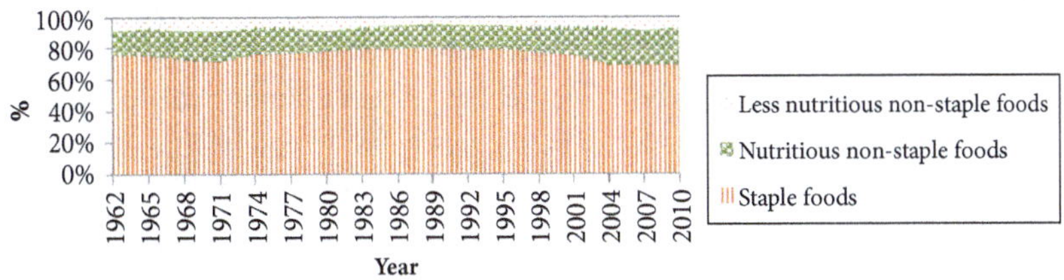

FIGURE 4.4C *Composition of food balance for Tanzania by staple foods (cereals, roots & tubers, plantains and pulses), nutritious and less-nutritious non-staple foods (% of total)*

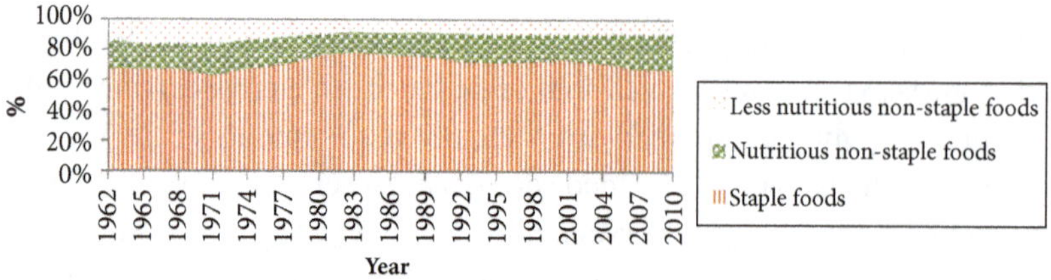

FIGURE 4.4D *Composition of food balance for Uganda by staple foods (cereals, roots & tubers, plantains and pulses), nutritious and less-nutritious non-staple foods (% of total)*

dietary energy supply for the four countries. Dietary diversity has increased in all the countries in the past decade. Nigeria has enjoyed higher levels of nutritious non-staple foods than the other countries all the time, while Kenya has seen the greatest improvement since independence.

Figures 4.5 a and b for Kenya give the breakdown of Figure 4.3 for staple-food and nutritious non-staple food categories. It shows that all individual staple foods except plantains participated in the decrease in staple foods and that most individual nutritious non-staple foods participated in a strong compensatory trend. Figure 4.6a for Nigeria shows a contraction in all types of food between 1976 and 1984, while Figure 4.6b shows how nutritious non-staple foods partly compensated for that dip. Nigeria is on its way to closing its serious food gap of the past (Figure 4.2), both in terms of staple foods and non-staple foods. It should be noted that the large share of edible oil is explained by its high energy density of 9 kcal/g, while the energy density of foods with a high water content and low fat (such as vegetables and fruits) may be as low as 1 kcal/g, so that the energy-based graphs underestimate their true importance for nutrition (i.e. their levels of fibre, antioxidants, vitamins and minerals). Total food availability in Tanzania started from the lowest level of the four countries in the decade after independence (Figure 4.2). Since the mid-1970s there has been a clear increase in the share of cereals and pulses (Figure 4.7a) and since 2000 in the share of nutritious non-staple foods, especially oils, oil-bearing crops and fruit (Figure 4.7b). In Uganda the food balance has fluctuated, with strong vagaries in the 1980s, but has not improved since the 1960s and is only slightly above minimum standards (Figure 4.2). That the modest availability for human consumption of cereals and roots and tubers (Figure 4.8a) is less than one might expect on the basis of production figures (cf. Chapter 3, this volume) can be explained by the fact that quite a lot of roots and tubers but also part of the cereal harvest are being used as livestock feed. The availability of quality foods rich in essential nutrients (fruit, vegetables, nuts, oils, fats and foods of animal origin) showed a strong dip between 1979 and 1987, mainly due to less calories from oil(crop)s, picking up gradually after the late 1980s before starting to rise above 1961 levels around 2003/2005 (Figure 4.8b).

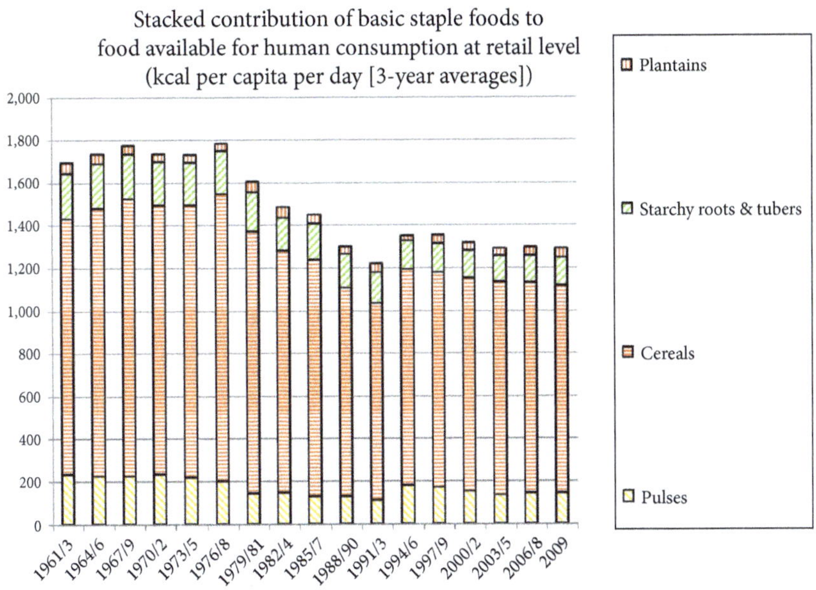

FIGURE 4.5A *Composition of Kenya's staple foods at retail level, 1961–2009*

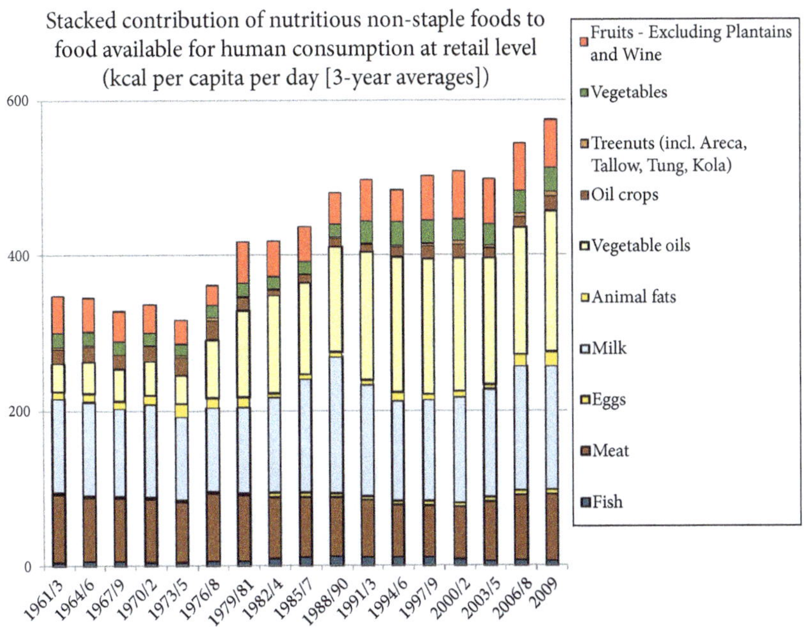

FIGURE 4.5B *Dietary diversity: contribution of nutritious non-staple foods to diets in Kenya*

FIGURE 4.5 *Food balance of Kenya (1961–2009) by food groups, within two categories: (A) basic staple foods and (B) nutritious non-staple foods*

SOURCE: FAOSTAT FOOD BALANCE SHEETS | © FAO STATISTICS DIVISION 2013. UPDATED: 29 JUNE 2012. HTTP://FAOSTAT.FAO.ORG/SITE/368/DESKTOPDEFAULT.ASPX?PAGEID=368. ACCESSED 24 JUNE 2013. [AS OF SEPTEMBER 2014: HTTP://FAOSTAT3.FAO.ORG/FAOSTAT-GATEWAY/GO/TO/DOWNLOAD/FB/FBS/E].

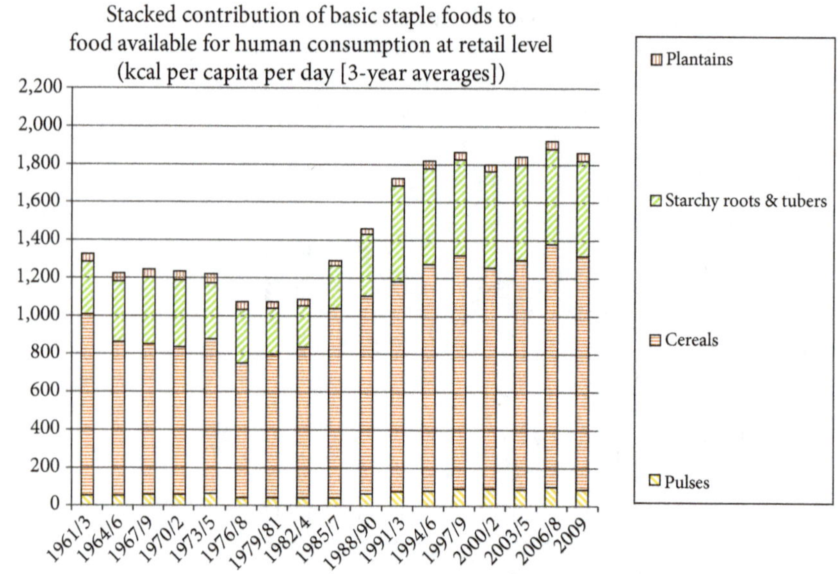

FIGURE 4.6A *Composition of Nigeria's staple foods at retail level, 1961–2009*

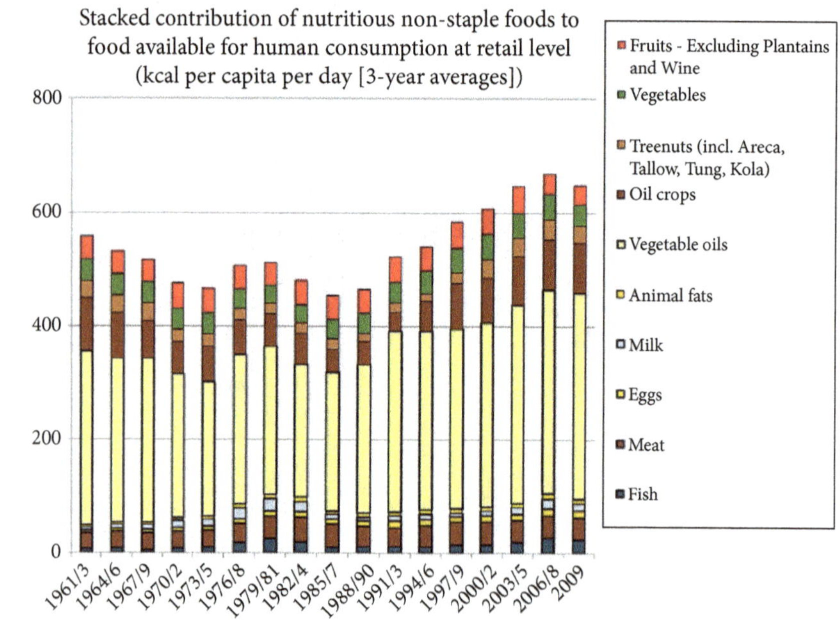

FIGURE 4.6B *Dietary diversity: contribution of nutritious non-staple foods to diets in Nigeria*

FIGURE 4.6 *Food balance of Nigeria (1961–2009) by food groups, within two categories: (A) basic staple foods and (B) nutritious non-staple foods*

SOURCE: FAOSTAT FOOD BALANCE SHEETS | © FAO STATISTICS DIVISION 2013. UPDATED: 29 JUNE 2012. HTTP://FAOSTAT.FAO.ORG/SITE/368/DESKTOPDEFAULT.ASPX?PAGEID=368. ACCESSED 23 APRIL 2013. [AS OF SEPTEMBER 2014: HTTP://FAOSTAT3.FAO.ORG/FAOSTAT-GATEWAY/GO/TO/DOWNLOAD/FB/FBS/E].

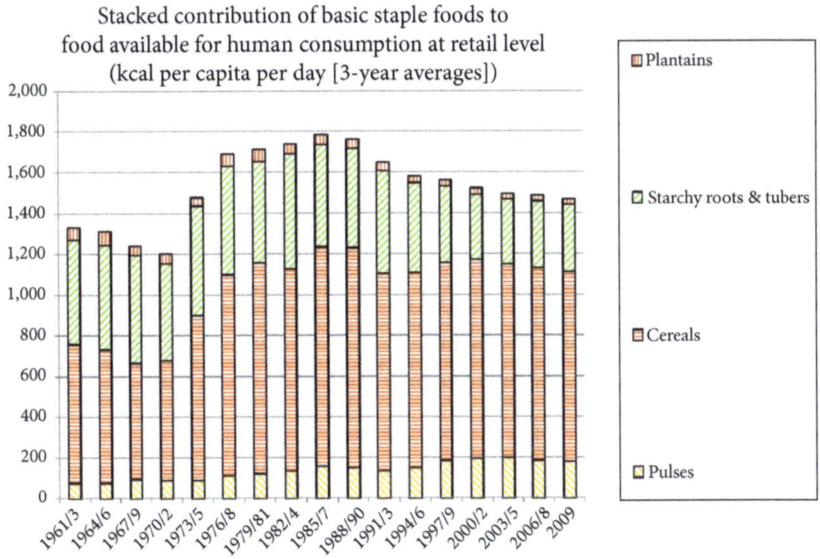

FIGURE 4.7A *Composition of Tanzania's staple foods at retail level, 1961–2009*

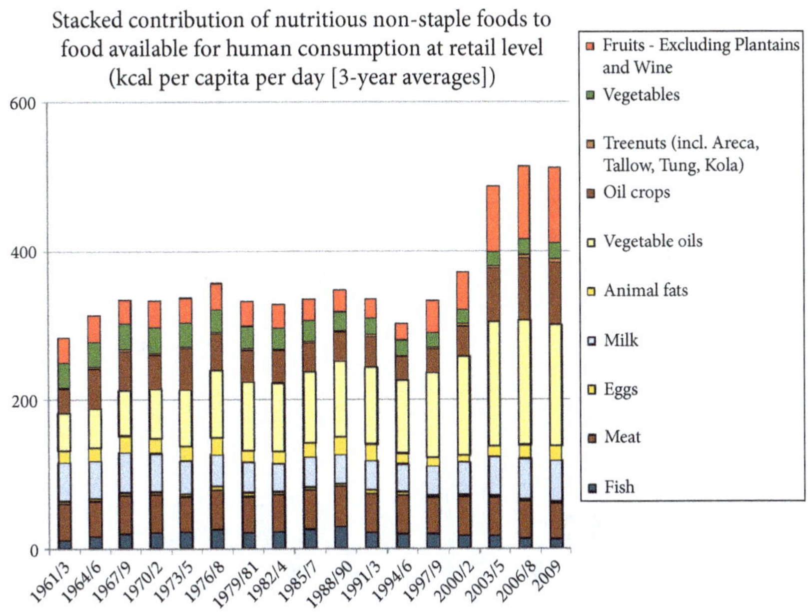

FIGURE 4.7B *Dietary diversity: contribution of nutritious non-staple foods to the diet in Tanzania*

FIGURE 4.7 *Food balance of Tanzania (1961–2009) by food groups, within two categories: (A) basic staple foods and (B) nutritious non-staple foods*

SOURCE: FAOSTAT FOOD BALANCE SHEETS | © FAO STATISTICS DIVISION 2013. UPDATED: 29 JUNE 2012. HTTP://FAOSTAT.FAO.ORG/SITE/368/DESKTOPDEFAULT.ASPX?PAGEID=368. ACCESSED 1 JULY 2013. [AS OF SEPTEMBER 2014: HTTP://FAOSTAT3.FAO.ORG/FAOSTAT-GATEWAY/GO/TO/DOWNLOAD/FB/FBS/E].

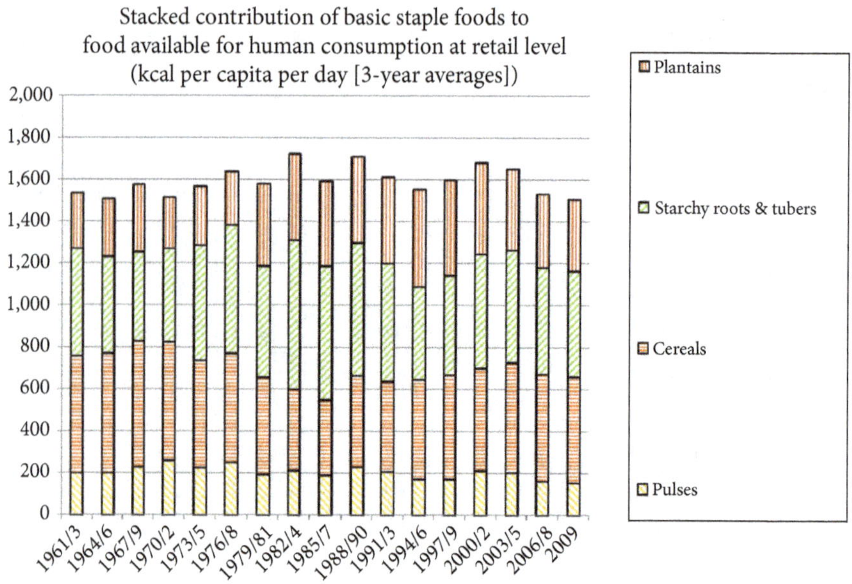

FIGURE 4.8A *Composition of Uganda's staple foods at retail level, 1961–2009*

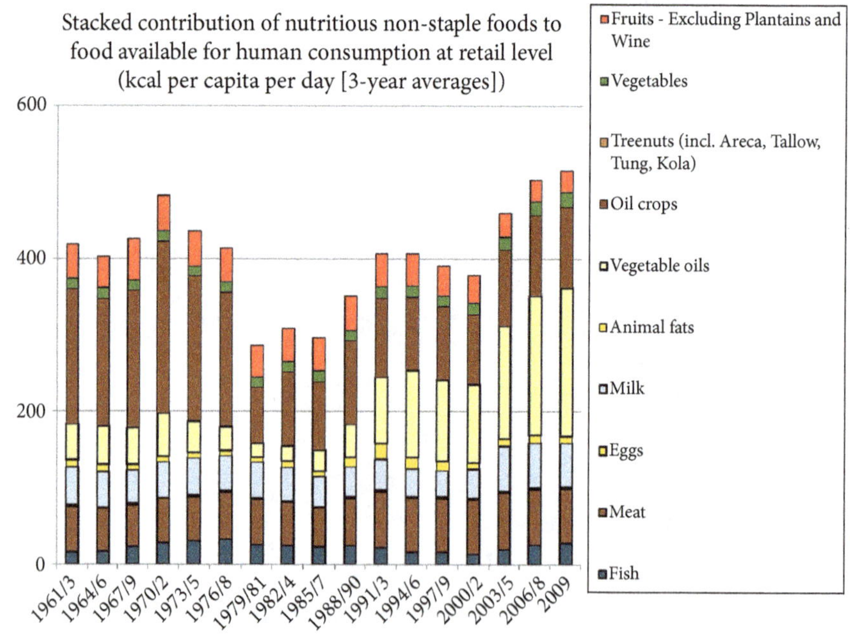

FIGURE 4.8B *Dietary diversity: contribution of nutritious non-staple foods to diets in Uganda*

FIGURE 4.8 *Food balance of Uganda (1961–2009) by food groups, within two categories: (A) basic staple foods and (B) nutritious non-staple foods*

SOURCE: FAOSTAT FOOD BALANCE SHEETS | © FAO STATISTICS DIVISION 2013. UPDATED: 29 JUNE 2012. HTTP://FAOSTAT.FAO.ORG/SITE/368/DESKTOPDEFAULT.ASPX?PAGEID=368. ACCESSED 3 APRIL 2013. [AS OF SEPTEMBER 2014: HTTP://FAOSTAT3.FAO.ORG/FAOSTAT-GATEWAY/GO/TO/DOWNLOAD/FB/FBS/E].

It has long been known (Périssé *et al.* 1969) that as incomes increase, the structure of people's diets changes, with increased intakes of refined foods, sugar, fat, nuts, vegetables & fruits and animal source foods and decreased intakes of unrefined starchy staple foods. As both staple and non-staple foods contribute to the fulfilment of dietary needs, this means that a relative decrease in basic foods (staple foods and legumes) does not necessarily mean a growing food gap as long as the availability of other foods (nutritious and less nutritious such as sugar) can compensate for the decreasing availability of basic foods. The question is whether there is an empirical way of breaking down the overall energy requirement of 800,000 kcal per capita per year (2250 kcal per day) into a partial requirement for staple foods and a partial requirement for non-staple foods, taking into account dietary patterns.

To address this issue, the ratio of staple to non-staple energy in the food balance data for Kenya was regressed over the past fifty years. The time trends showed three distinct periods: a higher ratio in the period immediately after independence and a transition between 1975 and 1985 to a lower ratio that has remained fairly stable ever since (Figure 4.9). The regression function in the first period can be

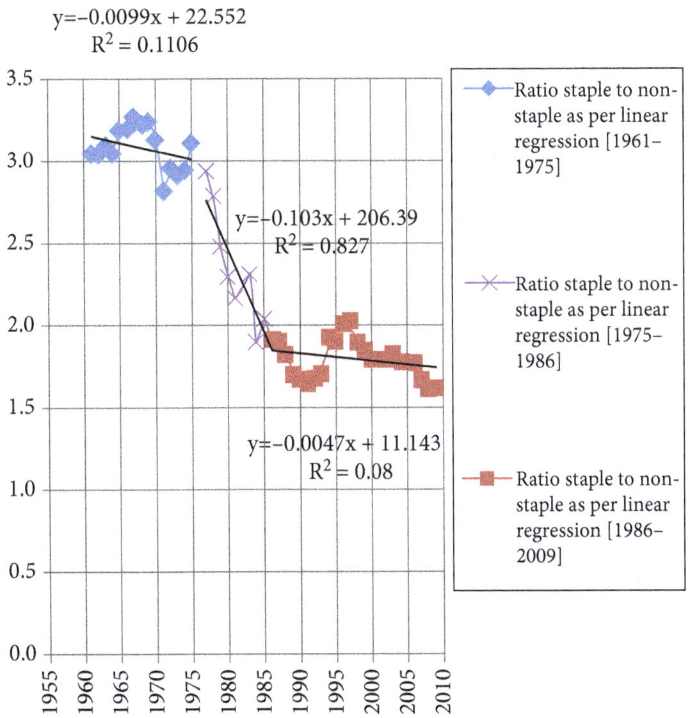

FIGURE 4.9 *Regression of ratio of staple to non-staple energy in Kenya in periods, 1961–2009*
SOURCE: CALCULATIONS BASED ON DATA FROM FAOSTAT FOOD BALANCE SHEETS | © FAO STATISTICS DIVISION 2013. UPDATED: 29 JUNE 2012. HTTP://FAOSTAT.FAO.ORG/SITE/368/DESKTOPDEFAULT.ASPX?PAGEID=368. ACCESSED 24 JUNE 2013. [AS OF SEPTEMBER 2014: HTTP://FAOSTAT3.FAO.ORG/FAOSTAT-GATEWAY/GO/TO/DOWNLOAD/FB/FBS/E].

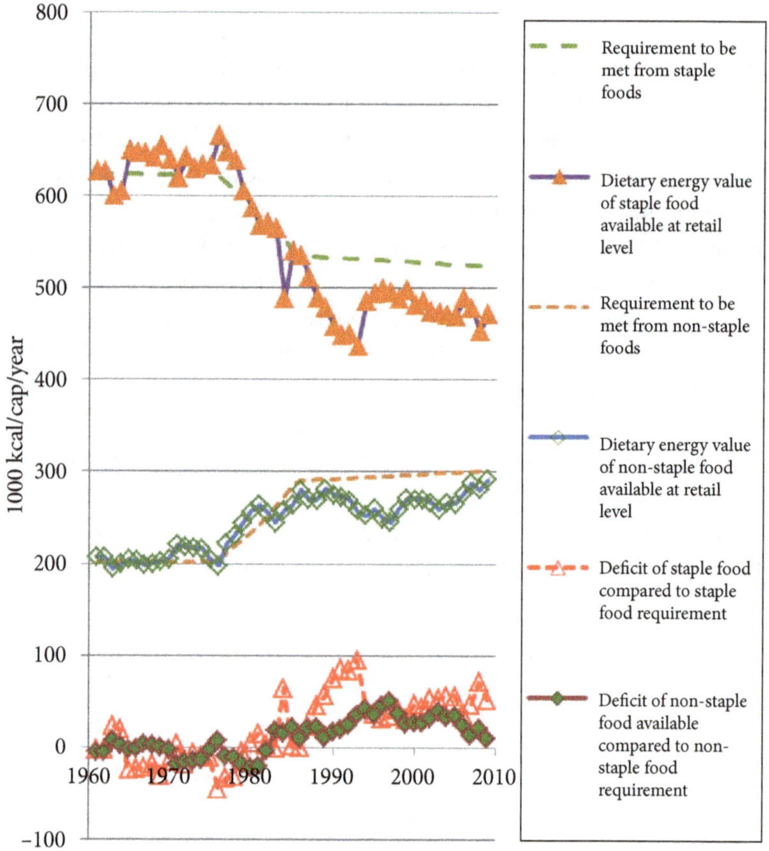

FIGURE 4.10A *Availability and requirements of staple and non-staple foods and resulting deficits, Kenya 1961–2009 (i.e. before, during and after the dietary transition between 1975 and 1986, based on linear regressions of staple to non-staple energy ratios)*
SOURCE: CALCULATIONS BASED ON DATA FROM FAOSTAT FOOD BALANCE SHEETS | © FAO STATISTICS DIVISION 2013. UPDATED: 29 JUNE 2012. HTTP://FAOSTAT.FAO.ORG/SITE/368/DESKTOPDEFAULT.ASPX?PAGEID=368. ACCESSED 24 JUNE 2013. [AS OF SEPTEMBER 2014: HTTP://FAOSTAT3.FAO.ORG/FAOSTAT-GATEWAY/GO/TO/DOWNLOAD/FB/FBS/E].

written as: y = 3.15–(0.01*the number of years after 1960) and the function after 1985 as y = 1.81–(0.005*the number of years after 1986). As the term with the number of years is negligible, even after 15 years, the ratios can be approximated by the intercept. The former *ratio* of 3.15 translates into 76% from staple foods and 24% from non-staple foods, while the latter *ratio* of 1.81 translates into 64% and 36%, respectively. These empirically derived percentages, which reflect a change in the structure of the diet in Kenya, were applied to the overall energy requirement to obtain the breakdown for the analysis shown in Figures 4.10a and 4.10b.

Figures 4.10a and 4.10b show how both staple and non-staple foods have failed to keep up with population growth from the mid-1980s onwards. Between

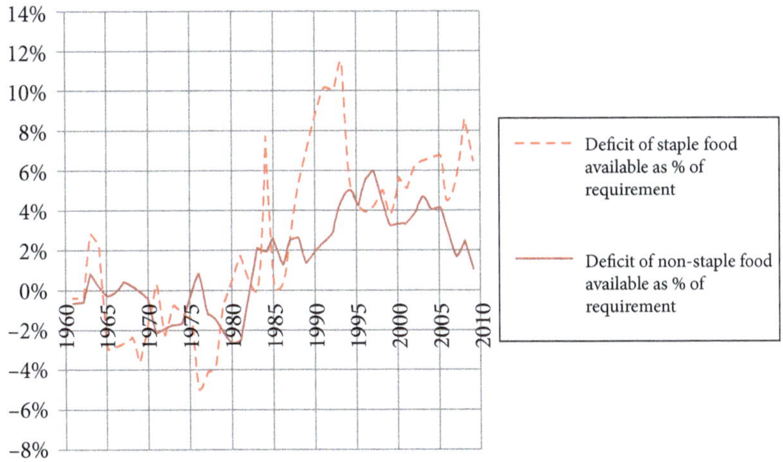

FIGURE 4.10B *Deficits of staple and non-staple foods, Kenya 1961–2009 (before, during and after the dietary transition between 1975 and 1986, based on linear regressions of staple to non-staple energy ratios) as a % of requirements allotted to each component of the total food basket energy*

SOURCE: CALCULATIONS BASED ON DATA FROM FAOSTAT FOOD BALANCE SHEETS | © FAO STATISTICS DIVISION 2013. UPDATED: 29 JUNE 2012. HTTP://FAOSTAT.FAO.ORG/SITE/368/DESKTOPDEFAULT.ASPX?PAGEID=368. ACCESSED 24 JUNE 2013. [AS OF SEPTEMBER 2014: HTTP://FAOSTAT3.FAO.ORG/FAOSTAT-GATEWAY/GO/TO/DOWNLOAD/FB/FBS/E].

2000 and 2010, the staple food deficit (assuming that about 64% of requirements are to be covered by staple foods) is approx. 6% of total requirements (i.e. 51,000 kcal per person per year or 140 kcal per day) and barely decreasing, with a peak of 8% in 2008, while the non-staple food deficit (assuming that 36% of requirements are to be covered by non-staple foods) is about 3% of requirements (i.e. 26,000 kcal per person per year or 70 kcal per day) and decreasing.

Undernourishment

The FAO estimates hunger (Prevalence of Undernourishment or PoU) based on (i) average aggregate food availability (according to annual Food Balance Sheets) and (ii) a statistical procedure based on budget-consumption survey data to generate a statistically generated 'distribution' of this food across income classes. The PoU is an estimate of the number of people (of all ages) who are chronically hungry in a country in a given year. Fluctuations in the PoU appear to depend mainly on the first component (i) considering the countervailing fluctuations of the adequacy of the food supply and the percentage of hungry people (PoU): Figure 4.11 shows that when the food balance decreases (i.e. the food gap increases), the PoU increases at the same time.

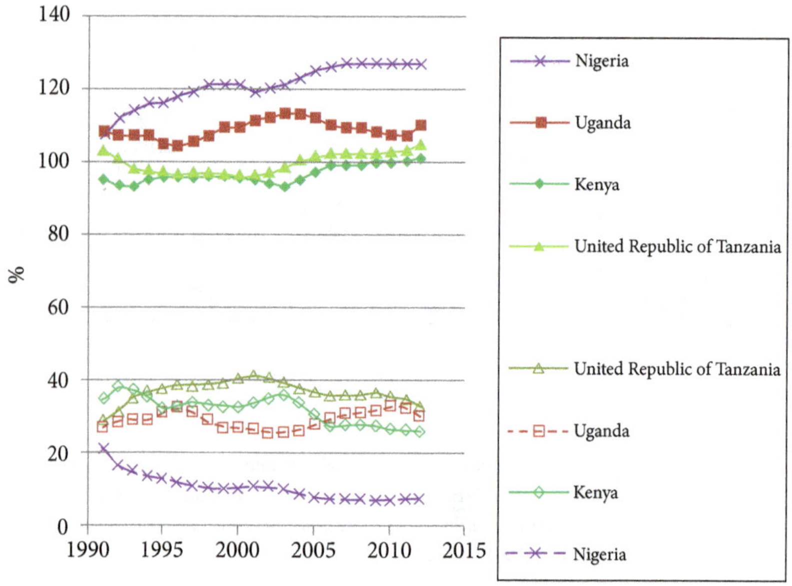

FIGURE 4.11 *Adequacy of the food supply (top: % of energy requirements) vs. the Prevalence of Undernourishment (bottom: % of the population)*
SOURCE: FAO (2013)

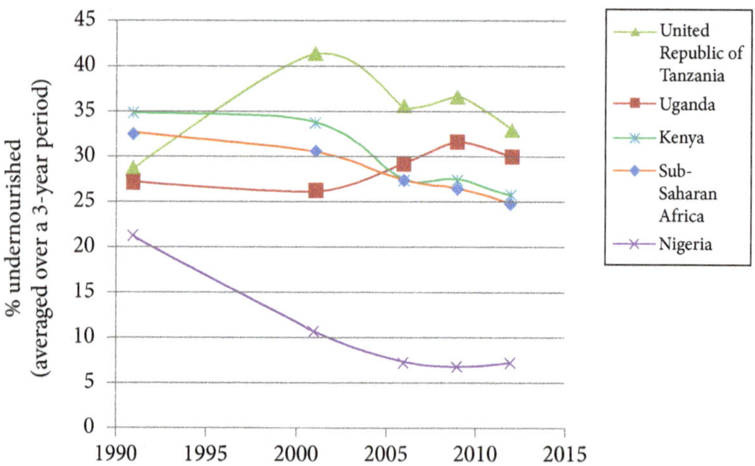

FIGURE 4.12 *Trend of the proportion of undernourished (PoU) in the four countries compared to the trend in Sub-Saharan Africa, 1991–2012*
SOURCE: CALCULATED FROM DATA IN FAO, IFAD & WFP (2013)

The statistical 'distribution' of food by income class (component ii) appears to be more stable over time than the food supply.

Figure 4.12 shows a gradual improvement in the food security situation in Nigeria between 1990 and 2005, although this seems to have slowed down more recently. It clearly demonstrates the impact of the improvement in the food situation since 1990 (cf. Figures 4.1b, 4.2 & 4.6). There was increased hunger in Tanzania during the 1990s, peaking shortly after 2000 to 10 percentage points (one third) higher than in Sub-Saharan Africa (SSA) as a whole. Levels subsided somewhat after this but the PoU remains about 8 percentage points (still one third) higher than the SSA average. The prevalence of hungry people in Kenya was above average in SSA although the trend shows an improvement after 2000–2002. In Uganda the PoU was more favourable before 2005, but rose afterwards to a peak in 2008. It has recently decreased but has remained about 5 percentage points (one fifth) above the average in SSA.

Undernutrition (Anthropometric Failure)

Demographic and Health Surveys are conducted every five years in many countries. They are sometimes complemented by results from the UNICEF-sponsored Multi Indicator Cluster Survey (MICS). These give an indication of child undernutrition and therefore the utilization pillar of food security (see page 84). Children grow both in weight and height and if a child's attained growth falls too far (i.e. 2 standard deviations or more) below the median value of an internationally recognized standard, s/he is considered to be suffering from undernutrition. A child's overall degree of undernutrition is determined by body weight for its age, while height allows an additional distinction to be made between being thin and being short in stature.

Figure 4.13 (using Tanzania as an example) presents the results of the three indicators compared with levels of public-health importance (immediately to the right of each indicator). Chronic undernutrition among under-fives in Tanzania in 2010 was an alarming 42% (i.e. above the cut-off value of 40% that indicates serious public-health concerns). The prevalence of underweight (15.8%) is also a reason for concern (medium prevalence).

Figure 4.14 illustrates the time dynamics of the three indicators of anthropometric failure for Tanzania. Levels may differ in the other countries but the following general tendencies are fairly universal. It highlights how acute undernutrition (wasting = thinness) is highest in the second semester of the first year of life when a baby is particularly vulnerable as new foods are introduced in addition to breastfeeding. While the occurrence of wasting subsides with increasing age, the

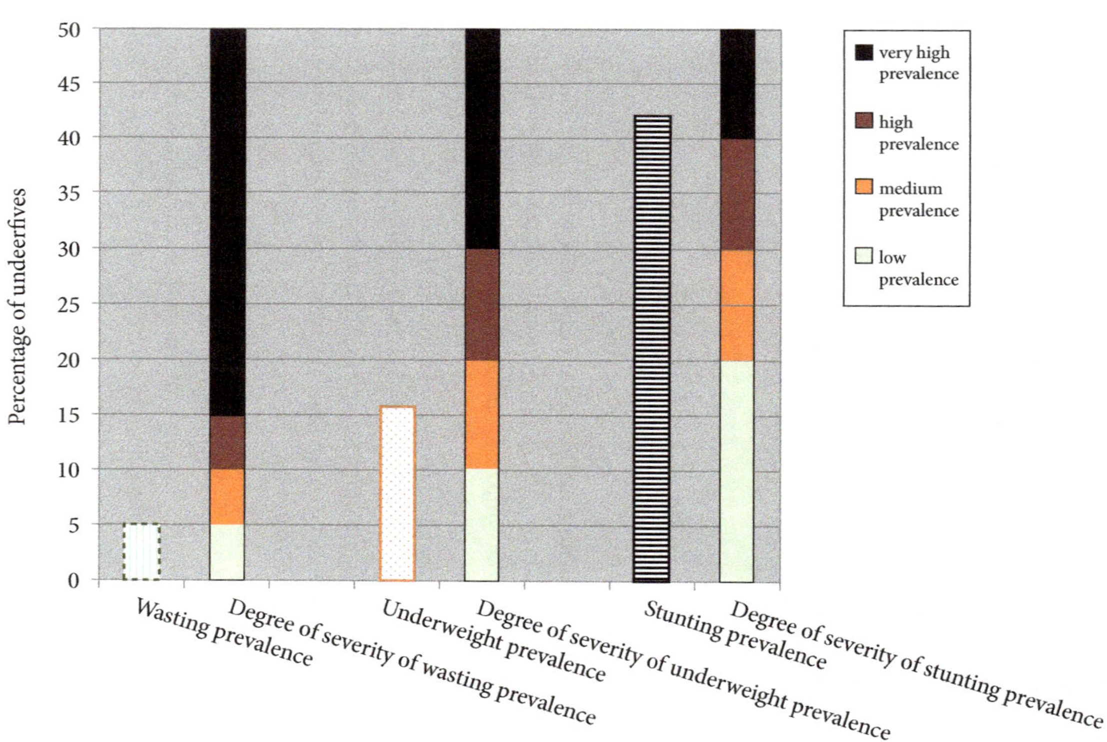

FIGURE 4.13 *Tanzania: Children's undernutrition in 2010: Three indicators of the prevalence of anthropometric failure compared to cut-off values for public-health significance*

SOURCE: BASED ON DATA FROM NATIONAL BUREAU OF STATISTICS (NBS) [TANZANIA] AND ICF MACRO (2011)

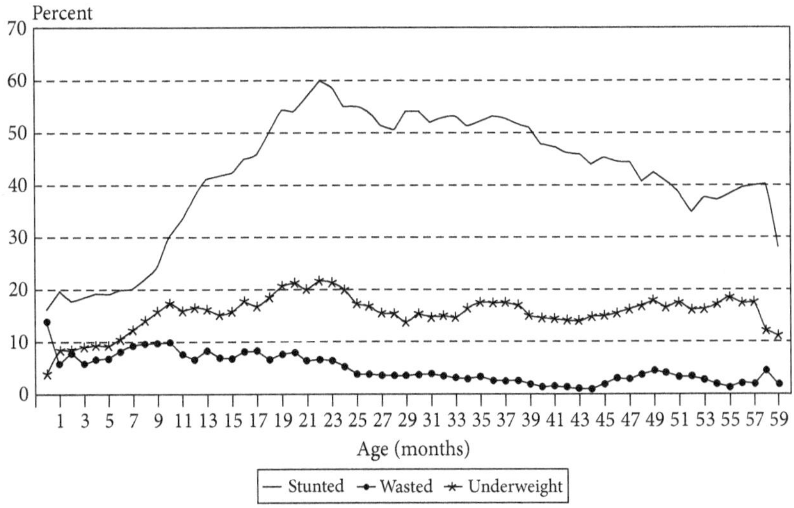

Note: *Stunting* reflects chronic malnutrition; *wasting* reflects acute malnutrition; *underweight* reflects chronic or acute malnutrition or a combination of both. Plotted values are smoothed by a five-month moving average.

TDHS 2010

FIGURE 4.14 *Tanzania: Children's undernutrition in 2010 (by age)*
SOURCE: NATIONAL BUREAU OF STATISTICS (NBS) [TANZANIA] AND ICF MACRO (2011: FIGURE 11.1)

chronic form (stunting = linear growth retardation) increases up to two years of age and remains alarming afterwards.

Low height-for-age (stunting, small stature) is a measure of chronic undernutrition and reflects the accumulative effects with age. In general, it is an irreversible condition that impacts negatively on health and on performance once the child reaches adulthood. The combined result of acute undernutrition (on the left-hand side in Figure 4.13) and chronic undernutrition (on the right-hand side) is reflected in the prevalence of underweight (in the middle). Stunting is regarded as the indicator of choice in nutrition insecurity. Yet it is interesting to note the prevalence of wasting as well. The following figures (4.15–4.18) thus combine these two distinct dimensions of anthropometric failure in one graph in such a way that a situation near zero is optimal (see the realistic 'ideal' situation indicated in Figure 4.18), while a situation in the upper right-hand corner would be disastrous with a very high prevalence of underweight (not shown). These graphs show reference values for severity. Prevalences of stunting are qualified as a 'serious problem' with public-health significance if they are between 30% and 40% and 'very serious' if they are above 40% (cf. Figure 4.13). Critical levels for wasting are lower (10% and 15% respectively) as its nature is generally a temporary (not accumulating) episode in a young child's life (see Figure 4.14).

The 2010 DHS results (Figures 4.15a–d) show that acute undernutrition in the four countries is high in the first two years of a child's life but drops later and that the negative effect on linear growth accumulates with age, reaching a high level among two- to three-year olds.

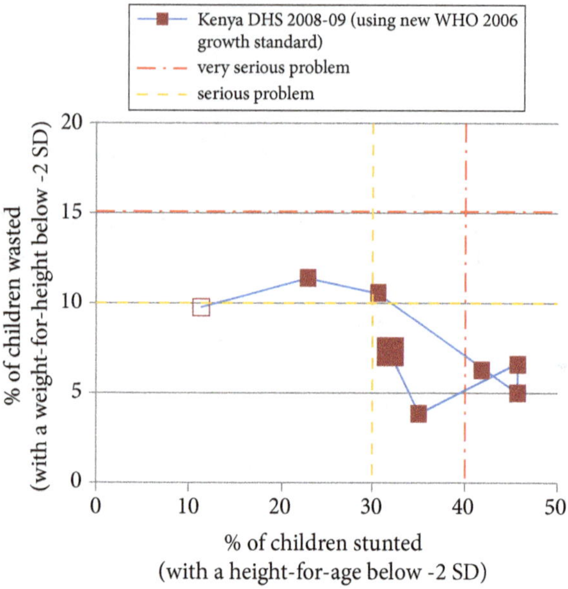

FIGURE 4.15A *Under-five undernutrition in Kenya, by age subgroup: <6 months (open icon), 6–8, 9–11, 12–17, 18–23, 24–35, 36–47 and 48–59 months (heavy icon)*

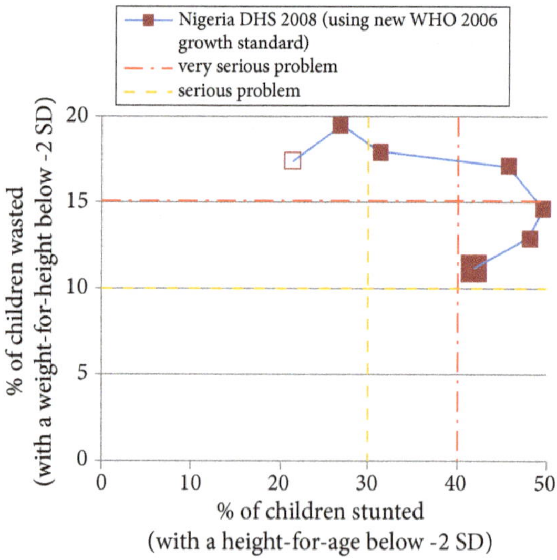

FIGURE 4.15B *Under-five undernutrition in Nigeria, by age subgroup: <6 months (open icon), 6–8, 9–11, 12–17, 18–23, 24–35, 36–47 and 48–59 months (heavy icon)*

Note: The age groups of 36–47 and 48–59 months overlap in this figure.

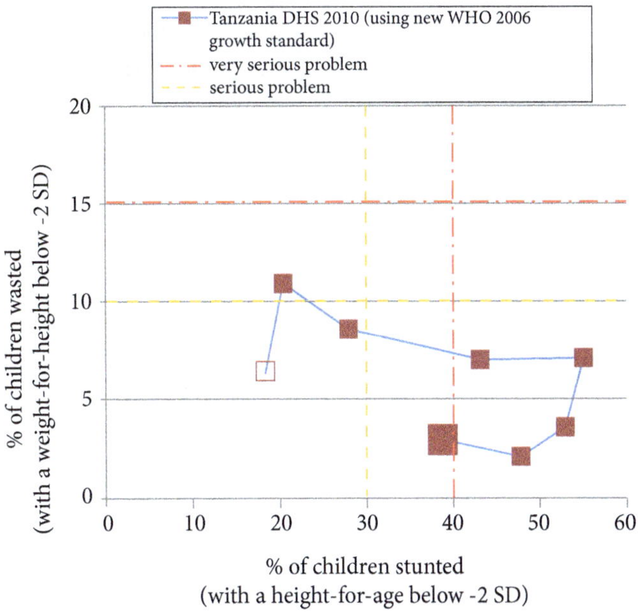

FIGURE 4.15C *Under-five undernutrition in Tanzania, by age subgroup: <6 months (open icon), 6–8, 9–11, 12–17, 18–23, 24–35, 36–47 and 48–59 months (heavy icon)*

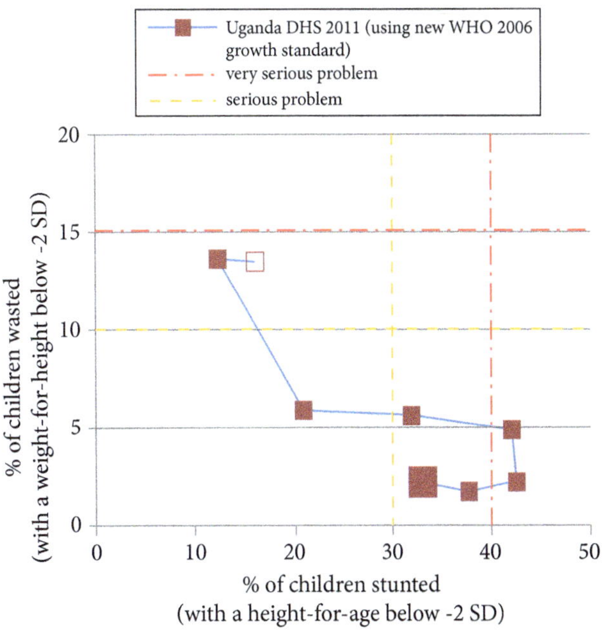

FIGURE 4.15D *Under-five undernutrition in Uganda, by age subgroup: <6 months (open icon), 6–8, 9–11, 12–17, 18–23, 24–35, 36–47 and 48–59 months (heavy icon)*

Boys under five years of age are more undernourished than girls of the same age; Figure 4.16a shows the differential in stunting for Tanzania, which is about the same in the other three countries (not shown), varying between 5–7 percentage points. Rural children under five are more undernourished than under-fives living in urban areas; Figure 4.16b illustrates the differential in stunting for Tanzania, which is about the same in the other three countries (not shown), varying between 10–15 percentage points.

Contrary to FBS data that are based on national accounts, anthropometric results are available for sub-samples at sub-national level. Of the four countries studied, Uganda has the largest discrepancy in levels of stunting between regions, ranging from a low 12% in the capital region to an extremely high 45% in Karamoja (Figure 4.17).

The DHS data are also analyzed by wealth. Wealth groups, which are measured on the basis of household possessions and characteristics, serve as a

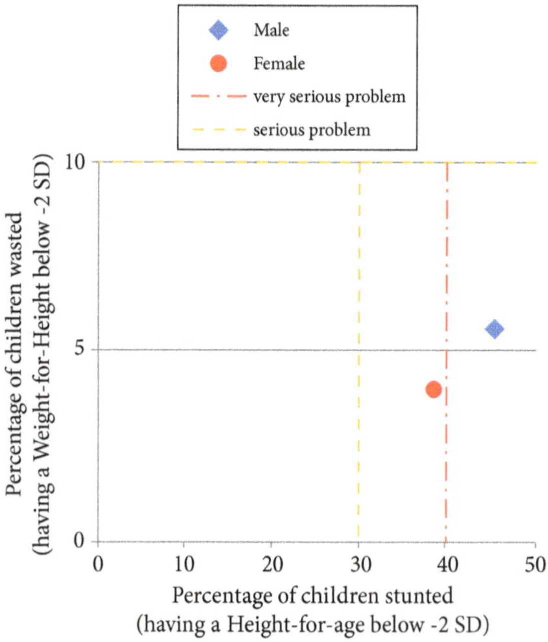

FIGURE 4.16A *Undernutrition in Tanzania: Differences between boys and girls, 2010*

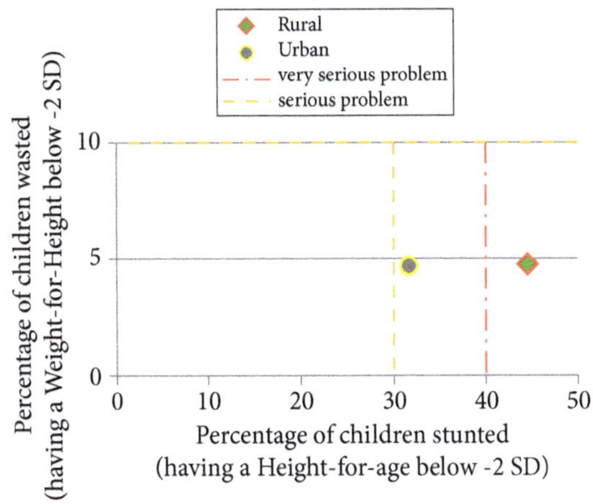

FIGURE 4.16B *Undernutrition in Tanzania: Differences between rural and urban areas, 2010*

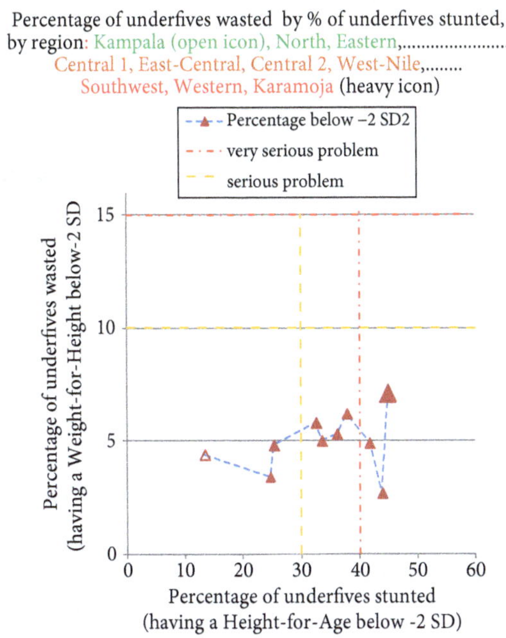

FIGURE 4.17 *Undernutrition: Differences between regions in Uganda, 2011*

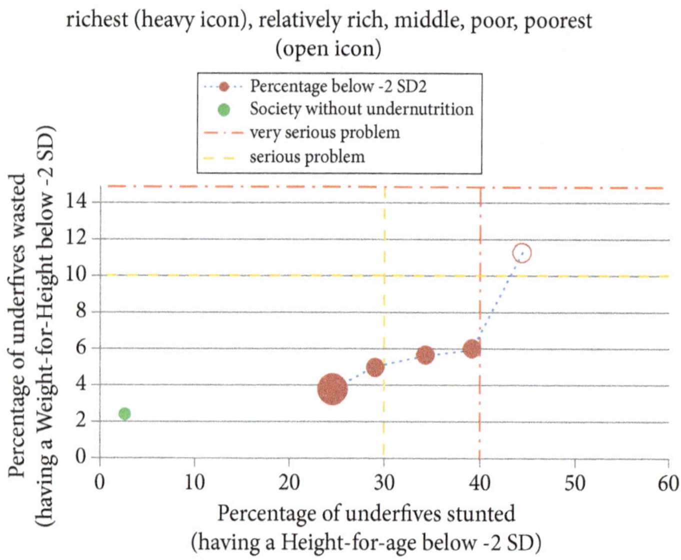

FIGURE 4.18 *Undernutrition by wealth group (income quintiles) in Kenya, 2008–2009*

proxy for income groups. Of the four countries studied, Kenya has the largest discrepancy in levels of wasting, ranging from a high 12% among the poorest to a low 4% amongst the richest children (Figure 4.18). The ideal situation (of absence of undernutrition) is set at 2.5% (not zero)[3] for each indicator.

Undernourishment and Undernutrition Compared: The Case of Kenya

Figure 4.19 shows that both the food-related and the anthropometrical indicators of food insecurity showed a slight downward trend in Kenya in the period under consideration (1986–2009). This points to an improvement in food access (fewer hungry people) and of food utilization (fewer undernourished young children) respectively.

When comparing the trend of hunger (percentage of people having a lack of food) on the one hand, and trends of childhood anthropometric failure on the other, the preferred indicator of chronic undernutrition is the prevalence of stunting, as it shows the effects of undernutrition that accumulated in the past over several years.

3 This is the result of the statistical consideration that under normal circumstances the majority of the reference group has a value above the mean minus 2 standard deviations and only an insignificant minority of 2.5% of the child population has a value below that cut-off point.

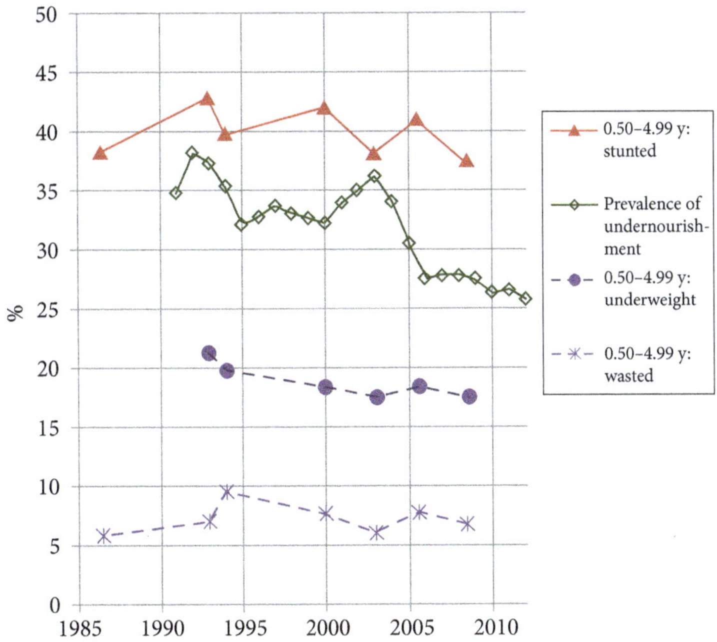

FIGURE 4.19 *Trends of hungry people[1] and of undernutrition among under-fives[2] in Kenya compared, 1986–2012*

SOURCE: FAO (2013). THIS DATABASE DERIVES THE PREVALENCES OF STUNTING, WASTING AND UNDERWEIGHT FROM THE WHO GLOBAL DATABASE ON CHILD GROWTH AND MALNUTRITION THAT USES THE NEW WHO CHILD GROWTH STANDARDS OF 2006. PREVALENCES OF UNDER-FIVES UNDERNUTRITION FROM OLDER SURVEYS (USING OLD NCHS GROWTH REFERENCE VALUES) WERE CONVERTED BY WHO INTO THE NEW SYSTEM USING A PUBLISHED ALGORITHM.

Notes:

[1] The prevalence of undernourishment (hungry people) is calculated annually as a three-year average, starting in 1990.

[2] The prevalence of undernutrition among young children is only available for the year in which a national nutrition survey was done.

As Figure 4.19 suggests, the two indicators at first sight appear to have their own dynamics and are seemingly unrelated: the peaks and dips in PoU and stunting do not coincide. Yet if one assumes that a change in stunting reflects a change in PoU from perhaps three to four years ago, the ups-and-downs in the modified graph, with PoU shifting to the right (not shown here) appear to coincide.

Figure 4.20 shows the clear internal dynamics at play between the two dimensions of child undernutrition: (i) there were more wasted children in 1994; (ii) more stunting concomitant with less wasting is noted between 1994 and 2000; (iii) an improvement on both accounts can be seen from 2000 to

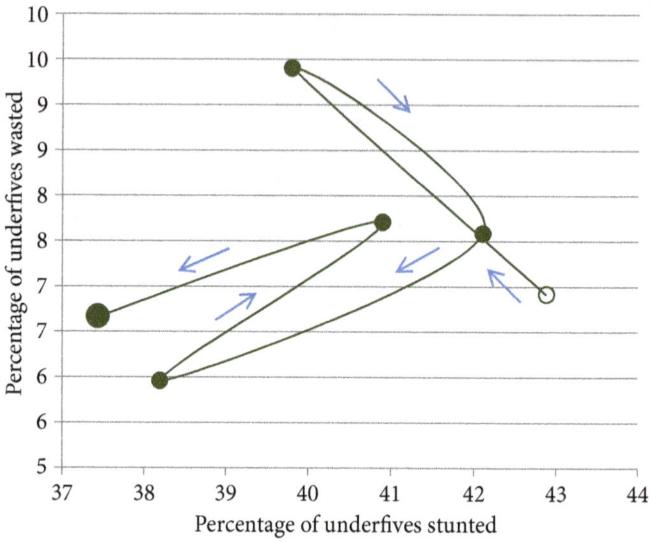

FIGURE 4.20 *Time trends in anthropometric failure among children 0.5–5 years of age in Kenya, 1993–2009*

SOURCE: WORLD HEALTH ORGANIZATION (2014). RESULTS OF NATIONAL NUTRITION SURVEYS OF FEB-AUG 1993 (OPEN ICON), JUN-AUG 1994, SEP-OCT 2000, APR-SEP 2003, YEAR ROUND MAY 2005–MAY 2006, NOV 2008–FEB 2009 (BOLD ICON)

2003; (iv) from 2003 to 2006 the reverse of (iii) was true; and (v) there was redress from 2006 to 2009. These internal dynamics, in particular when they have a spiralling characteristic, may reflect the influence of seasonality, when a season of stasis (stagnation) in length growth and relative accumulation of weight is followed by a season with a length growth spurt concomitant with some thinning of the body. In a longitudinal study at the Kenya Coast over five seasons in 1985–1986, growth in height was at a maximum in the period of December to February (in the dry season), while maximum weight growth took place in the subsequent period between March and May (during the long rains) (Hoorweg *et al.* 1995: 88; Klaver & Mwadime 2000: 288; Klaver & Nubé 2008: 298). The 1993 DHS survey was in March to May, while the 1994 DHS survey was done after the wet season and the DHS survey in 2008–2009 was carried out in the period before the wet season.

Conclusion

There are important differences in the degree and type of disaggregation of the three indicators of food (in)security presented in this chapter. The food-related indicators for food insecurity, namely food balance sheets and the derived prevalence of hungry people, are based on aggregated national accounts. So while food-related indicators are still limited to the national level,[4] they have the virtue of allowing any analysis to be broken down by food group and even by individual commodities. Insight is thus given into the composition of the national food supply. The PoU incorporates a distributional simulation and if it becomes available at subnational level in the future, it will allow a refined locally relevant analysis.

The most disaggregate indicators are the anthropometric ones drawn from the DHS. However, they cannot indicate what foods are lacking in the diet of the children and their mothers. The combined analysis of PoU and anthropometric failure thus looks promising, with the latter being considered as a late outcome of the former, with a time lag of a few years.

The FBS results show that food availability has kept pace with population growth to a large extent and that the quality of people's diets is improving. On the other hand, there is no reason for complacency, as the percentage of hungry people and of children with undernutrition has remained high and is still alarming in some cases.

References

Akinyoade, A., T. Dietz, D. Foeken & W. Klaver in collaboration with A. Leliveld, H. Smits, S. Soeters & M. van 't Wout (2013), Agricultural Dynamics and Food Security Trends in Nigeria. Developmental Regimes in Africa (DRA) Project. ASC-AFCA Collaborative Research Group: Agro-Food Clusters in Africa (AFCA). Research Report 2013-ASC-1. London/Leiden. https://openaccess.leidenuniv.nl/handle/1887/22942. Accessed 23 August 2014.

Dietz, T., D. Foeken, S. Soeters & W. Klaver in collaboration with A. Akinyoade, A. Leliveld, H. Smits & M. van't Wout (2013), Agricultural Dynamics and Food Security Trends in Kenya. Developmental Regimes in Africa (DRA) Project; ASC-AFCA Collaborative Research Group: Agro-Food Clusters in Africa (AFCA) Research Report 2013-ASC-4. London/Leiden. https://openaccess.leidenuniv.nl/handle/1887/23291. Accessed 23 August 2014.

FAO (2013), Food Security Indicators. http://www.fao.org/fileadmin/templates/ess/foodsecurity/Food_Security_Indicators.xlsx. Accessed 30 March 2014.

4 Only recently has the FAO started to collect disaggregated data by region in a number of countries (COUNTRYSTAT) but these data are as yet too fragmentary to allow trend analysis.

FAO, IFAD & WFP (2013), The State of Food Insecurity in the World 2013. The Multiple Dimensions of Food Security. Rome: FAO. http://www.fao.org/docrep/018/i3434e/i3434e00.htm. Accessed 1 October 2013.

FAO/WHO (2012) Draft: Country Nutrition Paper. Inputs from the Food and Agriculture Sector prepared for the Joint FAO/WHO 2nd International Conference on Nutrition ICN2, 12 June. Rome: FAO. http://www.fao.org/fileadmin/user_upload/agn/documents/Draft_country_paper_outline_for_regional_consultations_25March 2011_rev1.doc. Accessed 16 January 2014.

Hoorweg, J., D. Foeken & W. Klaver (1995), *Seasons and Nutrition at the Kenya Coast.* Leiden: African Studies Centre.

Kenya National Bureau of Statistics (KNBS) & ICF Macro (2010). Kenya Demographic and Health Survey 2008–2009. Calverton, MD: KNBS/ICF Macro.

Klaver, W. & R. Mwadime (2000), Food Consumption and Nutrition. In: J. Hoorweg, D. Foeken & R.A. Obudho (eds), *Kenya Coast Handbook: Culture, Resources and Development in the East African Littoral.* Münster: LIT Verlag, pp. 279–295. https://openaccess.leidenuniv.nl/handle/1887/9672. Accessed 25 August 2014.

Klaver, W. & M. Nubé (2008), The MDG on Poverty and Hunger: How Reliable Are the Hunger Estimates?. In: M. Rutten, A. Leliveld & D. Foeken (eds), *Inside Poverty and Development in Africa: Critical Reflections on Pro-poor Policies*. Leiden: Brill, pp. 273–302. https://openaccess.leidenuniv.nl/handle/1887/18529. Accessed 25 August 2014.

Leliveld, A., T. Dietz, W. Klaver, B. Kilama & D. Foeken in collaboration with A. Akinyoade, H. Smits, S. Soeters & M. van't Wout (2013a), Agricultural Dynamics and Food Security Trends in Tanzania. Developmental Regimes in Africa (DRA) Project. ASC-AFCA Collaborative Research Group: Agro-Food Clusters in Africa (AFCA). Research Report 2013-ASC-3. London/Leiden. https://openaccess.leidenuniv.nl/handle/1887/22944. Accessed 23 August 2014.

Leliveld, A., T. Dietz, D. Foeken & W. Klaver in collaboration with A. Akinyoade, H. Smits, S. Soeters & M. van't Wout (2013b), Agricultural Dynamics and Food Security Trends in Uganda. Developmental Regimes in Africa (DRA) Project. ASC-AFCA Collaborative Research Group: Agro-Food Clusters in Africa (AFCA). Research Report 2013-ASC-2. London/Leiden. https://openaccess.leidenuniv.nl/handle/1887/22943. Accessed 23 August 2014.

National Bureau of Statistics (NBS) (Tanzania) & ICF Macro (2011), Tanzania Demographic and Health Survey 2010. Dar es Salaam: NBS/ICF Macro.

Périssé, J., F. Sizaret & P. Francois (1969), The Effect of Income on the Structure of the Diet, (*FAO*) *Nutrition Newsletter* 7(3): 2.

World Health Organization (2014), Global Database on Child Growth and Malnutrition. http://www.who.int/nutgrowthdb/database/countries/ken/en. Accessed 6 June 2014.

SECTION 2

Agricultural Production and Effectiveness

CHAPTER 5

Dairy Clustering in Kenya

Diederik de Boer and Jackson Langat

Dairy is a booming industry in Kenya. This chapter presents a cluster analysis of lead firms/suppliers in the dairy-processing industry in the country's two main dairy regions: the Northern Rift and the Central Rift regions. Gaps and opportunities are discussed, with a focus on the dynamics and linkages in the cluster and a view to developing sustainable policies to increase productivity in dairy ventures. Results show that the linkages with the financing institutions in both regions are stronger than with research and training institutions. Coordination among the actors is better in the north but a lack of clarity remains on the division of responsibilities between government and the private sector. While the cooling-system infrastructure is more than sufficient in the Central Rift cluster, it is not adequate in the north where the roads are also poor. In general, the Northern Rift's potential is much higher in terms of milk processing. With improved linkages, the cluster in the northern region could become much stronger and this would result in higher levels of production and greater competitiveness. More emphasis on this cluster could, therefore, be an interesting route to economic development in Kenya.

Introduction

As the country's single largest agricultural sub-sector, dairy is an important growth sector in Kenya. It accounted for around 3.5% of GDP in 2008 (Wambugu *et al.* 2011) and the amount of dairy products consumed increased by 72% over the last ten years (Kenya Dairy Master Plan 2010). The demand for dairy products is growing in the wider region and a strong dairy sector in Kenya could be a sustainable source of economic growth over the next decade.

The dairy industry's productivity figures are, however, still low at 1500 litres of milk per cow per year when compared to cows in the Netherlands that give 10,000 litres, and South African cows produce 5000 litres per cow per year.[1]

1 See http://seasofchange.net/file/downloads/2012/04/31.03-SoC-dairy-fact-sheet-ah1.pdf. Accessed 15 March 2014.

 | DOI 10.1163/9789004282698_006

The low figures in Kenya are attributed to poor breeds and feeding regimes due to a combination of constraints, including poor access to information, the low and inconsistent quality of feed, inadequate extension services, a lack of adherence to standards and the high costs of inputs. The prices of inputs often limit the use of quality feeds and veterinary services, which then has a negative impact on the quality of any milk produced (Wambugu *et al.* 2011). These constraints vary according to the region[2] and so any strategies designed to overcome them and to increase productivity should therefore differ as well (Kenya Dairy Master Plan 2010).

The cluster analysis[3] of the two dairy regions that is reported in this chapter highlights the gaps and opportunities and also the dynamics and linkages in the clusters. The analysis will assist in developing sustainable policies aimed at increasing productivity (see Figure 5.1). The chapter assesses the dairy-processing industry in the two main dairy regions that were identified by the Kenya Dairy Board (KDB): the Northern Rift and the Central Rift regions (see Map 5.1). There is a clear difference between the regions, with the Central Rift region being dominated by small-scale intensive farming while the Northern Rift region has predominantly large-scale extensive farming.

Small-scale dairy farming is mostly found in the Central Rift Valley Province and on the coastal lowlands. A typical small-scale dairy farm consists of 2 to 3 cows per farm on about 1 ha of land. These farmers also engage in other agricultural activities and practise free grazing, which is often combined with zero-based grazing. Most of their cows are Friesians.[4] A typical large-scale dairy farm would have more than 20 cows. They are usually predominantly Friesian but Ayrshire and Channel Island cattle can be found too. These farms, which are mainly in the Northern Rift area (Majiwa *et al.* 2012), have zero-based grazing although free grazing is also extensively practised.

---

2 Three different dairy regions can be identified in Kenya: the East, Central and the North (Wambugu *et al.* 2011). Two regions were analyzed in this research: the Northern and Central regions. The Central region is characterized by zero-based grazing while this is practised less in the Northern region. But more land is available in the Northern region than in the Central region, which affects the amount of fodder available. The Eastern region was not included in this review due to time and budget constraints.

3 This study was conducted as part of Nuffic's Capacity for Competitiveness Programme in the Dairy Value Chain in Kenya (C4C-DVC). See also Footnote 7.

4 See http://www.fao.org/ag/AGP/AGPC/doc/counprof/kenya/Kenya.htm. Accessed 20 March 2014.

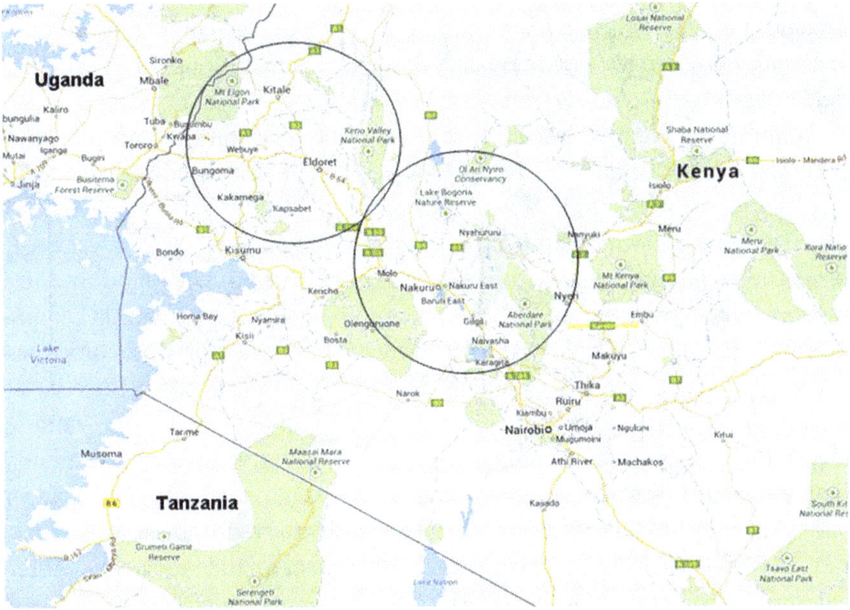

MAP 5.1 *Two of Kenya's dairy clusters*

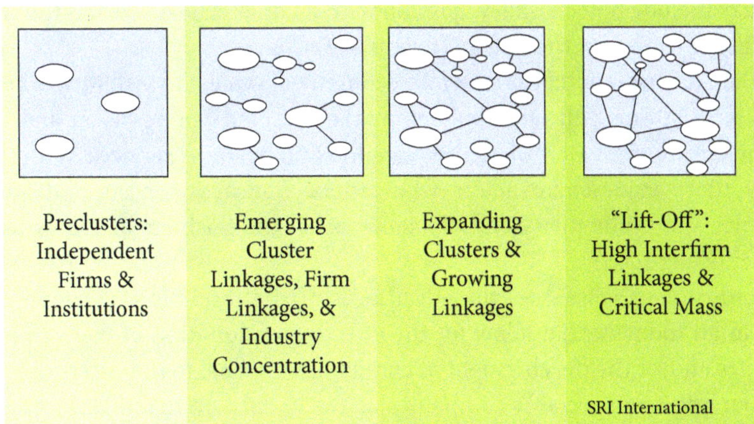

FIGURE 5.1 *Cluster formation*
SOURCE: SRI (IN USAID 2003)

Cluster Theory

The close interplay between firms, their suppliers and the business environment is the reason why competitiveness theorists and practitioners focus on 'clusters' as the locus of action, as opposed to individual firms or broader sectors.

Clusters are 'geographic concentrations of interconnected companies, specialized suppliers, service providers, and associated institutions in a particular field' (Porter 1998).

According to competitiveness theory and practice, individual firms cannot become competitive and stay competitive in global markets on their own. The vitality of an enterprise's cluster and its level of sophistication, productivity and the environment in which it operates all strongly influence its competitiveness. A firm's suppliers and supporting institutions have to continually improve their capabilities to provide the necessary inputs and services. This means that companies in the cluster constantly have to innovate and create new products, processes and ways of managing their operations if they want to remain competitive.

Supporting institutions might include educational and (vocational) training institutions that produce the industry's (future) workforce; research institutions that generate the scientific knowledge required for technological change; banking and financial institutions that ease access to credit; government institutions whose policies and practices have an impact on the industry; and providers of infrastructure for the industry.

Clusters evolve over time and develop from being independent firms and institutions to high inter-firm linkages (Figure 5.1). Well-known examples of clusters include Silicon Valley, London with its financial centre and the wine-producing regions in the Balkan (USAID 2003).

When a cluster achieves a certain economy of scale and visibility, it begins to exert a strong pull on suppliers, marketers, distributors, users and even financiers and providers of communication and business services. With them come more value-chain actors who reduce transaction costs and greatly enhance the productivity and efficiency of all the business activities in the cluster (Zeng 2008).

Nordin (2003) discusses the advantages of the formation of cooperation nets in an industry that allow for the effective exploitation of collective efficiencies and/or the development of companies external to the enterprises (i.e. outsourcing and specialist companies active in entrepreneurial cooperation, production specialization, collective infrastructure and service specialization). They make it possible for firms in the industry to benefit from an increased ability to negotiate collectively with suppliers of inputs and other components. In addition to these advantages, cooperation nets facilitate the development of new models, production processes and organization, the exchange of technical and market information, consortiums for buying and selling goods and services as well as joint marketing campaigns. The interaction and synergy arising from joint actions provide competitive advantages over the isolated actions of individual companies.

According to Zeng (2008), the benefits of clustering are:

- efficiency gains through cluster-based business value chains
- knowledge and technology acquisition, adaptation and dissemination by means of:
  - knowledge network and inter-firm links including subcontracting, market links with customers and suppliers, informal and formal collaborations, membership of professional and trade associations, and the movement of skilled staff from one firm to another
  - tapping into foreign knowledge and technology, including cooperative production agreements, franchising, licensing, reverse engineering and imitation
  - learning and training including apprenticeship, on-site training at suppliers' factories, on-the-job training, expert contracting, support mechanisms provided by public institutions, learning through transactions with local and external agents and by doing in the areas of production and maintenance
  - universities and institutes of technology including graduate programmes, technical assistance, training programmes, coordinated research and technology transfer
- the pooling of (relatively) well-educated entrepreneurs and workers since an educated, skilled labour force is at the core of innovation and competitiveness
- government and institutional support, which can be attracted in a number of areas including defining and enforcing sectoral policies, regulations and standards; creating a special agency or organization to promote, coordinate and facilitate cluster development; establishing various public institutions (such as councils, incubators[5] and technology centres) to provide technological and technical support, training and capacity development; providing incentives such as tax holidays, special funds, duty-free provisions, cheap land to promote technology imports and enterprise development; promoting alliances and partnerships through joint ventures and strategic alliances; and providing infrastructure such as roads, water, power, ports, warehouses and information technology

5 Business incubation is a flexible combination of business development processes, infrastructure and people that is designed in one location to nurture new and small businesses by helping them to survive and grow through the difficult and vulnerable early stages of development. See http://www.diogenes-incubator.com/incubation/business-incubation-definition. Accessed 4 March 2014.

- joint action and cooperation: the small size of many individual cluster enterprises allows firms to establish joint organizations and cooperate to protect their common interests; exchange information and ideas; form synergies and inter-firm links; provide technical, marketing and policy services; increase collective bargaining powers; and survive in the face of fierce competition[6]

In summary, clustering enhances firms' competitiveness through agglomeration economies. This happens as a result of the availability of a skilled labour force, access to specialized suppliers, having specialized service providers, improved market access and the circulation of information. The advantages of a division of labour and collective actions, according to UNIDO (2007) and Rodrik (2008), are that, if they work together, firms can pool their resources to purchase inputs, supply larger markets by increasing their volume of production, share costs related to technology purchase, investments in shared facilities and access to services, specialize in complementary functions, and exchange information and innovate.

The following section considers how firms could analyze clustering and the link between the formation of clusters and enhance their own competitiveness.

Cluster Development

The cluster pyramid developed by Stanford Research Institute (SRI) International (Figure 5.2) is a helpful tool to analyze the development of a cluster (USAID 2003). The top tier of the pyramid represents the 'core cluster firms' – the lead firms – that produce and export goods or services, while the second tier represents 'supplier firms' or those firms that provide inputs to the core cluster firms and are an essential part of the value chain. And the bottom tier of the pyramid includes the 'foundation factors' that provide the cluster's building blocks.

The model's most important dimension is perhaps the circle that surrounds the pyramid, i.e. the synergies and dynamism that result when the pyramid's three layers are all engaged and working towards a common goal (USAID 2003). It should be noted that the co-location of actors within geographic proximity does not automatically lead to interaction, learning and innovation (Zeng 2008).

6 Such collective gains are normally obtained through various industrial and professional associations.

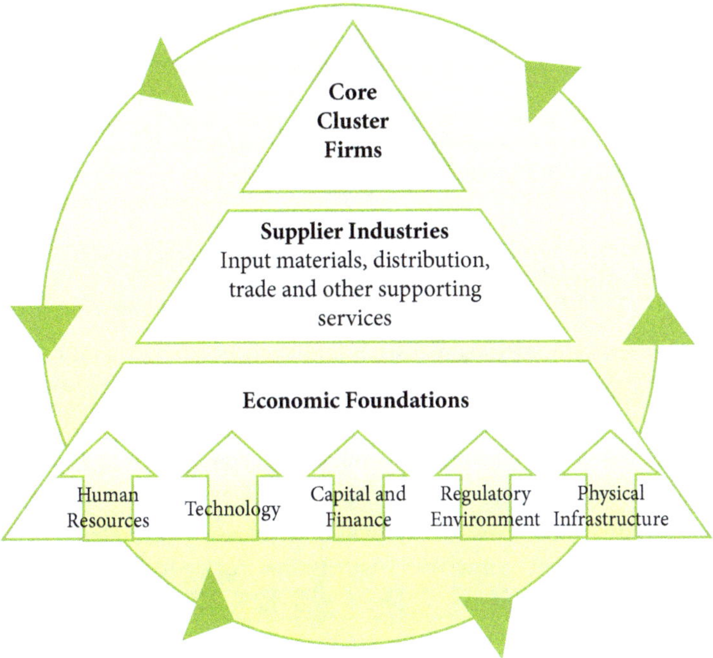

FIGURE 5.2 *Cluster pyramid*
SOURCE: SRI INTERNATIONAL IN USAID (2003)

Hargroves & Smith (2005) recommend a five-step process to advance cluster development, with industry preferably taking the lead but supported by government:

i. understanding the industry context
ii. setting objectives
iii. understanding customers' needs
iv. articulating strategic positioning
v. cluster leaders developing action guidelines for the industry

UNIDO (2013) has proposed a methodology to analyze cluster development based on the following steps:

- cluster selection that entails the identification of the cluster(s) to be assisted by a cluster development agent or cluster broker
- a diagnostic study that involves an action-oriented analysis of the cluster's strengths, weaknesses, opportunities and threats
- vision building and action planning to formulate a vision and a corresponding development strategy shared by the entire cluster

- ➢ implementation, i.e. the management and coordination of the activities outlined in the action plan, including the establishment of horizontal and vertical networks
- ➢ monitoring & evaluation (M&E) of the qualitative and quantitative outcomes of the project

Cluster Development in African Countries

Based on eleven case studies in Africa and their challenges and constraints, Zeng (2008) identified a number of elements that can lead to the spontaneous or constructed development of clusters. Their growth and sustainability in the long term will depend on how they cope with the challenges facing them. The challenges include: keeping up with the knowledge revolution and increasing global competition; the lack of a critical mass of skills and talent; weak links between businesses and knowledge institutions; weak governmental and institutional support; and resource depletion and the failure to meet international standards.

To compete in a globalized world, specialization in an identified cluster means pooling scarce resources in order to be excellent at a global level. Countries have to select sub-sectors that have growth potential as they cannot support and excel in all sectors. The dairy industry in Kenya is a cluster that is potentially of interest for Kenya at the level of the East African market. The Kenyan dairy sector is ahead of that in other countries in terms of having infrastructure already in place and in potentially having a certain critical mass of specialized businesses that would provide an initial basis for a competitive dairy cluster. These issues are, however, likely to arise as challenges for the dairy industry chosen for this study. An assessment will need to be made of the challenges that are currently being experienced and the efforts being undertaken to resolve them. The following section discusses how the cluster pyramid approach was used to identify the main actors in the dairy sector in Kenya and how they relate to each other.

Approach and Methodology

Two of the three milk regions identified by the Kenyan Dairy Board were selected, namely the North Rift region and the Central Rift region. They were assessed using UNIDO's cluster approach (described above). According to the World Development Report (2012), 'the cluster-based approach helps to identify policy

PHOTO 5.1 *Dairy farming at Marula farm in Naivasha (Central Rift region)*
PHOTO: ALBERTO GIANI

and institutional impediments to competitiveness and innovation'. Based on the SRI cluster pyramid, the impediments to competitiveness and innovation were assessed by looking at supportive ('economic foundation') actors in each of the following five dimensions in relation to lead firms and suppliers in the dairy cluster: training, policy, finance, research and infrastructure. The research objective was to develop sustainable policies to improve dairy-processing production. To develop these policies further, the government wanted to understand the existing institutional situation and how its institutions are accessed and linked with lead firms and suppliers.[7] The concept of linkages is of crucial importance to understanding competitiveness. Porter (2003) identified horizontal and vertical linkages in the value chain where vertical linkages focus on supplier/lead (core cluster) firms' relations and horizontal linkages on the relations within one tier. The notion of linkages is also used in the cluster theory and this study investigated the linkages between the supplier/lead firms and economic foundation

7 This Nuffic-funded project is being carried out in cooperation with Egerton University and the Dairy Training Institute in Naivasha, both of which are government or semi-government training institutions. Provision was made to meet the wishes of these organizations regarding a better understanding of the institutional dairy landscape in order to provide the training programmes required in the sector.

actors. The main aim of the research was to understand the conditions needed to optimize local dairy production. The following sub-questions were formulated:

- ➢ What kinds of relations exist between dairy-processing lead firms and suppliers and the five economic foundation actors in the two clusters?
- ➢ What are the gaps and opportunities per actor in the Kenyan Dairy Board clusters that are being researched regarding these relations?
- ➢ What are the differences between the Northern Rift Dairy Cluster and the Central Rift Dairy Cluster?

The interaction focuses on three kinds of linkages: information-knowledge provision; supportive actions; and coordination efforts (Rodrik 2004, 2008; Rabellotti, 2009). The strength of each interaction was operationalized in terms of their frequency as follows:

- ➢ Information-knowledge provided: less than twice in six months was classified as 'little', two times in six months as 'medium' and more than two times in six months as 'strong'
- ➢ Supportive actions: none to one supportive action per week was classified as 'weak', two supportive actions per week as 'medium' and three or more as 'strong'
- ➢ Coordination: none to one coordination effort with the private sector in six months was classified as 'weak', two efforts as 'medium' and three or more as 'strong'

To address these questions, a schematic cluster map was prepared. Then the strength of interaction (as perceived by the other actors) was assessed in relation to three kinds of linkages using a 3-point Likert scale. The Likert scale makes it possible to classify the opinions of the various actors and provides insight for policymakers on how much attention should be given to developing policy instruments to change the level of interaction per actor and/or per cluster. The three indicators for the strength of the linkages are not always applicable for all actors. If not applicable, the researchers used the qualification N/A. Two cluster maps were prepared based on the SRI model (see Figure 5.2). The top tier features lead firms, which are the processing firms that provide processed milk and yoghurt to different parts of the country. Representatives of the dairy associations were also interviewed to gain a sense of the issues affecting the industry more generally.

The second tier considers supplier industries that provide four services. Stakeholders from the following four types of supply firms were interviewed: farms that could be organized in cooperatives are the main supplier of 'raw material', i.e. milk for processors; the packaging industry; transporters who take the finished products to markets; and the markets themselves as they are important suppliers of 'market space' for the dairy products. In this tier, one or two actors from each of the four general areas were interviewed. The selection of actors was based on their visibility and how active they were in the industry. The main supplier industries are listed on the cluster map.

The third tier consists of the five main economic actors. These are: (i) educational institutions; (ii) policy institutions; (iii) financial institutions; (iv) research-related institutions; and (v) infrastructure-related actors. One or two institutions were assessed for each group that provides milk processing, not for dairy farming. Market and road infrastructure[8] were also investigated. In the research/technology section, institutions responsible for dairy market research, product development and marketing were analyzed. Financial actors are those who ease access to credit for dairy farmers or processors, and the educational institutions prepare the workforce for the industry.

The most relevant organizations in each sector were identified at an expert meeting and key representatives were interviewed using semi-structured questionnaires. The organizations considered to be relevant were the largest or most important in the cluster as perceived by the experts. By actor, and based on purposive sampling, key policy-strategy makers in the organizations were interviewed. Purposeful sampling was used to isolate the relevant members of the selected organizations and extend relationships and logic among constructs in the study (Eisenhardt & Graebner 2007).

Finally, the questionnaires were analyzed and the two clusters were compared according to the strength of the perceived linkages. Information was collected from 33 different decision makers (actors from all three tiers in the cluster pyramid) to determine the strength of the linkages of these economic actors in the two clusters being researched. Much of the information relied on the stakeholders' own experiences and views but factual information was verified by cross-checking amongst the stakeholders.

The data were classified by the researchers in a Likert scale and ranked as weak, medium or high. The information allowed the links to be assessed between the lead firms or the suppliers with the economic foundation actors for the entire circuit to be identified. Recommendations were then provided on cluster development.

8 Here the infrastructure itself was considered rather than the actors involved in it.

Mapping

As a result of fieldwork, the following actors were mapped (Figures 5.3 and 5.4).[9] The lead firms are defined as processing industries. As the processors set the price in the chain, the main processing association is listed here too. The final products range from processed milk, yoghurt, cheese, butter and milk powder to ghee (butter oil).

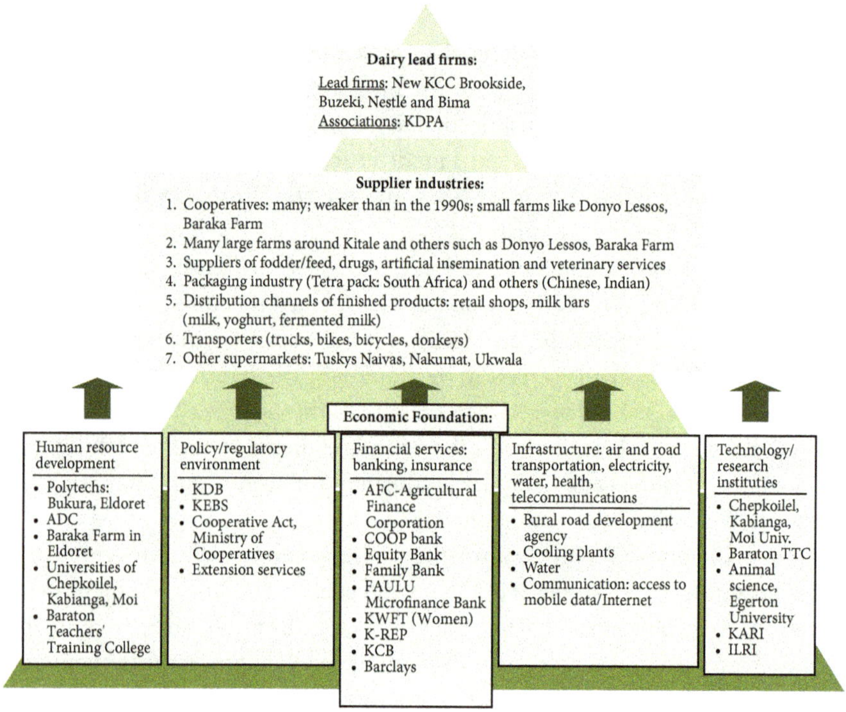

FIGURE 5.3 *Northern Rift dairy cluster actors*[10]

9 Special thanks go to Simone van Vugt at the Centre for Development Innovation, Wageningen UR for assisting with the compilation of part of this data.

10 Adapted from the SRI International cluster pyramid. NKCC = New Kenya Cooperative Creameries, KDPA = Kenya Dairy Producers Association, KDB = Kenya Dairy Board, KEBS = Kenya Bureau of Standards, KWFT = Kenya Women Finance Trust, K-REP = Kenya Rural Enterprise Programme Bank, KCB = Kenya Commercial Bank, ADC = Agricultural Development Corporation, KARI = Kenya Agricultural Research Institute, ILRI = International Livestock Research Institute.

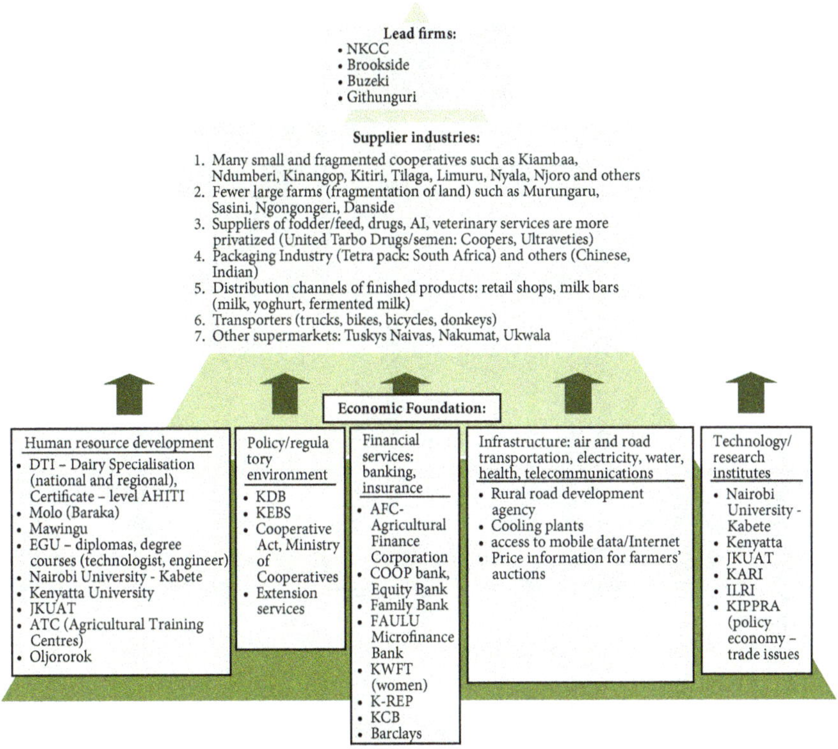

FIGURE 5.4 *Central Rift dairy cluster actors*[11]

Findings

The data were clarified. When the linkages between the lead firm or the supply firms with the five economic foundation actors were assessed, the following linkages were noted based on the respondents' perception. They were ranked: weak, medium or strong by the researchers. The outcome in terms of the strength of the linkages were almost the same for suppliers and lead firms. It was, therefore, decided to show the results in one table. An interpretation of the data in Table 5.1 is given in Table 5.2. The findings in Table 5.1 are visualized in Figure 5.5.

11 NKCC = New Kenya Cooperative Creameries, DTI = Dairy Training Institute, AHITI = Animal Health Training Institute, EGU = Egerton University, JKUAT = Jomo Kenyatta University of Agriculture and Technology, KEBS = Kenya Bureau of Standards, KWFT = Kenya Women Finance Trust, KDB = Kenya Dairy Board, KARI = Kenya Agricultural Research Institute, ILRI = International Livestock Research Institute, KIPPRA = Kenya Institute of Public Policy Research and Advocacy.

TABLE 5.1 *Linkages between lead firms/suppliers and economic foundation actors in the Northern and Central Rift regions*

	Dimensions of economic foundation				
Kind of linkages	**Policy**	**Financing**	**Research**	**Training**	**Infrastructure roads/ cooling plants**
Information & knowledge provision (North)	Medium: two information sessions were provided in six months	Strong: banks generally provide information on a regular basis	Weak: information is provided to the private sector less than once a year	Weak: information on training programmes provided is less than once a year	N/A
Supportive actions (North)	Medium: extension services limited	Medium: coverage by banks is limited to urban centres	Weak: hardly any research actors available and there are few supportive actions provided for the private sector	Weak: hardly any training programmes available each year	Weak/medium: roads are in a poor state and some feeder roads not passable the whole year; cooling systems available but no complete coverage
Coordination (North)	Medium/strong: the government is taking a coordinating role	Medium: efforts to coordinate every six months	Weak: hardly any coordination with other actors	Weak: hardly any coordination with other actors	Medium: disputes over the coordination of maintenance about once a year

Information & knowledge provision (Central)	Weak/medium: less than two information/knowledge sessions conducted in six months	Strong: regular and adequate information provided	Weak/medium: information provided about once a year	Weak/medium: information about training provided about once a year	N/A
Supportive actions (Central)	Medium: extension services are limited	Medium/high: present in about 10 locations	Medium: limited number of research actors available and assistance about once a year	Medium: limited number of actors available and about 1–5 training programmes a year	Medium/high: roads in a reasonable state; cooling structure more than enough but 'white elephant' risk
Coordination (Central)	Weak/medium: less than two coordination efforts every six months	Medium: efforts to coordinate with other actors in six months	High: with IFAD's Smallholder Dairy Commercialisation Programme (SDCP), coordination linkages are provided on a regular basis	Weak: only starting	Medium: disputes about who is responsible for types of roads/feeder roads/highways; cooling plants are fine but it is not always clear which farmers the cooling plants are intended for

TABLE 5.2 *Key to Table 5.1: Scoring on the strength of linkages on a 3-point Likert scale*

		Dimensions of economic foundation					
	Score	**Policy**	**Financing**	**Research**	**Training**	**Infrastructure: roads**	**Infrastructure: cooling plants**
Information – knowledge provision	Weak	Less than twice in six months	Provided but not on a regular basis and not available (or limited) when requested	Information provided less than once a year	Information on training programmes provided less than once a year	N/A	N/A
	Medium	About twice every six months	Provided on a regular basis *or* when requested	Information provided about once a year	Information about training programmes is provided about once a year	N/A	N/A
	Strong	More than two times in six months	Provided on a regular basis and when requested	Information provided more than once a year	Information on training programmes is provided more than once a year on a regular basis	N/A	N/A

Supportive actions	Weak	None to one supportive action per week	Presence in less than 5 urban areas in the region	Research actors are rarely available and assist less than once a year	Adult training almost never available	Roads: feeder and main roads perceived to be in a poor state, not passable the whole year	Milk collection structure: no cooling facilities
	Medium	About two supportive actions per week	Presence in about 5–10 urban areas in the region	Research actors assist about once a year and are available on a limited scale	1–5 programmes available for adult training	Roads: feeder and main roads in a reasonable state, passable all year but not paved	Collection centres: cooling facilities available but no complete coverage
	Strong	More than two supportive actions per week	Presence in more than 10 urban areas in the region	Research actors available and assist more than once a year	A comprehensive training programme is available	Roads: mainly paved	Collection centres: complete coverage of cold cooling facilities
Coordinative efforts	Weak	None to one every six months	Almost non-existent	Almost non-existent	Almost non-existent	Many disputes (more than once a year) about who is responsible for road maintenance and cooling structures	
	Medium	About two every six months	About once every six months	About once every six months	About once every six months	Limited disputes (about once a year)	
	Strong	More than two in six months	More than once in six months	More than once in six months	More than once in six months	No disputes	

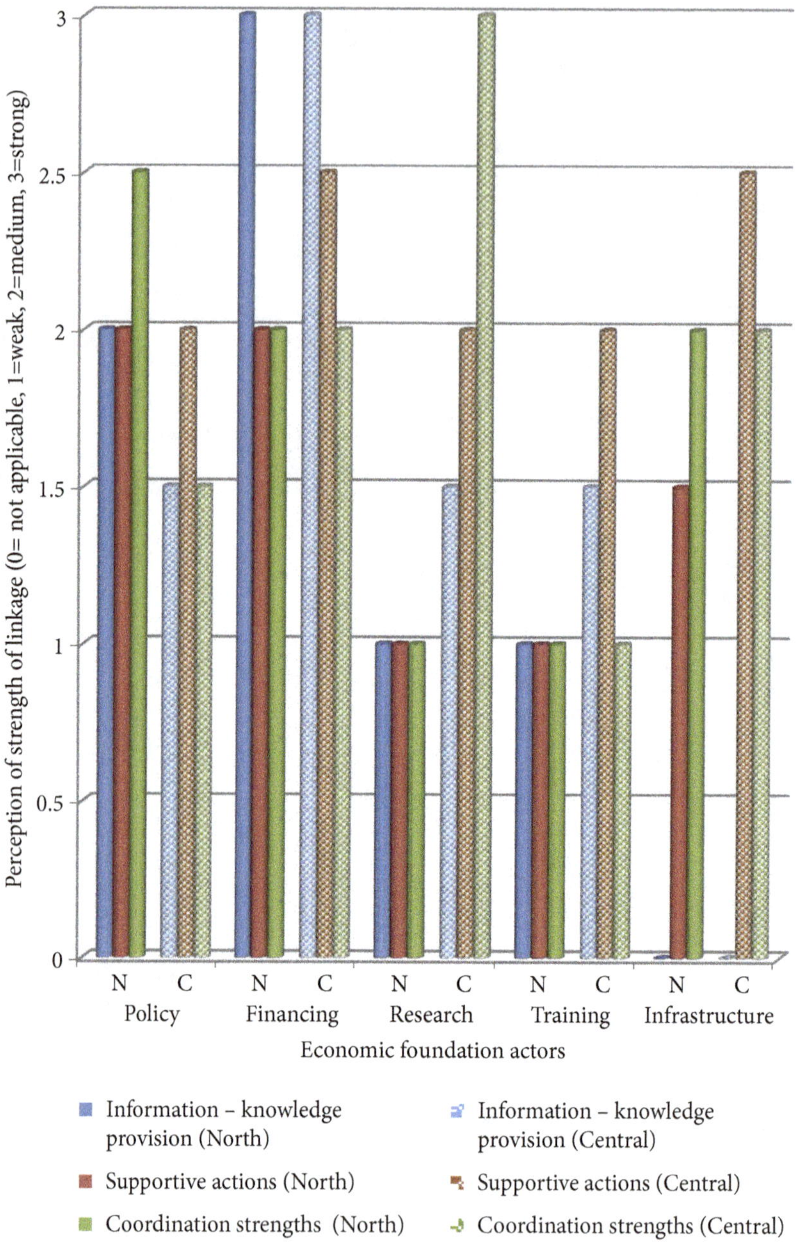

FIGURE 5.5 *Perceptions of the strength of linkages between lead firms/suppliers and five types of economic foundation actors in three domains for two dairy clusters (N = Northern Rift, C = Central Rift)*

The government regulator concerned with dairy, i.e. the Kenya Dairy Board and its extension officers, were considered as policy actors, as were its linkages with lead firms and suppliers. Table 5.1 shows that the coordination efforts of the Dairy Board in the Northern Rift region were stronger than in the Central Rift region in the sense that it had more regular coordination meetings and provided more information-exchange programmes than there were in the Central Rift. However, the transportation budget can be seen to be insufficient in both clusters and the extension officers did not, therefore, go into the field as often as they otherwise might have. It is also interesting to note that, with the new Constitution since 2010, the role of the government is not clear and this is hampering coordination in the dairy sector. This was particularly true in the Central Rift.

The role of the financial players is also interesting. In general, they were perceived as being aggressive towards the industry in the way they offered their services. They were perceived as being knowledgeable about the dairy industry and supportive in their role as professional advisor to the business. There were sufficient banks and branches to serve the dairy industry but not all the larger villages had banks focusing on the dairy industry. This was especially the case in the Northern Rift. In terms of coordination, it was found that they were active but were not doing much regarding coordinating actions in the cluster.

When it came to research, the linkages between the lead firms/suppliers and research were considered to be weak overall. Very little knowledge was being transferred from the research institutions to the lead firms. In the Northern Rift region, hardly any research institutions were found, although there were some in the Central Rift. Outreach to the private sector remains limited and is weaker in the Northern Rift region because of the distances involved. It was noted that only IFAD's Smallholder Dairy Commercialisation Programme was addressing this linkage in the Central Rift and the strength of coordination was therefore considered a little higher than in the Northern Rift.

The situation concerning the linkages between the lead firms/suppliers and the training institutions was similar. In the Northern Rift region, hardly any training institutions existed, which affected knowledge provision and coordination strengths. The situation in the Central Rift was a bit better simply because there were more training institutions there. However, the recent start-up of an outreach training programme for adults might change this situation.

As for infrastructure, the project considered the state of the roads and to what extent cooling plants were available across the region. Roads are important for the improvement of linkages between the actors in the dairy cluster. The roads in the Northern Rift are much worse than those in the Central Rift. This is, therefore, an important factor that hampers the cluster dynamics in the Northern Rift. In terms of coordination, it is important to mention that it is not always clear who is responsible for road maintenance, which does not help to improve road infrastructure in general. The development of a cooling plant network is a hot topic for donors and private investors/businesses. In the North, the network distribution of cooling plants in the north has been organized quite well but, in the Central Rift, the number of cooling plants is too high, which has resulted in some 'white elephants' emerging.

Conclusion

Dairy is a booming industry in Kenya and the dynamics of the clusters are increasing in both the Northern Rift and in the Central Rift in the sense that linkages are getting stronger. However, the dynamics differ in each cluster. Considering the three functions that were operationalized – coordination, information sharing and support – it can be concluded that overall coordination is an issue that is conducted by the Dairy Board. But, with the new Constitution now in place, this role is no longer evident. The coordination role seems to have been picked up better in the Northern Rift than in the Central Rift. However, more donors in the Central Rift region are working with the various dairy actors, such as IFAD and the Dutch government, which is cooperating with the Kenyan Ministry of Agriculture. As a result, more multi-stakeholder meetings are being organized in the Central Rift region, although none have yet been structural in nature.

Overall Information/knowledge provision between the actors in both the Northern Rift region and the Central Rift region is poorly developed. Only the financial institutions perform well on this component and are actively involved in showing what they can offer. Information/knowledge provision is better provided by the different actors in the Central Rift region but improvements are still required there too. As to the information linkages between the research and training institutions with the private sector, these are perceived to be weak and a sound framework in which such linkages can be provided is still lacking.

The financial actors do well in supporting private-sector actors with the help they require. But this supportive action is not perceived to be high for the other actors involved. In relation to policy, extension officers are often not supportive as they do not have much budget for transport. Training support by the institutions is often provided for younger students but not for existing managers and farmers. And research support is rarely available as the research actors and the private-sector actors have few points of contact.

Finally, the infrastructure is supportive in terms of cooling systems in the Central Rift region and to a lesser extent in the Northern Rift region, but the quality of the roads, especially in the Northern Rift region, remains poor. Overall, the supportive actions of the actors in the cluster are classified as 'weak to medium', although the Central Rift cluster is doing slightly better than the Northern Rift cluster. This may be because there are more economic actors in the Central Rift and the condition of the roads there is better, but also possibly because this region is closer to Nairobi and financial support is easier to access.

In summary, the linkages in both clusters are still weak. The level of organization in the Central Rift seems better than in the Northern Rift. This may be because more actors are attempting to focus on the linkages in the cluster, such as research and training institutions. The Northern Rift region needs more knowledge and support, which could be provided by setting up training and research programmes and/or institutions in the region.

If the linkages amongst the lead firms/suppliers are considered, it is striking that the coordination role has not been fully adopted by the Dairy Association for processors as it does not feel responsible for the coordination of the whole cluster. Sharing knowledge is not common and the firms see each other more as competitors than as team actors that could assist each other's cause. This notion still needs to be developed in both the Northern Rift and the Central Rift regions.

The potential of the Northern Rift region is generally much higher in terms of the quantity of milk processed. Clear statistics are not available but, according to those interviewed, the Northern region, provides 75% of all the milk produced in Kenya. With better linkages, the cluster in the Northern region would become much stronger and production levels and the competitiveness of this cluster would increase. More emphasis on this cluster is, therefore, an interesting route to economic development in Kenya.

References

Eisenhardt, K.M. & M.E. Graebner (2007), Theory Building from Cases: Opportunities and Challenges. *Academy of Management Journal* 50(1): 25–32.

Hargroves, K. & M. Smith (2005), *The Natural Advantage of Nations, Business Opportunities, Innovation and Governance in the 21st Century*. London: Earthscan.

Kenya Dairy Master Plan (2010), Nairobi: Government of Kenya.

Majiwa, E.B., M.M. Kavoi & H. Murage (2012), Smallholder Dairying in Kenya. *Journal of Agriculture, Science and Technology* (*JAGST*) 14(2).

Nordin, S. (2003), Tourism Clustering and Innovation: Paths to Economic Growth and Development. *Etour Analysis and Statistics*. European Tourism Research Institute, Mid Sweden University.

Porter, M.E. (1998), Clusters and the New Economics of Competition, *Harvard Business Review* (Nov–Dec): 77–90.

Porter, M.E. (2003), The Economic Performance of Regions. *Regional Studies* 37: 6–7.

Rabellotti, R. (2009), Knowledge and Information Networks in an Italian Wine Cluster. *European and Planning Studies* 17(7): 983–1006.

Rodrik, D. (2004), Industrial Policy for the Twenty First Century. http://myweb.rollins.edu/tlairson/pek/rodrikindpolicy.pdf. Accessed 28 February 2014.

Rodrik, D. (2008), Economic Research Forum, Industrial Policy: Don't ask why, ask how. *Middle East Development Journal.* Demo Issue (2008) 1–29.

UNIDO (2007), Introduction to the UNIDO Cluster Development Methodology. Presentation for workshop on 'SME Cluster and Network Development: Principles and Practice'. Port of Spain, Trinidad & Tobago.

UNIDO (2013), *The UNIDO Approach to Cluster Development, Key Principles and Project Experiences for Inclusive Growth.* Vienna: UNIDO.

USAID (2003), *Promoting Competitiveness in Practice. An Assessment of Cluster-Based Approaches.* Washington, DC: US Agency for International Development.

Wambugu, S., L. Kirimi & J. Opiyo (2011). Productivity Trends and Performance of Dairy Farming in Kenya. Nairobi: Tegemeo Institute of Agricultural Policy and Development.

World Development Report (2012), *Gender Equality and Development.* Washington, DC: World Bank.

Zeng, D.Z. (ed.) (2008), *Knowledge, Technology and Cluster-based Growth in Africa.* Washington, DC: World Bank.

CHAPTER 6

Biofuel Feedstock Production in Ethiopia

Status, Challenges and Contributions

Maru Shete and Marcel Rutten

Climate change and a rise in fuel prices over the last decade have led to increased demand for land worldwide, including in the developing South, for the production of biofuel feedstock. In Ethiopia, considerable amounts of land has been transferred to investors for biofuel feedstock production. Depending on the type of land conversion and the feedstock produced, this contributes either positively or negatively to countries' green economies and food-security efforts. Based on data collected from secondary sources, this chapter examines the status, challenges and contributions of biofuel feedstock production in Ethiopia.

Introduction

The production and use of first- and second-generation[1] liquid biofuel as a solution for the twin challenges of energy security and climate change have been widely promoted since the 2007/2008 fuel-price crisis (FAO 2008; Fargione *et al.* 2008; Peters & Thielmann 2008). The European Union Renewable Energy Directive (Directive 2009/28/EC), which was adopted in 2009, requires EU member states to meet at least 10% of their energy demands for transport from renewable sources by 2020 (EU 2009). This measure was followed up by interested parties acquiring large tracts of what was seen as cheap, underdeveloped land for biofuel feedstock production, particularly in African countries. The increased global demand for land for biofuel feedstock production is expected to quadruple in the next fifteen to twenty years (Fairless 2007) and is projected to account for 20% of the world's agricultural land by 2050 (White & Dasgupta 2010). This figure is probably too high. Van Eijck (2007: xiii) has shown some of the challenges linked to transferring to jatropha biofuels in Tanzania. 'Production of biofuel could help to stop soil erosion, create additional income for the rural poor and provide a source of energy both locally and internationally'. She noted several obstacles and uncertainties but concluded

1 'First-generation fuels are produced from food crops, including palm oil, rapeseed, sugarcane, and corn. Second-generation fuels are non-food crops, like switch grass and *Jatropha curcas*, or the residual inedible parts of food crops, such as the husks and stems of corn. Third-generation biofuels are derived from algae' (Dauvergne & Neville 2010: 635).

 | DOI 10.1163/9789004282698_007

that, overall, the transition to jatropha biofuels could be made although it might take a long time. This conclusion was reached before food prices rose in 2008. A 2007 publication by a team at Stanford and Nebraska universities concluded that: (i) rapid growth in the bioethanol and biodiesel markets is placing increasing demands on key agricultural commodities and it is likely that the demand for biofuels and the related effects on agricultural prices will continue as long as petroleum prices remain at more than US$55–60 per barrel; (ii) political economy interests in some important countries and regions, such as the US, China, Brazil, Indonesia and the EU, will likely perpetuate growth in biofuel production capacity in the medium-term regardless of short-lived fluctuations in petroleum prices; and (iii) biofuel growth will rely primarily on agricultural commodities in the coming decade and be constrained largely by food-crop production capacity. The expansion of land for biofuel will come at a cost to the environment due to forest-clearance schemes, a loss of biodiversity, reduced water and impaired soil quality, which could offset the benefits of biofuels (Naylor *et al.* 2007). The team stated that research should determine whether the growth in biofuels could be sustainable but expressed the hope that second-generation technologies would become available. They also questioned whether food security would benefit more from biofuel investment by specific companies or groups or investments in agriculture at a regional and/or national level. And they called for sustainability audits to understand the effects of biofuels on food security and the environment.

Several groups have voiced their concerns about biofuel production, especially its impact on crop prices for (poor) net consumers of food. Between 2003 and 2007, two-thirds of the global increase in maize production went to biofuels (World Bank 2009: 4).[2] Pressure on arable land due to biofuel feedstock production is argued to challenge food security over the next 20 years (FAO 2011). Opponents have also stated that biofuels are not necessarily a panacea

2 Initially only the maize market was affected but as farmers switched land from wheat and soybean production to growing maize, the price of these commodities also began to rise. Higher oil and fertilizer prices increased food production costs, especially in high-income countries where they can account for as much as 30% of overall costs. This, plus the demand of biofuels for grains, has made the price of these products much more sensitive to changes in oil prices. Above US$50 a barrel, every percentage-point increase in the barrel price causes maize prices to rise by 0.9% (World Bank 2009: 7). Finally, a series of poor wheat harvests in Australia resulted in lowered stocks and contributed to the price rise. In addition to these fundamental drivers, agricultural prices have been influenced both by increased investor interest in these commodities as an asset class and by government policies, including the decision by several countries to impose export bans. All of these factors are driven by forward-looking expectations and may have exacerbated the price rises in 2007–2008 and their subsequent decline (*Ibid.*: 4).

for the twin crises of energy security and climate change that initially served as the justification for investing in biofuels.

Using biofuels as a source of energy and minimizing dependence on fossil fuels was generally expected to curb the challenges of climate change. However, this has not been the case. Levels of greenhouse gas (GHG) emissions per unit of energy from biofuels depend on how and where the biofuel crops are produced (UNIDO 2010; Pacheco *et al.* 2010) and the conversion technology used (Alder *et al.* 2007; Pacheco *et al.* 2012). For instance, Fargione *et al.* (2008) estimated that biofuel crop expansion on to natural landscapes could release 17 to 420 times more CO_2 than the annual greenhouse gas reductions gained by using liquid biofuels. Gibbs *et al.* (2008) noted that biofuel feedstock production is carbon-saving when produced on degraded or cultivated plots and carbon-emitting when produced as a replacement for tropical forests. In the more specific case of palm oil expansion in South East Asia, Danielsen *et al.* (2009) and German *et al.* (2010) noted that widespread deforestation has already occurred and reversing the carbon emissions that have been generated as a result of the conversion will require several decades or more of carbon-payback periods (Gibbs *et al.* 2008; Achten & Verchot 2011). Some also argue that land conversion might be a threat to livelihoods and/or food security among local populations if high-quality lands that were sources of food and animal-feed production are used for biofuel feedstock production (Dauvergne & Neville 2010; German *et al.* 2010; FAO 2011; Rahmato 2011) and to the natural ecosystem (Dauvergne & Neville 2010; German *et al.* 2010). A UK study concluded that, without strict policies, biofuel production would 'result in net greenhouse emissions and loss of biodiversity through habitat destruction' (Sweney 2009: 1). The Gallagher Review (2008: 64) also reported that while it should be possible to develop a long-term global sustainable biofuel industry, 'the introduction of biofuels should be significantly slowed until adequate controls were in place to address the indirect impact of production'. But as Pickmeier & Rutten (2013) show for Kenya, local people strongly oppose growing sugarcane for ethanol plantations that would destroy crucial dry-season grazing areas, although many welcomed a jatropha plantation on underused livestock ranches for a variety of reasons including the paying-off of land rent debts, keeping out foreign pastoralists and the destruction of an invasive shrub (*prosopis juniflora*). It is clear that detailed studies examining various economic and ecological issues are needed for each specific case while general statements should be rejected.[3] Let us first examine the biofuel situation in Ethiopia in more detail.

3 A Canadian investor successfully developed a 25 ha trial but then pulled out due to a lack of financial means (personal communication, Bedford official 2013).

When interest in biofuels started, Ethiopia saw that domestic, foreign and state investors were keen to acquire land for biofuel production. The government thought it would contribute to employment creation, fiscal revenue generation, their Clean Development Mechanism (CDM)[4] and the strategy of saving[5] and generating foreign currency (Ministry of Mines and Energy 2007). In 2006, the government developed a biofuel strategy aimed at tapping this opportunity and identified land for biofuel production by private-sector and state actors. The strategy emphasized the availability of suitable land and a good climate for the production of biodiesel crops, such as castor seed, jatropha (*Jatropha curcas*) and palm oil, and of bioethanol from sugarcane plantations (*Ibid.*). The target is to produce 1.8 bn litres of liquid biofuel by 2015, of which 1.6 bn litres are expected to come from biodiesel and 195 m litres from bioethanol (MoFED 2010). Accordingly, the government has embarked on a strategy of state-sponsored large-scale sugarcane plantations in order to meet part of the country's energy demands from bioethanol production. The Ethiopian government has justified its intervention in the sector because of the limited involvement of the private sector in sugar development as it requires enormous capital investment (Davison 2011). In its five-year Growth and Transformation Plan (2010/11-2014/15), Ethiopia planned to build eleven[6] new sugar factories and expected to generate US$1 bn per annum from biofuel exports (MoFED 2010). The four existing sugar mills and factories are currently being expanded.

Based on secondary data collected from different sources, this chapter examines the status of biofuel feedstock production in Ethiopia and the amount of land being allocated to state and private actors for biofuel production. It also assesses the amount of land developed to date and the challenges of promoting biofuel production. The contribution that biofuel production has on local people's lives and the country's energy security are then discussed. The concluding section provides some key policy implications and suggests

4 The Clean Development Mechanism (CDM) is one of the flexibility mechanisms defined in the Kyoto Protocol that provides for emissions reduction projects that generate Certified Emission Reduction units, which may be traded in emissions trading schemes: http://en.wikipedia.org/wiki/Clean_Development_Mechanism.

5 Ethiopia spends 87% of the foreign currency it earns on financing fuel imports. Biofuel production is being adopted as a strategy to save foreign currency by substituting fossil-fuel imports (Ministry of Mines and Energy 2007).

6 While ten are going to be developed by the state at a cost of US$4.6 bn, one sugar factory is being established by a private Ethiopian-owned company in Amhara Regional State. According to Global Trade Alert (2013), the Indian government provided the Ethiopian government with US$90 m to develop the sugar industry.

the way forward for biofuel production. The chapter thus adds to the scant literature on land acquisition for large-scale biofuel production in Ethiopia and serves as a useful source of information for further research.

Status of Biodiesel and Bioethanol Production in Ethiopia

Ethiopia is estimated to have 23.3 m ha of land suitable for biodiesel (Forum for Environment 2011) and 333,500 ha of land for irrigated and rain fed sugarcane and associated bioethanol production (Ethiopia Investment Agency 2008). This section considers the amount of land licenced and transferred for biofuel production and discusses the status of some investment projects in different regional states based on specific case studies. A joint report by a team of experts from the Ministry of Agriculture and the Ministry of Energy and Water Resources indicated investment licences were offered to 83 investors in Ethiopia to produce biodiesel feedstock in various regions. Although 22.9% of these investors were reported to be in the pre-implementation phase, the team found that only 7.2% of them had started activities and then just on a limited scale (MoA *et al.* 2011). Table 6.1 shows the major investors that acquired land parcels greater than 2000 ha. A total of 1.5 million ha of land were leased to 34 investors for biodiesel feedstock production in Ethiopia. Based on this figure, only 6.5% of what has been estimated as being suitable for biodiesel feedstock production has so far been transferred to investors and the average amount of land acquired by the investors was 45,115 ha.

TABLE 6.1 *Major companies with land acquired for biodiesel production in Ethiopia*

Company name	Country of origin	Region	Land size (ha)	Year
Adventure Ethiopia Agric. Development PLC	China/South Africa/Ethiopia	Amhara	50,000	2007
Biomassive AB	Sweden	Amhara	100,000	2007
BDFC Ethiopia Industry	N/A	Amhara	18,000	N/A
Ambasel Trading	Ethiopia	Benshanguel	20,000	2007
S&P Energy Solution	India	Benshanguel	50,000	2010
Sun Biofuel Eth	UK	Benshanguel	80,000	N/A
Jatropha Biofuels Agro Industry	N/A	Benshanguel	100,000	N/A
IDC investment	Denmark	Benshanguel	15,000	2007
Sisay Yohannes	Ethiopia	Gambella	100,000	2008

Company name	Country of origin	Region	Land size (ha)	Year
African Climate Exchange PLC	US/Ethiopia	Multiregional	100,000	2005
Africa Sustainable Energy Corporation PLC	Netherlands/US	Multiregional	20,000	2006
Ertale Bio Diesel PLC	Britain/Ethiopia	Multiregional	50,000	2007
A.B.S.A. Bio-Fuels PLC	South Africa China/Ethiopia	Multiregional	30,020	2007
Horizon Plantation PLC	Saudi Arabia/Ethiopia	Multiregional	300,000	2007
OBM Ethio Renewable Energies PLC	Italy/Ethiopia	Multiregional	50,000	2008
Ardent Energy Group	US	Multiregional	10,000	2008
Petropalm Corp-Ethiopia PLC	Austria/US	Oromiya	50,000	2009
Yehuda Hayun	Israel	Oromiya	8000	2007
Green Energy plc	Ethiopia	Oromiya	50,000	2007
Flora Ecopower PLC	Germany	Oromiya	6148	2008
J.M.B.O. Bio Fuel Production PLC	US/Ethiopia	Oromiya	2000	2008
Soubra Abdallah Khalid	Lebanon	Oromiya	10,000	2008
Emami Bioteck	India	Oromiya	11,000	2009
Petro Palm Corporation Ethiopia	N/A	Oromiya	50,000	N/A
VATIC International Business	N/A	Oromiya	20,000	N/A
Etan Biofuels	Ethiopia	SNNPR	5550	2007
Getachew Mulugeta	Ethiopia	SNNPR	25,000	2007
2A 2S International Business PLC	Ethiopia	SNNPR	60,000	2007
F.E.P.E. Amaro Bio-Oil PLC	Cyprus/Israel	SNNPR	50,000	2008
Gereth Modular Fuel & Energy PLC	Germany	SNNPR	5000	2009
Global Energy Ethiopia	N/A	SNNPR	2700	N/A
Omo Sheleko Agro Industry	N/A	SNNPR	5500	N/A
Fri-el Green	Italy	SNNPR	30,000	2006
Agro-Peace Bio Ethiopia PLC	Israel	Somalia	50,000	2009
Total			1,533,918	

SOURCE: DOCUMENTS FROM THE MINISTRY OF AGRICULTURE AND ETHIOPIAN INVESTMENT AUTHORITY (2013)

Biodiesel Investment Case Studies and their Status

Various cases of biofuel investment that started on a small scale are discussed here. At least one case study from each regional state was considered to provide a picture of the status of biofuel development in the country. The cases were selected based on the availability of information and do not necessarily represent the whole regional situation.

Ambasel Trading

Ambasel Trading is a domestic investment firm that acquired 20,000 ha of land for jatropha cultivation in Benshanguel Gumuz Regional State in 2007. The company developed close to 200 ha of land and planted jatropha on 15 ha. After acquiring the land, the company changed its licence from one for 'production of biofuel' to 'other ends' without sufficient justification.

S&P Energy Solution PLC

S&P Energy Solution is part of the large Indian construction conglomerate Shapoorji Pallonji that was founded by Pallonji Mistry. In 2010, the company leased 50,000 ha of land in the Dangur and Guba Districts of Benshanguel Gumuz Regional State to cultivate *Milletia pinnata* (Pongamia) as feedstock for biodiesel production and other edible food and oil crops. The soil is a combination of Nitisols and Acrisols. The lease runs for 50 years at ETB 143.4 per ha (€5.4/ha)[7] but only 3500 ha has been developed to date. The land is inhabited by the Gumuz indigenous people whose livelihoods are based on shifting cultivation, gold mining, hunting and gathering wild foods from the forest. The company has no prior experience in agriculture and the farm, the first in its kind in Ethiopia, is jointly owned by Mr Cyrus Pallonji Mistry and his brother Mr Shapoorji. The former has a 30% share in Tata Share Company and served as the CEO of S&P Energy Solution PLC until recently when the leadership transferred to his brother, Mr Shapoorji.

Agro-Peace Bio Ethiopia

Agro-Peace Bio Ethiopia is owned by an Israeli national who acquired an investment licence for the production of caster seed for biodiesel production in Somalia Regional State in Ethiopia. The company started screening 40 caster-seed varieties on 10 ha of land in 2009. Irrigation water from boreholes using a drip irrigation system has been put in place and the company has been providing extension services to pastoralists who engage in crop production.

7 €1 = ETB 26.44 (Ethiopian Birr) on 7 January 2014. In July 2010, €1 was the equivalent of ETB 16.41.

The company's plan is to grow caster bean on 50,000 ha of land and produce biodiesel for the export market. However, the land it was allocated was affected by salinity and was not suitable for caster-bean production. This has hampered the company's progress.

Fri-el Green

Fri-el Green is owned by an Italian citizen who acquired 30,000 ha of land in the south of Ethiopia for palm-oil production in 2006. It developed less than 1000 ha of its concession and, as a result, the government revoked 15,000 ha due to the company's sluggish performance. While Ethiopia's biofuel strategy stated that land allocation for biofuel development does not affect land uses for food production and pastoral grazing, the land leased to Fri-el Green has been used by Dassanech pastoralists for the summer-season cultivation of sorghum, maize, millet and grazing of livestock. The company had planned to cultivate jatropha but this ran counter to government strategy and a large proportion of the land has consequently been left idle.

Flora Eco-Power

Flora Eco-Power is owned by a German national who leased 6148 ha of land in Oromiya Regional State in 2008 and started producing caster bean in 2009. The company set up an oil-processing plant in Fedis District in Oromiya Regional State that has a daily processing capacity of 250 tons of oil seed for the export market. It subcontracted its activities to an Israeli-owned foreign firm and had an outgrower scheme on 12,000 ha of land and produced 5500 tons of caster bean for processing and export to China and Germany. However, following disputes between the company owner and the Israeli operator, operations ceased soon afterwards and the company has not been able to pay for 120 tons of caster bean produced by outgrower farmers according to its contractual agreement. This has created conflict with surrounding farmers. The company took out a ETB 7 m loan from the Nib International Bank Share Company and received fertilizer from Afersa Cooperatives Union for ETB 5 m on credit, but did not pay for either the loan or the fertilizer credit. Later it subcontracted the activities to a French-operated firm but has been cultivating caster bean only on its 5000 ha concession since the outgrower farmers were not compensated for the earlier agreement.

Emami Biotech

Emami is an Indian-owned farm that leased 11,000 ha of land in Oromiya Regional State for the cultivation of jatropha feedstock for biodiesel production in 2009. The company started to develop its land and started plantation activities using flooding irrigation from the Areba River. The company faced challenges from local people as a result of land-related disputes and the low wage rates it offered (ETB 15 per day), which has negatively affected its progress because of conflicts. The local administration and village elders tried to mediate between the local people and the company to ease tensions but progress has not been what was expected.

Status of State and Privately Owned Sugarcane Production

In addition to biodiesel feedstock production, huge tracts of land have been allocated to privately owned and state-run sugarcane and bioethanol developers in Ethiopia's regional states. This is aimed at minimizing the negative effects of climate change by using energy from bioethanol to generate/save foreign currency by substituting fossil-fuel imports for the transport subsector by blending gasoline with ethanol, the creation of employment opportunities on plantations and the processing of sugarcane (Ministry of Water Resources & National Meteorological Agency 2007; MoFED 2010; Ministry of Mines and Energy 2007). A total of 294,000 ha of land has been allocated for eleven new sugar factories that are planned in three regional states[8] (Ethiopia Sugar Corporation 2013a). Data from the Ethiopian Sugar Corporation showed that 330,000 ha has already been allocated to the state-owned Sugar Corporation and private investors for the production of sugarcane and bioethanol. Compared to the amount of land identified as suitable for sugar development, the land transferred so far accounts for 98.9% of land potentially suitable for this purpose (Table 6.2). Many of the projects are well underway in 2014 and should be completed towards the end of the Growth and Transformation Plan (GTP) in 2015 (Ethiopia Sugar Corporation 2013b). Ethanol for blending with gasoline is one of the products targeted in the projects.

Ethiopia's sugar-production capacity before the expansion of the existing sugar mills and the establishment of new ones was 300,000 tons annually and, the country imported 150,000 tons of sugar to meet domestic demands in 2013. With the expansion of old sugar mills, the country expected to meet its domestic requirements by the end of 2013. And when the new mills start crushing sugarcane, production is expected to increase to 2.3 m tons a year making the country a net exporter of sugar (Davison 2011).

8 A total of seven new sugarcane farms are being set up on 175,000 ha of land in Southern Nations, Nationalities and People's Regional State; three new sugarcane farms will be established in Amhara Regional State on 75,000 ha of land; and, a new sugar factory is being opened on 44,000 ha of land in Tigray Regional State as well (Ethiopia Sugar Corporation 2013a).

TABLE 6.2 *Status of state and privately owned sugar factories and bioethanol targets in Ethiopia*

Sugar factories	Region	Size (ha)	Water source	Production capacity		Electric power generating capacity (megawatt)[1]	Expected employment	Status
				Sugar (tons/ year)	Ethanol (m^3/year)			
Fincha[2]	Oromiya	21,000	Finchaa River	270,000	20,000	31	Data not available (N/A)	Completed in 2013
Metehara	Oromiya	11,000	Awash River	130,000	12,500	9	12,500	Completed in 2013
Wonji/Shoa	Oromiya	25,022[3]	Awash River	95,000	20,775		N/A	Completed in 2013
Tendaho	Afar	50,000[4]	Tendaho River	619,000	55,405	120	50,000	2015
Omo Kuraz[5]	South Omo, SNNPR	175,000	Omo River	278,000	26,162	405	117,131	Kuraz is expected to be completed in 2014
Tana Beles[6]	Amhara and Benshanguel	75,000	Beles River	726,000	62,481	150	50,199	First two factories are expected to be completed in 2014
Wolkayit	Tigray	11,000	Zarema & Kalema Rivers	284,000	26,750	60	N/A	2015
Kessem	Afar	20,000	Kebena River	153,000	12,500	26	N/A	To be completed in 2014

Arjo-Dedessa[7]	Oromiya	28,000	Dedissa River	155,520*	N/A	N/A	N/A	90 percent completed
Hiber Sugar*	Amhara	6000	Tana Beles	N/A	N/A	N/A	N/A	Implementation started
Karuturi[8]	Gambella	15,000	Baro River	N/A	N/A		N/A	In its initial stages
Total		330,000						

SOURCES: ETHIOPIA SUGAR CORPORATION (2013A, 2013B & 2013C); MINISTRY OF FOREIGN AFFAIRS (2013); * HIBER SUGAR SHARE COMPANY (2009)

NOTES: 1 *Business Times*, 24 May 2014: Addis Ababa, Ethiopia.

2 Fincha sugar factory was initiated in 1977 but sugarcane plantation only started in 1991/92 on 55.74 ha and the factory was completed in 1999. It has expanded its ethanol production capacity to 45,000 litres a day, i.e. yields of 20,000 m^3 of ethanol per annum and it has increased its sugar production to 270,000 tons annually.

3 The factory was set up by the Dutch Company, HVA, in 1952 and Shoa sugar factory was incorporated in 1962. An expansion plan was recently implemented on 16,000 ha of land and 1000 ha is being developed by outgrowers.

4 The factory planned to have local farmers grow sugarcane on 25,000 ha using outgrowers.

5 Five different sugar factories are planned for the Omo Kuraz sugar plantation.

6 Three factories are being planned for the Tana Beles sugar plantation. While 50,000 ha of land is in the Awi zone of Amhara Regional State, the remaining 25,000 ha are being developed in Dangur District of Benshanguel Gumuz Regional State (Ethiopia Sugar Corporation 2013c).

7 Arjo-Dedessa was owned by a private Pakistani investor called Al-Habesha PLC before being transferred to the state in 2012. The company went bankrupt after taking out huge loans with the Ethiopian Development Bank.

8 Karuturi Agro Products PLC acquired 100,000 ha in Gambella Regional State for the production of sugarcane, palm oil, rice and maize. It planned to produce sugarcane on 15,000 ha but there were financial problems even though the company applied for US$ 100 m in loans from different sources.

Challenges of Biofuel Production in Ethiopia

Biofuel production was considered a solution for the twin challenges of climate change and energy security in Ethiopia. However, the sector's performance has not been encouraging. As mentioned above, the implementation of projects to produce biofuels have shown sluggish progress. While this is generally true for biofuel projects planned by private agribusiness companies, state-operated sugarcane plantations and processing factories are functioning in Ethiopia regardless of the massive capital requirements.

The common challenges faced by private agribusinesses that acquired land for biofuels in Ethiopia are as follows:

- Investors often do not have a clear business plan, annual activity plan and/or Environmental Impact Assessment (EIA) to guide biofuel development operations.
- Companies own or use farm machinery and employ numbers of farm workers that are incompatible with the size of the land available. For example, S&P Energy Solution acquired only two dozers to develop land covered in trees and bushes and only six tractor-mounted ploughs to cultivate 50,000 ha of land. It is thus difficult to develop land in a few years and cultivate the entire parcel in a single planting season. This is related to the farms' (limited) financial capacity and affects their sustained and smooth operations. Due to financial constraints, S&P Energy Solution PLC adopted a strategy of reducing operations. As explained by the Human Resource Manager at S&P Energy Solution PLC, the company has invested close to US$16 m to date and there has recently been a serious push from head of the company (Shapoorji) to cover the costs of the farm from own revenue. As a result, the farm has reduced its major operations (e.g. land development/clearing land) and it is operating at reduced capacity. It has fired 23 Indian expatriates since June 2013 and reduced the number of permanent local staff from 200 to 81 in 2014. The farm's Operations Manager, Mr M.V. Sira Reddy, explained the challenges faced by S&P and the unsuccessful efforts it made to get a loan from domestic banks in Ethiopia. He explained that the cost of land development is very high and on average 70 big trees, in addition to several small trees and bamboo, have to be cleared to develop each ha of land. This has resulted in the company incurring huge costs by developing close to 3500 ha of land to dater. Wages are seldom paid on time and are lower than the going rate locally. During the survey period, S&P delayed paying wages by two weeks. The company pays ETB 39 for casual workers and interviews

with labourers indicated that the pay scale is too low for them to cope with the high costs of living in the area.

- Biofuel plantations use limited farm inputs (fertilizer, pesticides) and few improved technologies to enhance their farms' productivity. In the case of S&P Energy Solution, where the crop – pongamia – is new to Ethiopia, the company was unable to find planting materials locally. Importing pongamia seeds from India was the only option available but the germination percentage of these seeds was low because they remained in the Ethiopian port awaiting customs clearance for several months. As explained by the management, inputs such as urea and improved varieties for maize and sesame are also not readily available. Investors faced challenges due to pests, weeds and diseases and support from the national agricultural research institute in terms of appropriate farm technologies has been very limited. In addition, the availability of technology that processes and converts biofuel feedstock to liquid fuel is a challenge that is also faced by domestic private investors. In none of the cases were the biofuel projects able to set up a biofuel processing plant to convert feedstock into liquid fuel and start to earn revenue.
- Companies have failed to collaborate closely with local, regional and federal administration and they rarely pay land rent fees. This has decreased the level of support that they could expect to receive if they face problems of different types. The management of S&P Energy Solution PLC revealed that they have a poor relationship with the district administration, which hampers their smooth operation. But, they indicated to have better relationship with leaders at regional and federal level.
- Very weak linkages often exist between companies and the local community. Some of them did not live up to expectations and honour the promises they made to the local people, which resulted in conflict and hampered the smooth running of operations. For example, land transferred to S&P Energy Solution had previously been used by the local people for shifting cultivation and collecting different non-timber forest products such as wild honey, firewood, forest fruits and roots. The local people did not support the biofuel project's arrival. This is evidenced by the theft of maize plantation upon harvesting.
- Availability of both skilled and unskilled labour is a major problem for farms operating in remote regions, such as Benshanguel Gumuz and Gambella regional states. The management of S&P Energy Solution PLC stated that the farm was dependent on Indian expatriates paying huge outlays since knowledge of pongamia plantations is limited in Ethiopia. Unskilled labour availability is a also a problem and the farm is forced to

> bring in labour from other areas and thus to pay additional transportation costs. Due to malaria and poisonous snakes around the farm, which are common challenges for labourers in the area, labour productivity is low. The health situation is worsened by the limited availability of medicine. The management of S&P Energy Solution PLC also revealed that there is unhealthy competition for labour between them and other domestic farms operating in the area. After they brought labour in from other regions, labourers moved to other farms that paid slightly higher wages. Since these farms in the vicinity had not incurred any transport costs to bring labourers into the area, they could afford to pay a little over the going wage.

While countries like Ethiopia claim that land allocated for biofuel production was not previously in use and the plantation of biofuel crops has contributed to the Clean Development Mechanism, it frequently emerges that land used for biofuel production is suitable for crop cultivation and livestock grazing, as was the case with S&P Energy Solution. This has been a source of conflict between local populations and investors and has hampered the progress of project implementation. In some areas, the jatropha planted on communal lands has been damaged by livestock that were grazing openly in the field. While In other cases, investors have stopped operating because they failed to be productive as the land allocated was not suitable for the production of biofuels.

The slowing down of the global economy and the subsequent drop in crude oil prices as well as the use of shale gas in the US have negatively affected demand for biofuels. The financial crisis has been a major factor in investors not gaining access to financial capital to develop the land they have acquired for biofuel production. The challenge presented by limited access to financial capital, coupled with Ethiopia's poor road, electricity and water infrastructure and the government allocating land for biofuel investment has affected the sector's development. It is also important to note that biofuel production is relatively new to Ethiopia. There is, therefore, very little expertise, few trained and skilled staff and limited technology supporting (domestic) investors. The absence of improved seed for biofuel crops and the weak institutional structure from federal to district level to provide the needed technical and administrative support to investors in the sector were mentioned as ongoing challenges (MoA *et al.* 2012).

Opinions in the North have changed regarding the wisdom of large-scale biofuel production, as was evidenced by the amendment of the European Union's Renewable Energy Directive. This decreased the mandate for member countries having to meet the energy demands of the transport sector from renewable energy sources from 10% to 5%. Following critiques and the poor progress made, many investors have pulled out of Ethiopia.

Contributions of Biofuel Production in Ethiopia

Employment creation is one of the positive contributions expected from investment in biofuel production in Ethiopia. Due to the relative incipiency and sluggish implementation of many of the biodiesel feedstock-producing and processing projects, the level of employment generated is very small. For example, Agro-Peace Bio Ethiopia created jobs for 50 people while many more had been expected. Similarly, Fri-el Green only created work for 51 permanent and up to 100 casual workers, who were paid at a daily rate of ETB 35. Data collected in 2014 for S&P Energy Solution PLC indicated that jobs were created for 187 individuals on a temporary/casual basis. During the picking season, this increased to 700 wage workers.

Expectations that S&P Energy Solution would create jobs for local people were high in the households interviewed. People hoped this would allow them to generate a decent income from employment and contribute to other community-development activities. However, these expectations were not met, as is illustrated in Table 6.3. The company contributed to community development by launching a school-feeding programme for 199 students at Kota Junior School. It spent ETB 160,000 for three consecutive years and this resulted in improved school enrolment among the indigenous Gumuz population. The company also built a delivery room/labour ward for mothers from Kota village although it has not started providing services since the building has not been furnished with the necessary facilities. Nevertheless, interviews with local people indicated that these contributions are insignificant compared to the initial promises made. These included: upgrading the junior school to a high school; supplying electricity to Kota village; training Gumuz youth with skills and recruiting them on the farm in different capacities; and providing a furnished labour ward. The household-level perception analysis also confirmed the significant difference between ex-ante expectations shaped by earlier promises and the ex-post reality of the company's efforts regarding community development. In terms of the project's contribution to technology transfer to the local people and to the local availability of food supplies, households had very low prior expectations. Actual contributions were close to the households' initial expectations, a fact that was confirmed by the non-significant difference result (Table 6.3).[9] Key local informants also explained that, due to the company's

9 It is worth noting that households had a prior perception of losing key forest-based livelihood resources as a result of the S&P project and this was confirmed by the paired *t*-test analysis that showed that there is no statistically significance difference between the ex-ante expectations and the ex-post reality regarding deforestation and loss of associated livelihood sources.

TABLE 6.3 *Households' perceptions of expected and realized outcomes from S&P Energy Solution in Dangur District, Benshanguel Gumuz Regional State*

Outcome variables	Score for ex-ante expectations		Score for ex-post reality		Mean difference	St. Err.	t-value
	Mean	Mode	Mean	Mode			
Technology transfer	0.71	1	0.69	0	0.02	0.04	0.58
Employment generation	4.17	4	2.01	2	2.16	0.08	26.68*
Decent income generation from employment	4.13	4	1.88	2	2.25	0.07	31.9*
Contributes to community-development activities	4.11	4	1.36	1	2.75	0.06	45.96*
Increase in domestic availability of food supply	0.24	1	0.23	0	0.01	0.02	0.59
Deforestation and loss of forest-based livelihood sources	4.14	4	4.1	4	0.04	0.06	0.54

SOURCE: AUTHORS' OWN SURVEY DATA

Note: * Significant (2-tailed) at $p < 0.001$

operations, forest-based livelihood sources had been lost, fallow periods were shorter and crop productivity had declined. Cultivation of sorghum (locally known as *Quanch*) through shifting cultivation, followed by the gathering of foods from the forest (forest honey, a root crop locally called *Aecha* and hunting wild animals) are the dominant food sources for the Gumuz population, but all have declined. As key informants explained, before the land clearing by S&P, a typical Gumuz collected up to 90 kg of forest honey but this has now declined to 10–24 kg. These factors have increasingly undermined the community's food security.

Another company, Flora Eco-Power, would have performed relatively better in terms of value addition and creating employment in a short period of time had it not ceased operations due to a dispute between the owner and the operator. During its short period of operations, it generated 820 permanent jobs and 3500 jobs for casual workers. By contrast, job creation at the state-owned sugarcane plantation and processing factories is enormous (Table 6.2). It was reported in 2012 that jobs were created for 28,000 citizens by the different sugarcane plantation and processing factories. This is projected to increase to 200,000 when some of the factories move to full operation (Ethiopian Ministry of Foreign Affairs 2013). Improving the local population's food security through employment in both private and state-owned biofuel projects is also a justification for promoting the sector.

The other benefit expected from biofuel feedstock-producing farms is the production of liquid biodiesel and bioethanol. This is presented in the national biofuel development strategy as part of the country's Clean Development Mechanism that minimizes GHG emissions. Biofuel production is also claimed to be a way of increasing foreign currency by substituting fossil-fuel imports and generating foreign currency through exports. Some examples of contributions of this sort in Ethiopia from biodiesel developers in Tigray and Amhara Regional States should be mentioned. A company called Africa Eco-Power Initiative established a biodiesel processing plant in Mekelle Town in Tigray Regional State with a daily capacity of 2000 litres of biodiesel. Similarly in Bati District in Amhara Regional State, a plant that processes jatropha into biodiesel with a daily capacity of 200 litres has already been set up. The processing plants in both Tigray and Amhara Regional States have a very small capacity and cannot be considered as big contributors. Yet the Ministry of Agriculture felt that it was important to have a good start that could serve as valuable experience for this emerging sector (MoA *et al.* 2011). None of the biofuel projects presented here have, however, set up a biofuel feedstock-processing plant.

The government plans to increase the amount of ethanol production from sugar factories to 181,604 m^3 by the end of the GTP period (Sugar Corporation 2013a). So far Ethiopia has blended about 39 m litres of ethanol with gasoline since it started blending activities in 2009, saving over US$30 m. The government intends to increase the ethanol content to 25% (up from the current 10%) by the end of its five-year Growth and Transformation Plan.

In his annual report, the Director General of the Ethiopian Sugar Corporation stated that the country is importing more than 2 m tons of sugar every year and spending vast sums of scarce foreign currency. The setting-up and the expansion of new and existing sugar plantations and processing factories is expected to meet domestic demands for sugar in 2014, which would save much-needed foreign currency (Ethiopia Sugar Corporation 2013b; Kebede 2013). By the end of the GTP period, the production of sugar in Ethiopia is expected to have increased eightfold from the current level of 300,000 tons annually. This would provide the country with the opportunity to export 1.25 m tons of sugar that would, in turn, generate much-needed foreign currency (Davison 2011). In addition, the sugar mill factories aim to generate a total of 832.15 MW of electricity, of which it is expected that 530.1 MW will be contributed to the national power grid, with the balance being used by the factories themselves (*Business Times* 2014).

Controversy, however, surrounds state-sponsored sugar development in Ethiopia's regional states. While the government claims that the intervention has benefited the local people by providing employment and social services, including improved roads, potable water, schools, health stations and flour mills (Ethiopia Sugar Corporation 2013d), human rights groups have criticized it as 'land grabbing' sponsored by the state that has displaced the local people from their ancestral land for little benefit (Human Rights Watch 2012; Oakland Institute 2011). The Ethiopian Wild Life Conservation Authority (EWCA) has also raised its concerns about the possible negative effects of the large-scale sugarcane plantation in the Lower Omo Valley on the wildlife resources of the Omo-Tama-Mago complex. EWCA is concerned because no environmental impact assessment was carried out on the possible effects of the Kuraz sugar project on the distribution, movement and habitat requirements of the area's wildlife resources before the project started (EWCA 2011).

Conclusion and Policy Implications

The global trend of meeting part of the world's energy demands through liquid biofuels encouraged the Ethiopian authorities to embark on the development of a new biofuel strategy that resulted in the allocation of (cultivable) land to both private and state actors.

The land identification process for biofuel development seems to have been carried out hastily and the allocation of land suitable for biofuel production that does not compete with other objectives of food production and environmental goals was not done carefully. Private investors were not screened for their capacity and commitment and the government did not establish a well-coordinated institutional framework or provide the necessary support to private biofuel producers. Different government institutions, such as the Ministry of Agriculture, the Ministry of Water and Energy, the Ethiopian Investment Agency, the Ethiopia Sugar Corporation and regional governments were involved in the land allocation process, monitoring and the provision of policy support for investors engaged in biofuel development. But little coordinated effort was observed among them. The availability of technology that allows the processing and conversion of biofuel feedstock to liquid biodiesel and a supply of farm inputs, better yielding crop varieties, fertilizers and pesticides were limited. Support from national agricultural research centres by providing the technical know-how required for biofuel development was limited as the sector is still new.

In contrast with state-owned sugarcane development projects, privately owned biofuel development projects showed discouraging performances. Many investors pulled out before the implementation phase and those that started operations showed very sluggish progress. This raised questions about whether biofuel development by private agribusinesses is a viable strategy in Ethiopia, something that requires revisiting the approach.

References

Achten, W. & L. Verchot (2011), Implications of Biodiesel Induced Land Use Changes for CO_2 Emissions: Case Studies in Tropical America, Africa, and Southeast Asia. *Ecology and Society* 16(4): 14.

Alder, P.R., S.D. Grosso & W.J. Parton (2007), Life Cycle Assessment of Net Greenhouse-gas Flux for Bioenergy Cropping Systems. *Ecological Applications* 17: 675–691.

Danielsen, F., H. Beukema, N.D. Burgess, F. Parish, C.A. Brühl, P.F. Donald, D. Murdiyarso, B. Phalan, L. Reijnders, M. Struebig & E.B. Fitzherbert (2009), Biofuels Plantations on Forested Lands: Double Jeopardy for Biodiversity and Climate. *Conservation Biology* 23(2): 348–358.

Dauvergne, P. & K.J. Neville (2010), Forests, Food, and Fuel in the Tropics: The Uneven Social and Ecological Consequences of the Emerging Political Economy of Biofuels. *Journal of Peasant Studies* 37(4): 631–660.

Davison, W. (2011), Ethiopia Building Ten New Sugar Plants to Become Leading Exporter. http://www.bloomberg.com/news/2011-09-13/ethiopia-builds-10-new-sugar-plants-as-it-aims-to-become-leading-exporter.html. Accessed 12 September 2014.

Ethiopia Sugar Corporation. (2013a), Sugar Corporation & Ethiopian Sugar Industry Profile. http://www.etsugar.gov.et/en/about.html. Accessed 12 December 2013.

Ethiopia Sugar Corporation. (2013b), Implementation of Sugar Development Plan Encouraging. http://www.etsugar.gov.et/en/news/item/108-implementation-of-sugar-development-plan-encouraging.html. Accessed 1 January 2014.

Ethiopia Sugar Corporation. (2013c), Tana Beles Sugar Project Makes A Stride By Completing Main Irrigation Canal. http://www.etsugar.gov.et/en/news/item/103-tana-beles-sugar-project-makes-a-stride-by-completing-main-irrigation-canal.html. Accessed 4 January 2014.

Ethiopia Sugar Corporation. (2013d), Residents around Omo Kuraz Happy with Sugar Development Project, Villagization Program. http://www.etsugar.gov.et/en/news/item/101-residents-around-omo-kuraz-happy-with-sugar-dev%E2%80%99t-project-villagization-program.html. Accessed 4 January 2014.

Ethiopia Investment Agency (2008), Investment Opportunity Profile for Sugar Cane Plantation and Processing in Ethiopia, Addis Ababa, Ethiopia. Unpublished document.

European Union (EU) (2009), Directive 2009/28/EC of the European Parliament and the Council of the European Union on the Promotion of the Use of Energy from Renewable Sources and Amending and Subsequently Repealing Directives 2001/77/EC and 2003/30/EC. *Journal of the European Union* L140/16-L140/62.

EWCA (2011), Existing Challenges: Plantation Development versus Wildlife Conservation in the Omo-Tama-Mago Complex. Addis Ababa: Ethiopian Wildlife Conservation Authority.

Fairless, D. (2007), Biofuel: The Little Shrub That Could—Maybe. *Nature* 449: 652–655.

FAO (2008), Climate Change Adaptation and Mitigation in the Food and Agriculture Sector. Paper presented at the conference on 'Food Security: The Challenges of Climate Change and Bioenergy', Rome: 3–5 June.

FAO (2011), Land Tenure and International Investment in Agriculture: A Report by the High Level Panel of Experts on food Security and Nutrition. Rome: FAO.

Fargione, J., J. Hill, D. Tilman, S. Polasky & P. Hawthorne (2008), Land Clearing and the Biofuel Carbon Debt. *Science* 319: 1235–1238.

Forum for Environment (2011), Assessment of Biofuel Development Activities in Ethiopia, Addis Ababa: Unpublished Report.

Gallagher Review (2008), The Gallagher Review of the Indirect Effects of Biofuels Production. http://www.bioenergy.org.nz/documents/liquidbiofuels/Report_of_the_Gallagher_review.pdf. Accessed 26 May 2014.

German, L., G. Schoneveld, M. Skutch, R. Andriani, K. Obidzinski, P. Pacheco, H. Komarudin, A. Andrianto, M. Lima & A.D. Norwana (2010), The Local, Social and Environmental Impacts of Biofuel Feedstock Expansion: A Synthesis of Case Studies from Asia, Africa and Latin America. *Brief Info* No. 34. CIFOR:

Gibbs, H.K., M. Johnston, J.A. Foley, T. Holloway, C. Monfreda, N. Ramankutty & D. Zaks (2008), Carbon Payback Times for Crop-Based Biofuel Expansion in the Tropics: The Effects of Changing Yield and Technology. *Environmental Research Letters* 3(2008): 1–10. http://iopscience.iop.org/1748-9326/3/3/034001/pdf/1748-9326_3_3_034001.pdf. Accessed 2 September 2013.

Global Trade Alert (2013), India: Loan to Ethiopia to Finance Development of Sugar Industry with the Condition to Source Inputs from India. http://www.globaltradealert.org/measure/india-loan-ethiopia-finance-development-sugar-industry-condition-source-inputs-india-1. Accessed 5 January 2014.

Hiber Sugar Share Company (2009), Hiber Sugar Share Company: Prospectus No. 01/01. Addis Ababa.

Human Rights Watch (2012), 'What Will Happen if Hunger Comes?': Abuses against the Indigenous Peoples of Ethiopia's Lower Omo Valley. http://www.hrw.org/sites/default/files/reports/ethiopia0612textonly_0.pdf. Accessed 12 October 2013.

Kebede, S. (2013), Sugar Corporation Plans to Satisfy Domestic Demand by 2006. http://www.etsugar.gov.et/en/news/item/109-sugar-corporation-plans-to-satisfy-domestic-demand-by-2006.html. Accessed 25 December 2013.

Ministry of Agriculture (MoA), Ministry of Water and Energy & Ethiopia Agricultural Research Institute. (2011), National Level Evaluation of Biofuel Development in Ethiopia. Addis Ababa: Unpublished Report [translated from Amharic].

Ministry of Agriculture, Ministry of Water and Energy & Ethiopian Agricultural Research Institute (2012), Minutes of the National Workshop held on Biofuel Production in Ethiopia. Addis Ababa, 5–6 April.

Ministry of Finance and Economic Development (2010), Growth and Transformation Plan (GTP) 2010/11-2014/15. The Federal Democratic Republic of Ethiopia, Ministry of Finance and Economic Development [MoFED], Addis Ababa.

Ministry of Foreign Affairs (2013), Ethiopia's Sugar Development Progressing Steadily: A Week in the Horn of Africa. http://www.mfa.gov.et/weekHornAfrica/morewha.php?wi=869#871. Accessed 7 January 2014.

Ministry of Mines and Energy (2007), Biofuel Development and Utilization Strategy. Addis Ababa: Unpublished Report.

Ministry of Water Resources and National Metrology Agency (2007), Climate Change Technology Needs Assessment Report of Ethiopia. Addis Ababa. Unpublished Report.

Naylor, R., A. Liska, M. Burke, W. Falcon, J. Gaskell, S. Rozelle & K. Cassman (2007), The Ripple Effect: Biofuels, Food Security, the Environment, Agronomy and Horticulture. Faculty Publications. Paper 386. http://digitalcommons.unl.edu/cgi/viewcontent.cgi?article=1386&context=agronomyfacpub. Accessed 29 May 2014.

Oakland Institute (2011), Understanding Land Investment Deals in Africa: Half a Million Lives Threatened by Land Development for Sugar Plantations in Ethiopia's Lower Omo Valley. http://www.oaklandinstitute.org/sites/oaklandinstitute.org/files/Land_Deal_Brief_Ethiopia_Omo_Valley.pdf. Accessed 10 November 2012.

Pacheco, P., D.A. Wardell, L. German, F.X. Johnson, N. Bird, J.W. van Gelder, H. Schwaiger, G. Schoneveld, K. Obidzinski, M. Guariguata, M. Skutsch, Y. Gao, G. von Maltitz, W. Achten, L. Verchot, H. Komarudin, O. Masera & R. Andriani (2012), Bioenergy, Sustainability and Trade-offs: Can We Avoid Deforestation While Promoting Biofuels? *Brief Info* No. 54. Indonesia: CIFOR.

Peters, J. & S. Thielmann (2008), Promoting Biofuels: Implications for Developing Countries. Ruhr Economic Papers No. 38. Ruhr-Universität Bochum (RUB).

Pickmeier, U. & M. Rutten (2013), Biofuels or Grazing Lands? Heterogeneous Interests in the Tana Delta, Kenya: A Cross-community Perspective. *ASC Infosheet* No. 16. http://hdl.handle.net/1887/20876. Accessed 2 July 2013.

Rahmato, D. (2011), Land to Investors: Large-scale Land Transfer in Ethiopia. Addis Ababa: Forum for Social Studies.

Sweney, M. (2009), Biofuels Ad Banned by ASA After George Monbiot Complaint. *The Guardian*, 14 January. http://www.theguardian.com/media/2009/jan/14/guardian-writer-george-monbiot-gets. Accessed 22 May 2014.

UNIDO (2010), Making It Industry for Development: Wind of Change? *Quarterly Magazine* No. 2. Vienna International Centre, Austria.

van Eijck, J. (2007), *Transition towards Jatropha Biofuels in Tanzania? An Analysis with Strategic Niche Management*. Leiden: African Studies Centre.

White, B. & A. Dasgupta (2010), Agrofuels Capitalism: A View from Political Economy. *Journal of Peasant Studies* 37(4): 593–607.

World Bank (2009), *Global Economic Prospects: Commodities at the Crossroads*. Washington DC: World Bank.

CHAPTER 7

Local Careers and Mixed Fortunes in Africa's Globalizing Food Exports

The Case of Nile Perch from Lake Victoria, Uganda

Joost Beuving[1]

New high-value food export crops have been emerging across Africa since the 1990s and are replacing traditional cash crops as the mainstay of African economies. Taking the macroscopic view, these global food exports appear promising as they can lift ordinary Africans out of poverty by creating new career opportunities, but a closer look at the local level unveils a situation of mixed fortunes. By looking at fishermen, traders and other economic actors making a living in the Nile perch export sector at Lake Victoria, Uganda, the chapter shows how local frames of meaning underpin their fortunes. Ethnographic analysis of their professional trajectories uncovers a tendency to perceive the business and other people in individualistic terms, and this compromises successful local careers. The chapter thus raises critical questions about conventional interpretations of Africa's globalizing food exports that emphasize institutions/structures over African agency.

The Problem: Understanding Local Careers in Africa's Globalizing Food Exports

New food exports have been emerging across the African continent since the 1990s: fresh fish, fruit and vegetables. Sometimes also referred to as non-traditional exports, these include high-value products that have penetrated the global market and thus connect Africa to overseas (mainly Western) destinations (Ouma 2012). These new global exports are making a significant contribution to national economies, often rivalling traditional cash-crop revenues (Swinnen 2007). At a local level, they are also associated with businesses, jobs and other economic activities that hitherto did not exist and are thus creating

1 The author would like to thank Moses Nsereko and Michael Ssali for their research assistance, the Makerere Institute for Social Research, the Ugandan National Council for Science and Technology in Kampala, the VU University Amsterdam, the Radboud University Nijmegen and Wenner-Gren for their financial andinstitutional support. The insightful suggestions made by participants at the authors' workshop at the African Studies Centre in Leiden and Andrea Downing were also greatly appreciated.

 | DOI 10.1163/9789004282698_008

new career opportunities. This is a significant development as these new exports are often emerging in parts of Africa where (i) possibilities for economic diversification were limited in the past; (ii) multiplier effects are virtually non-existent; and (iii) the opportunities for scaling-up into more profitable avenues are restricted because of global political-economic realities (Amanor 2009). This chapter hopes to provide a better understanding of the emergence and/or functioning of these new local careers as it is expected that they can throw more light on the causes and consequences of Africa's insertion into broader, global economic frameworks. It focuses on one particular example, namely the booming export business of Nile perch from Lake Victoria, Uganda.

Neoliberal interpretations of emerging local careers in Africa's global (food) export sector highlight the success of recent economic reforms (Diop & Jaffee 2005). Put in place since the 1980s, African economies have received an impetus by introducing policy packages consisting of: (i) currency devaluations; (ii) the abandoning of foreign-exchange controls; and (iii) the levelling of barriers to entry, such as taxes and trade licensing (Rimmer 2003). This led to a supply response, whereby hitherto subterranean, local African entrepreneurs, suppressed by decades of a hostile business climate, suddenly began to invest in new economic activities, thus also creating employment (Dijkstra & van Donge 2001). Once the supply response met with foreign investment and overseas demands, this process accelerated rapidly and led to a proliferation of new professional career opportunities and trajectories.

This chapter agrees that in the case of Uganda, economic reforms adopted by the Museveni administration promoted development after decades of economic mismanagement by the Amin and Obote regimes (Fulgencio 2009). Yet it challenges the optimistic notion that greater possibilities for exporting Nile perch pushed ordinary Ugandans into new careers. It will be shown how capital accumulation and a pattern of upward economic mobility are not the automatic outcomes of reform and questions will be raised about how a framework emphasizing economic opportunity can explain this.

More critical observers agree that Africa's new global exports may have boosted the careers of some Africans but they stress that these are usually those of people from the ranks of a small economically privileged class and the countries' administrative and political urban elites (Bayart 2000). Proponents of this view argue that these elites have seen a new occasion to create barriers to trade (or 'rents') in the emerging high-value food-export chains in order to benefit from it personally (Bernstein & Campling 2006). For instance, rapid access to essential airlifting facilities, needed to ensure the smooth functioning of the cold chain along which perishable food exports flow from source to overseas destinations, have evolved as a key economic resource over which

African entrepreneurs and powerful actors are struggling in the administration (Ponte & Gibbon 2005). Such struggles are seen as being detrimental to local development because they limit possibilities for capital accumulation and profit-making by underprivileged actors. While this may generally be true, this interpretation in the case of Ugandan Nile perch is of limited value as it focuses on what goes on at the higher end of the export chain and does not offer a satisfactory explanation for the diversity observed in career outcomes at lower levels. In particular, it fails to account for the local existence of both business success and failure at close range.

A major reason why the conventional interpretations outlined above have difficulties in appreciating the mixed fortunes of local Nile perch actors is that their careers are the result of contingencies associated with economic reform and/or economic globalization. From what might appear to be similar institutional/structural conditions (such as neoliberal economic policies and the creation of new export possibilities), local actors can pursue different courses of action. Previous research suggests that this may depend on (i) the web of social relations actors are a part of; (ii) the symbolic capital of authority and prestige that they have managed to build up; and (iii) the ambitions and expectations they have for the future (Long 2001).

This chapter takes issue with the reductionism characterizing the previous two interpretations, essentially seeing African initiatives as the by-product of the structures of class and/or economic opportunity (Dolan 2001). Instead it adopts an actor-oriented perspective that views these political-economic structures as important inasmuch as they are filtered through the life world of local actors and put into social action (Hobson & Seabrooke 2007).[2] It therefore considers these structures not as realities *sui generis* but as the outcome of human action. How this works in practice cannot be inferred from a study of political-economic structures but, in line with an actor-oriented perspective, must be appreciated situationally, i.e. with reference to concrete actors and the particular dilemmas they face and the solutions they find.

An ethnographic analysis of career histories can illuminate essential aspects of the agency of local actors (Guyer 1997), in this case of those working/living under conditions of economic globalization. To appreciate this approach, the chapter first explores the emergence of Nile perch exports from Uganda, looking in particular at how political-economic reforms and foreign investments have contributed to the creation of new career opportunities. To see

2 To some, the term 'actor-oriented methodology' may conjure up an image of methodological individualism but this is not the message of this chapter. It will be shown how careers are an emergent structural aspect of social life that can set real boundaries to human agency.

how local actors make use of these, the chapter then discusses the career histories of three ordinary Ugandans who are all engaged in income-generating activities related to Nile perch. Analysis of their social relations will receive particular attention in the analysis of these histories, as the economic decision-making underpinning professional careers is structured in ongoing networks of social ties (cf. van Donge 1992). A close look at the decision-making process will demonstrate how local Nile perch actors perceive their businesses in highly individualistic terms. This promotes the understanding of potential collaborators as competitors and shows how this particular value orientation is having a profound impact on local Nile perch careers.

The Emergence of Global Nile Perch Exports from Lake Victoria, Uganda

British colonial administrators introduced Nile perch into Lake Victoria, Kenya in the 1950s with the aim of boosting local animal protein production although this ambition did not materialize. The bucket of juvenile fish they emptied into the lake in an attempt to introduce them disappeared in the depths of the waters for many decades and these fish were not seen again until the early to mid-1980s (Pringle 2005). Along the Ugandan coast, there appeared to be little interest as local Ugandans did not consider Nile perch a delicacy and, due to its high fat content, the fish was reported to go off quickly. With other preferred fish species in abundance, in particular tilapia and *mukene*, which is a sardine-like fish and an important source of animal protein for ordinary Ugandans, local demand for Nile perch was limited. Many old-timers still remember the surprise they felt when the *bazungu* ('white men', Europeans) began to look for the fish in the late 1980s. They initially gave them away for free but when the local fishermen later began to realize the money they could make from selling Nile perch, they started to sell the fish to ambulant traders and a market for it quickly emerged.

With the market taking off in the mid-1990s (see Figure 7.1), the organization of the fishing and trading of Nile perch began to change. Initially it was considered a by-catch in a multi-species fishery that was connected to local and regional consumption markets. Many of the fishermen in those days were local villagers and sold the fish themselves either at local markets or at markets in town if they had a bicycle or some other form of affordable transportation. Today Nile perch are caught predominantly for the export market, with over 90% going to European destinations (Geheb *et al.* 2007). A distinction is usually made between those owning boats used for fishing and who often double

PHOTO 7.1 *View of a large landing site, Kasensero, central Uganda*
PHOTO: JOOST BEUVING

up as traders ('boat owners'), those who catch the fish ('crews') and those working in the service activities that support the fishing and trading sector (Medard 2012).[3]

This distinction coincides with a socio-ethnic segmentation. Boat owners are usually of local origin but the majority of the crews and those in the service industry are migrants from across the region who arrived at the lake in the wake of the Nile perch 'boom' of the 1990s (Nunan 2010). Though many of these new migrants settled at well-established landing sites, many more ended up at newly created sites (over 85% of the total landing sites in Uganda today). What was relevant for the development of the Nile perch sector is that these are places with a large, multi-ethnic and mainly transient or floating population (Balirwa 2007). They are usually engaged with Nile perch fishing and trading or, increasingly, some ancillary service activity connected to it, for example, providing accommodation and food for migrant fishermen, repairing fishing equipment such as nets or outboard engines, or selling mobile phone cards. The larger village landings are known for their vibrant leisure industry that offers pool tables, dances and video halls, and project an atmosphere of city rather rural life (Beuving 2010).

3 The boats are locally made canoes constructed from wooden planks and are suitable for beaching on the lakeshore and do not require any port-type infrastructure. Commercial trawling has been banned from the lake since the 1990s. Nile perch fishing is practised with (imported) monofilament nets and locally manufactured long-lines (Beuving 2014).

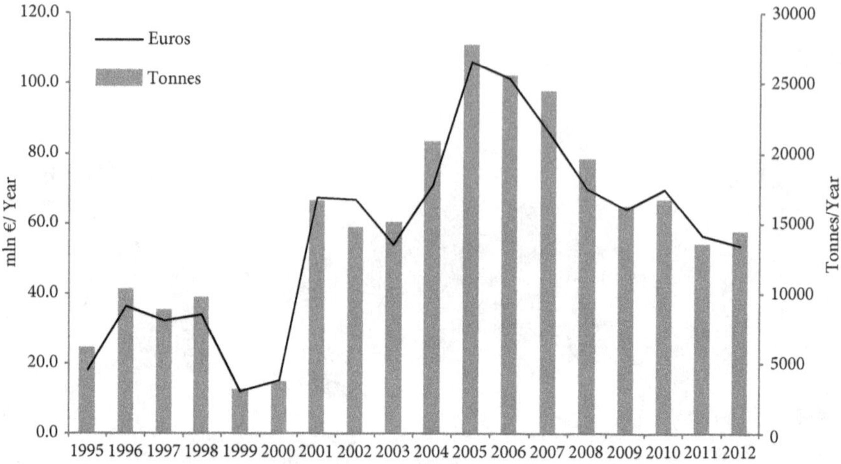

FIGURE 7.1 *Nile perch exports from Uganda to European countries, 1995–2012*[4]
SOURCE: COMPILED FROM EUROSTAT 2014

Making a distinction between boat owners/traders, their crews and others working in the service industry is to some extent an abstraction. Smaller-scale boat owners still often fish with their crews, while the large ones work as traders too. Successful crews invest in boats and become boat owners. Many of the bigger boat owners also own hotels, restaurants and video halls and the owners are occasionally seen to go out fishing. These positions are not neutral but function as part of a social hierarchy where increasing economic opportunity is being seen. Crews are usually those without sufficient means to procure a boat and, at the other end of the scale, the boat owners have the ambition to eventually retire from fishing and become fully fledged traders, preferably in close association with those further up the chain, such as the owners of fish-processing factories and exporters. It is the fluidity of these careers and their embedding in local frames of meaning that are the focus of this chapter.

Three Local Career Histories in Uganda's Nile Perch Sector

The findings presented here were collected during a twelve-month anthropological field study in central Uganda between 2006 and 2009. Subsequent field visits

4 It is difficult to compare Nile perch exports with Nile perch production (i.e. to gauge the proportion of local/regional trade) as catch statistics are hotly debated (cf. Kolding *et al.* 2008; Msuku *et al.* 2011). Nile perch fishing in the research area is predominantly export-oriented and the majority of the lorries used to transport the fish go directly to Entebbe/Kampala.

have since been made. The research started in Lambu, a prominent hub in Uganda's Nile perch export sector where large quantities of the commercial fish have been landed since the mid-1990s and then trucked to export points, particularly Entebbe and Kampala (Figure 7.2). Contacts were slowly developed in the field with a host of different actors – local boat owners, crews and traders – and this gradually made it possible to become closer to them and observe their social practices. Although Lambu was the starting point, local actors have tended to migrate so some of them were followed and the fieldwork site moved to other village landings, several of which were on the Ssese Islands in the lake. An attempt was thus made to build up an image of the social life of the landings sites 'from within'. By applying the principles of grounded theory, major themes were identified that are explored in more depth in this chapter (Corbin & Strauss 1998).

The career histories presented here are case studies and were selected with the aim of exploring the diversity in the career trajectories that can be observed at the local level i.e. boat owners/traders, crews, and leisure-industry entrepreneurs/workers (van Velsen 1967). These distinctions were adopted in the sampling process because they were believed to illuminate the essential dynamics that drive local careers. Adopting a purposeful sampling procedure clearly has methodological limitations, in particular because the findings of the study are not representative in the way surveys are (cf. Flyvbjerg 2006). It is, however, envisaged that the career studies selected could contribute to an empirically informed interpretation of the new career patterns emerging under the new conditions of globalization and/or reform and shed light on everyday economic decision-making processes.

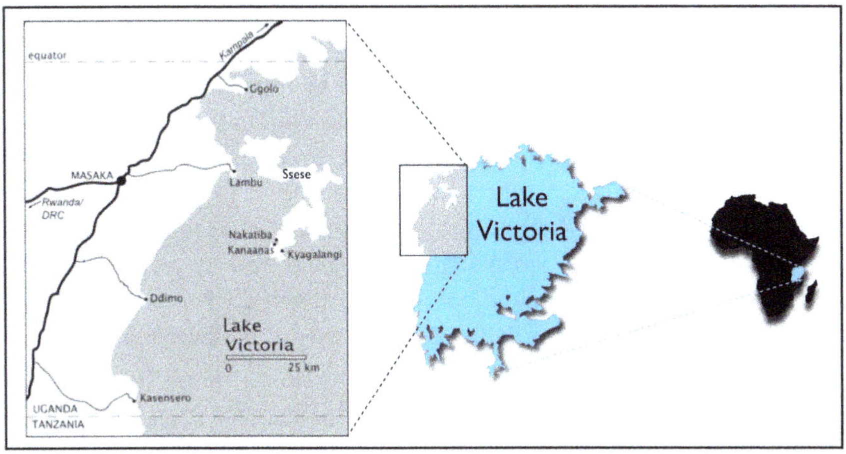

FIGURE 7.2 *Lake Victoria and the research area in Masaka District and the Ssese Islands in Central Uganda*

Case 1: Robert Muwanga: A (Former) Fisherman and Video-Hall Owner

Robert (born 1971), who originates from the Jinja area northeast of Kampala, is the only son in a Lusoga-speaking family of seven. Both his parents died of AIDS when Robert was still young. Cared for by poor relatives, he dropped out of school and began to work in town. His various jobs brought him into contact with a friend who was working at a nearby fish-processing factory and Robert was employed there too after some time. The factory was a Nile perch processing plant, set up a year earlier by enterprising Indian-Ugandan businessmen. Robert remembers the work as tough: cutting fish into fillets is smelly and noisy because of the refrigerating equipment. It is also a dangerous job: filleting knifes are kept razor sharp to ensure export-quality fillets and minimize losses during the cutting process. But at least it brought him some much-needed money.

The factory closed down in 1995 when Robert was 24. The EU issued an export ban on fresh fish from Lake Victoria around this time following several cases of salmonella contamination. This resulted in the EU's food hygiene inspectors closing its borders to African fish exports[5] and many factories had to lay off employees. For a while, Robert tried his hand at construction in the capital Kampala, which was booming in the 1990s as Uganda was experiencing rapid economic growth following decades of stagnation. But he was not happy: the masonry work he was assigned to was back-breaking and at times dangerous too so when an occasion presented itself to quit, he was ready to take it.

The opportunity arrived in the form of a former colleague at the fish-processing factory. He told Robert that the owner had begun to extend his business activities by sourcing fish directly. (Previously the factory depended on arrangements with traders.) In one of the villages designated for fishing, the owner was looking for an overseer who would be responsible for purchasing fish. Robert applied for the job and then started work in the fishing village of Lambu that was hidden deep in the countryside in Masaka District on the shore of Lake Victoria. Robert remembers that he was puzzled by life there at first as he had been used to city life and did not expect to last very long in Lambu.

While in his overseer's job, Robert noted that there were fishermen in Lambu who had money to spend but nothing to spend it on as Lambu is far from any town and leisure activities for fishermen were therefore extremely

5 Due to its export-oriented nature, the excess Nile perch could not be absorbed bv local/regional trade networks in the 1994–1997 period (Geheb *et al.* 2007) and fishermen saw their incomes drop dramatically. Figure 7.1 shows the dip in this period, which is still remembered vividly by veterans in the sector.

limited. So Robert, again through a friend, got the idea of showing videos – mostly Nigerian 'Nollywood' ones but also Chuck Norris fighting movies – to the fishermen for a small fee. He did this initially from his small room in Lambu but later built a shed-liked construction to accommodate his visitors. Initially he combined this with his official job and when he got tired of the locals' resistance to paying to see movies, he tried his hand at fishing for Nile perch. However the video business picked up unexpectedly and as he could not combine it with fishing, he stopped this. The idea was not particularly original as many people around this period in the late 1990s were starting to run video halls. Robert was the first in Lambu though.

With the boom in Ugandan fish exports to global markets, Robert's business fared well. A lot of fishermen moved to Lambu and even though other businessmen began video halls, they did not affect Robert's profits and he succeeded in building up a loyal circle of customers, all young crews. Robert started to grow dreadlocks and listen to reggae music, which was a common practice among young landing-site dwellers. After a while, he was able to afford a satellite dish and a subscription to a satellite television company, DSTV, that broadcast internationally. This allowed him to show Brazilian *telenovelas* (soap operas) that were hugely popular in Uganda. But his major breakthrough came when DSTV started broadcasting UK Premier League soccer games in real time.

From then on, Robert's hall was full of soccer-loving fishermen. This, it turned out, was the vast majority of them and the original video hall made of iron sheets (a customary construction material in Uganda's fishing villages) quickly made way for a stone building. Recently, in 2012, he managed to construct an even-larger building where he lives and which is one of the few stone houses in Lambu today. He has plans to purchase an outboard engine and boat, which are essential equipment for setting up a fishing enterprise, and expects to get the money for these soon.

Case 2: Bosco Kawuki: An Established Farmer/Boat Owner

Bosco Kawuki (born 1962) owned ten boats at one point, evidence of the business success that many have come to see as typical in the Nile perch sector in Uganda. He was born at the Ggolo landing site into a Luganda-speaking family with a long history of fishing on Lake Victoria. Kawuki finished Senior Six at a private school and then joined his father's *magendo* business, which was the profitable yet risky smuggling of essential commodities on Lake Victoria in the period when Uganda's economy was tightly regulated in the 1980s. In addition to running a *magendo* business, Kawuki's father also pioneered Nile perch fishing back in the 1980s and, with the profits from his *magendo* business, he was

PHOTO 7.2 *Robert, the owner of the video hall with its satellite dish* (*Lambu, Masaka District*)
PHOTO: JOOST BEUVING

among the first to purchase the expensive monofilament nets needed to land the massive Nile perch.[6]

Like many others from Ggolo, Kawuki's father supported Museveni's rebellion in the 1980s and helped insurgents to manoeuvre across the swampy terrain around Ggolo, which is strategically located close to the interstate road that connects Kampala with the Tanzanian border. Many of the soldiers he helped later acquired prominent positions in the government when Museveni took over power. Through his father's extensive political network in the early 1990s, Kawuki came in contact with Concern International, an American NGO working with the region's growing number of HIV/AIDS orphans. As a result of his well-developed local contacts, through his father but later also through his

6 An individual adult fish can weigh up to 200 kg and is very strong.

own trading activities in the area, and his excellent command of English, they appointed him as a field officer.

Kawuki enjoyed the work as it opened up an international circle of contacts for him but also provided him with an opportunity to travel extensively in the area and observe new developments. 'I saw this Nile perch business coming and many fishermen left *mukene* and *tilapia* and the other local fish and began to look for *mputa*', he remembers. Kawiki himself was hesitant, remembering his father's stories about the hardship of fishing in the past. While on a training course at Concern International's headquarters in Kampala, Kawuki befriended a local businessman, Michael Kyobe, who was looking for a local supplier of high-quality *mukene*. Making use of Concern International's transportation facilities, Kawuki secured access to affordable and dependable transport from Ggolo to Kampala, and the men went into business together.

With Kampala's population increasing rapidly in the early 1990s, the business went well and Kawuki succeeded in purchasing his own Nissan pick-up truck, locally called a 'Sahara', so that he no longer had to depend on Concern International's routes and schedules. About this time, the Indian-Ugandan business tycoon Mahmood Thobani, who had invested in a fish-processing plant near Kampala a few years earlier, acquired a permit to set up a facility in Ggolo to land fish for export. To boost catches of Nile perch there, which were subsequently trucked to the factory, Thobani began to sell monofilament fishing nets on credit to local fishermen in 1996. This initiative was a resounding success and one of the many fishermen who procured a set of nets was Kawuki. Through his contacts in Kampala, he purchased an outboard engine and also a second-hand boat, both of which were financed by selling his Sahara. In the following years, business went exceptionally well, with good catches and Nile perch fetching increasingly higher prices. Kawuki, with sadness in his voice, remembers that 100 kg a day was not exceptional.

In the early 2000s, at around the time that Kawuki bought a second boat that was fitted with a set of nets and an outboard engine and was roughly equivalent to the price of a second-hand car, catches started to drop (see Figure 7.1). He also had several brushes with the local police who accused him of using nets will mesh sizes smaller than legally permitted. This resulted in several of his expensive nets being seized and subsequently burned. Kawuki remembers that a few wealthy boat owners (who were mostly from the same privileged social circles that he was born into) quickly expanded their fleets in those days and switched to using iceboxes to keep their catches fresh, having what he surmised were better contacts in police circles that allowed them to avoid police harassment. However Kawuki lacked the financial means to do the same as he had invested most of his

PHOTO 7.3 *A wealthy Nile perch owner (Lake Victoria region, Uganda)*
PHOTO: JOOST BEUVING

profits in land, at the encouragement of his now-ageing father. He also lacked the right contacts in Kampala. Looking back, he regrets ever having sold his Sahara, as this could have been a ticket to trading Nile perch instead of fishing.

This all happened around 2001 and, ever since, business has been difficult. He continued fishing Nile perch and expanded his fleet but this could not compensate for steadily declining catches, especially in the waters off Ggolo. He finally sent his oldest son with a boat to Kasensero, a major landing site close to the Tanzanian border where catches were supposedly still good but the boat and its equipment were stolen, which is a common occurrence at larger landing sites. Nowadays, Kawuki resorts mostly to farming coffee, which saw a modest recovery after prices dropped throughout the 1990s and he is even planting new coffee trees. He looks enviously at those who have made it, including distant relatives who live in large and beautified houses that are visible signs of their newly accrued wealth. And he occasionally still fishes when he hears stories of good catches.

Case 3: Kyavira Mpwedde: A Young Crewman

Kyavira is a young man in his early twenties from a Banyankole village in West Uganda. His parents were illiterate so he does not know his precise age and he left school at an early age. With a much older brother who has tended his parents' small banana and coffee farm since they passed away when Kyavira was in his teens, he felt he faced limited economic possibilities at home and decided to follow a village friend, Twinne, to the lakeshore. This was in the early 2000s and for years there had been stories about the area's Nile perch fishing and how the *bazungu* paid good money for it. Though Kyavira had no practical fishing experience, Twinne impressed Kyavira with stories of an older brother who had gone to Lambu a few months earlier, convincing him that Nile perch fishing required no special skills. With little else to do, Kyavira decided to join Twinne and move to the lake.

Twinne introduced Kyavira to his first job: fishing for *nkejjei*, which are juvenile cyclids that are caught in coastal waters and require little but a fishing net improvised from cheap, locally available mosquito nets. As Nile perch fishermen with monofilament nets were gradually replacing these with (much cheaper) long lines fitted with hooks at the time of Kyavira's arrival, demand for bait was just picking up when Kyavira began to work in Lambu. He disliked standing in waist-deep water all day but the money was decent and, perhaps even more importantly, he was paid in cash. Kyavira was certainly not alone. The newly emerging *nkejje* business attracted many young men like Kyavira who came to the lake to make a living. Through this work, Kyavira quickly expanded his network of social contacts that would benefit him later on.

One of those he befriended was Soumani, a young man who had arrived in Lambu a few years previously and was known locally to be a dependable fisherman. As is common in this part of Lake Victoria, more seasoned fishers with a good reputation usually get promoted to overseeing the boats as captains. In addition, boat owners typically allow captains to handpick their crews and Kyavira, often with Twinne, were among those that Soumani liked to work with. This gave Kyavira an important entry into the more lucrative world of Nile perch fishing for two reasons. If catches are good, crews usually get to keep part of the catch, which they then distribute among themselves. Some of it is prepared on board during the long fishing trips, but most is sold. With prices still good, selling one's share of the catch can fetch more money than they would ordinarily receive in salary. In addition, the crews are handed petrol to power the outboard engines that drive the boats. Though most of it is used for propulsion, the crews trade some to ambulant traders too, mostly at night and/or far offshore where it can be done unnoticed by the boat owner.

Over the next few months, Kyavira worked for a succession of boat owners, mostly under Soumani. When pressed, Kyavira is not capable of reconstructing the sequence of the bosses he worked for. Work relations are experienced as problematic: crews generally complain about the low pay they receive, whereas boat owners commonly complain about their crews. Few boat owners, especially the larger ones, are able to give names or other details of those working for them. Relations with peers have presented a more stable domain of social relations for Kyavira. In fact, both he and Twinne have now become part of a group of young men in Lambu who have dreadlocks, listen to reggae music and refer to themselves as 'rasta' – as in 'rasta Kyavira'. This is common in the research area: having established themselves, young fishermen shift to a new, alternative identity, in this case, drawing on the symbols of Rastafarianism.[7]

Kyavira's fortunes turned when Lambu began to lose its prominent position in Uganda's Nile perch export network around 2006/2007. Several indicators offer evidence of this. First, the volume of Nile perch landed in Lambu has dropped over the past decade at a faster rate than national catches. Second, several leading boat owners left Lambu and moved to the Ssese islands, allegedly in response to falling catches in the vicinity of the landing site. Third, the establishment of a fish-processing plant in nearby Bukakate led to a reorientation of the Nile perch trade networks away from Lambu. Kyavira saw his support network of loose contacts with boat owners, often mediated through his informal network of contacts with like-minded peers (the Rastas), dwindle and was occasionally compelled to resort to trapping *nkejje* – the very activity he had initially tried to get out of.

Not long afterwards, Kyavira left Lambu and moved to Kanaanas, a medium-sized landing site across the narrow, unnamed strait that separates the Ssese Islands from the mainland. Earlier, in 2009, a Nile perch businessman from a prominent local trading centre (Lukaya) had begun to expand his fleet by hiring a crew of carpenters specialized in building boats for Nile perch fishing. Word had it that he would soon begin to recruit fishermen and Kyavira was hoping that he could somehow secure a position aboard one of the vessels, and perhaps even make it to the rank of captain himself. Meanwhile, he spends most of his days at or near the beach where the fishermen moor their boats and where last-minute recruitment takes place, often offering openings to those without fixed arrangements with boat owners. To make a living, Kyavira took on a job as a night watchman, overseeing the plastic containers filled with live *nkejje* that float on the water just offshore. He makes little money but at least it gives him a spot close to where the action is.

---

7 Rastafarianism in Africa is sometimes interpreted as a religious practice. However religious life in the world of Nile perch fishing is not as important as elsewhere in Uganda. In Lambu, for instance, there are churches and mosques to be found but these are small, inconspicuous buildings tucked away behind shops and residential dwellings. Few of the respondents in this research ever engaged in religious talk with this researcher, in striking contrast with encounters in places away from the lake.

PHOTO 7.4 *Crews waiting for work*
PHOTO: JOOST BEUVING

The Social Organization of Local Careers in Uganda's Nile Perch Sector

The material presented here highlights a striking diversity in local careers, centred on but not limited to Nile-perch-related economic activities. In a local understanding of professional success, Robert and Kawuki are examples of those who have made it: they have succeeded in establishing independent businesses. Kyavira seems to have been less fortunate, however, as he has not yet landed an established position. However he claimed not to be unhappy with his fate and evaluates his position favourably in relation to the rural life he left behind. And Robert talks about his ambition to invest in boats once the video hall has yielded sufficiently stable returns, whereas Kawuki is looking at the Nile perch business and regretting the opportunities that he missed. These narratives are not self-explanatory. Following the adopted actor-oriented perspective, they must be deconstructed to find the deeper layer of meaning that informs economic decision-making. The material suggests paying special attention to three points: (i) the political-economic context of local Nile perch careers; (ii) the role of social relations in the accumulation of capital and the mobilization of labour; and (iii) local actors' valuation of Nile-perch-related economic activities.

Political-Economic Context of Local Careers

The careers discussed above are all at the lower end of a complicated export chain. One result is that there are considerable fluctuations in demand and prices that are difficult to predict. Nile perch competes on the global market for affordable table fish, which has seen a rise in cheap alternatives over the last few years, particularly tilapia from South East Asian countries. Another result is that participation in the Nile perch chain is increasingly expensive as the supermarkets place stringent food-safety demands on the processing and trading up the chain. The smallest deviation from the norm can end exports, as was seen with the EU's 1995 trade ban. To comply with supermarket requirements, increasing investments are needed in processing and transporting practices of the cold chain, including the administrative procedures supporting it, to ensure high-quality fresh fish. Processing factories along the lake can, therefore, no longer procure fresh fish from local traders that has not been collected in an icebox. But few traders possess an ice-making facility and, for many, these new food-safety measures have meant having to procure ice from factories, which makes it more difficult for boat owners to move up the chain.

Government measures to regulate fishing add to the challenges that local Nile perch actors face, especially the boat owners and their crews. A discourse of overfishing permeates Uganda's fishery policies (cf. Kolding *et al.* 2008) and appears in the life world of local Nile perch actors as restrictions on the type of fishing gear that can be used. Certain traps are banned, as are fishing nets with holes greater than a certain size. Though the Ugandan government lacks the means to systematically enforce these measures due to state retrenchments because of economic reforms, larger landing sites are often the targets of police raids. Any arrests made are subsequently displayed in the media, often with full-colour pictures of the suspects, to warn fishermen harbouring any ambitions about using 'illegal' fishing gear. Arrests and the seizure of equipment can have dramatic consequences, as the case of Kawuki shows. Whereas the fine is usually modest (tens of € only), the loss of expensive nets constitutes the real financial damage. In Kawuki's case, it took him several months of hard work before discovering that the competition had moved on at a pace that made it difficult for him to catch up.

Social Relations: Capital and Labour

Access to external capital is essential for a successful local career. Kawuki, for example, had acquired it through an earlier *magendo* business, whereas Kyavira and Robert were in a less advantaged position. Capital is important to expand fleets to compensate for catches per boat once the Nile perch sector took off. As a rough rule of thumb, fishermen today consider that three to five boats are needed to balance out the catch fluctuations of individual boats. With profits under pressure as a result of increased local competition, building up a sizeable fleet increasingly depends on external capital. Capital is also important to move out of fishing and into trading, which is seen as a more stable business. This is also capital intensive because trading requires a vehicle to transport the fish and cooling equipment to keep it chilled. Trading additionally requires a healthy cash flow to purchase fish. Nile perch is bought on the spot but traders prefer to tie boat owners/fishermen to them in debt relations. There is a risk here too as debts may not necessarily be cleared by catches. And because the lake is vast and difficult to survey, heavily indebted boat operators/owners have been known to sometimes disappear, effectively transforming their debt into a financial loss for the trader.

The research has further shown that capital is not an input factor in itself (i.e. something that one either has or lacks) but is one that is socially constructed. As the case of Kawuki demonstrates, few Nile perch businessmen can mobilize sufficient funds to procure a fully equipped fleet of boats, which one needs to reduce the uncertainties of fishing and trading below manageable levels. Numerous informants agree that pooling capital could, in principle, be a solution here because banks rarely issue formal loans as they anticipate people defaulting on loans. Few of them do actually seem to resort to this. For example, Kawuki waited to purchase a boat until he could afford to buy one himself and the same logic underpins Robert's stated choice of waiting to buy a boat until he had accumulated sufficient capital from the video hall. A shareholder system could likewise reduce the cash that boat owners are required to pay their crews in wages. Lacking capital, crews could then invest their labour and, in return, receive cash related to the size of the catches they bring in. These options were discussed with a wide range of informants, however none of them considered this a serious option. Robert saliently summarized the main objection: 'This is impossible because many of them are *muyaye* [i.e. cannot be trusted]'.

A similar social pattern can be seen in labour relations. Nile perch fishing depends on collaboration between a captain and two men setting and hauling in the long lines or nets. Boat owners commanding two or three boats can still oversee their crews' activities themselves but this becomes impossible with more boats. Larger boat owners are also expected to stay on shore and manage their fleets from a distance, which places high demands on trust relations. However the cases of Kyavira and Kawuki suggest that relations between boat owners and their crews are usually limited because boat owners try to economize by underpaying their crew and routinely replace them, i.e. they are constantly firing them. Crews respond by compensating their uncertain position with whatever means they can find. The shallowness of these relations is readily observable at the village landings. While boat owners tend to trek from one landing to the next, the crews do not usually follow them. The case of Kyavira is typical. Crews build networks of solidarity, which manifest themselves for instance in their shared rasta identity, at one particular landing and this discourages them from leaving. Established boat owners, like Kawuki, on the other hand, command relations in a broad zone along Lake Victoria by virtue of their socialization into existing fishing networks and this promotes a more mobile existence.

The structure of these relations is important. With a few exceptions, the majority of crews work for boat owners who they have little in common with socially. Kinship ties between them are exceptional and, as the case of Kawuki's son demonstrates, do not necessarily promote business stability/success. Shared ethnic and/or religious ties are limited due to migration practices. The larger boat owners are like Kawuki and they socialize in mono-ethnic fishing/trading networks that, in this part of Uganda, means Baganda. Most of the crews are from the countryside and from diverse socio-ethnic backgrounds – in his case, Munyankore. This means that sentiments of solidarity between boat owners and crews, which were often thought to be fostered by a shared ethnic background, are not automatic. In addition, crews come to the lake with expectations of individual autonomy and making it economically, and are then confronted with boat owners who show little engagement with their fate. It is not surprising that they behave opportunistically.[8]

To better understand this seemingly self-deceiving social behaviour, the chapter now turns to a short exploration of how local actors value the Nile perch business, thereby distinguishing between boat owners and their crews.

Valuation: The Social World of Nile Perch Fishing

A major reason why few boat owners pool capital is that they regard one another as competitors rather than potential business partners. This has been considered a self-fulfilling prophecy of anticipated deceit as Nile perch businessmen expected to be deceived by prospective business partners. They preempt this by limiting common projects to capital-extensive ones requiring little interpersonal trust as these are low risk. This, in turn, discourages possible

8 This problem appears to be broader than fishing it self. A major reason why Robert failed to combine fishing with running a video hall is that he failed to build up a dependable circle of contacts to run the hall in his absence. Robert's narative is replete with stories about how he entrusted it to someone he knew from before but when he returned from the lake, he found that little real money had been made, even though he could tell that there had been several busy nights with paying customers.

partners from cultivating relations with them, thus setting a cycle of negative expectations in motion (Beuving 2013). This tendency explains, for instance, why one sees a limited concentration of boats in the study area. Boat ownership is considered prestigious, as is illustrated above by Kawuki's admiration of those commanding a sizeable fleet and Robert who hoped to own a boat. However, it is prestige that is perceived in highly individualistic, antagonistic terms and this limits the number of boats per boat owner (as few have access to the external cash needed to build up a fleet). This reduces the profitability of the business. The chequered fortunes of Kawuki, who was once a fairly established boat owner, are indicative of how such perceptions can impact negatively on a Nile perch career.

The relationship between boat owners and crews is affected by this value orientation. Owners are inclined to assume that their crew will cheat on them (which many in fact do) and, to pre-empt this, they construe them as antagonists. The crew themselves draw on yet another set of values for going to the village landings. The case of Kyavira suggests how migration from the countryside holds the key to understanding their trek to the lake. What is salient is that many of the crew go to village landings in the company of friends. Field research found few kinship ties between crew members. This is often the result of a premeditated decision, namely to move away from what are seen and felt to be constraining kinship ties. In traditional rural central Uganda, fathers or other direct patrikin such as grandfathers and/or uncles can exercise formidable control over a young man's actions which they can endlessly complain about. Orphans are even worse off because they are viewed as a burden on an already overcrowded relative's household and are yet another mouth to feed. The landing site provides an opportunity for them to expand their personal autonomy unobserved by their patrikin (cf. Beuving 2010). This presents such a strong force that many young men decide to stay on even though the prospects of moving up the Nile perch chain remain limited.

Concluding Remarks

This chapter has presented some new evidence and ideas in an attempt to understand the mixed fortunes of local actors in the rapidly globalizing export sector of Nile perch from Lake Victoria, Uganda. Economic reforms in Uganda have opened up new economic activities that could push large numbers of Ugandans into new careers. However ethnographic analysis of career histories highlights some of the challenges that they still face. Nile perch competes on the world market with other commercial fish, making local business dependent on unpredictable competitive forces higher up the chain. In addition, supermarkets are enforcing increasingly strict product criteria lower down the chain, which contributes to the crowding out of capital-poor boat owners and/or traders and compromises the upgrading of crew members to better established positions. These challenges place a premium on the cultivation of social relations down the chain, but the world of Nile perch fishing is highly individualistic. Others tend to be taken as adversaries and a quest for personal autonomy prevails, preventing people from building up social capital. Factors within the Nile perch fishing sector thus make and break careers. This chapter has not offered definitive conclusions on this matter but hopes to have contributed to the discussion on Africa's rapidly globalizing food exports and to have moved this beyond an analysis of institutional/structural factors.

References

Amanor, K. (2009), Food Chains, African Smallholders and World Bank Governance. *Journal of Agrarian Change* 9(2): 247–262.

Balirwa, J. (2007), Ecological, Environmental and Socioeconomic Aspects of the Lake Victoria's Introduced Nile Perch Fishery in Relation to the Native Fisheries and the Species Culture Potential: Lessons to Learn. *African Journal of Ecology* 45(2): 120–129.

Bayart, J-F. (2000), Africa in the World: A History of Extraversion. *African Affairs* 99(395): 217–267.

Bernstein, H. & L. Campling (2006), Commodity Studies and Commodity Fetishism I: Trading Down. *Journal of Agrarian Change* 6(2): 239–264.

Beuving, J. (2010), Playing Pool along the Shores of Lake Victoria. Fishermen, Careers and Capital Accumulation in the Ugandan Nile Perch Business. *Africa* 80(2):224–248.

Beuving, J. (2013), Chequered Fortunes in Global Exports: The Sociogenesis of African Entrepreneurship in the Nile Perch Business at Lake Victoria, Uganda. *European Journal of Development Research* 25(4): 501–517.

Beuving, J. (2014), Spatial Diversity in Small-Scale Fishing: A Socio-cultural Interpretation of the Nile Perch Sector on Lake Victoria, Uganda. *Journal of Economic and Social Geography* 105(3): 1–23.

Corbin, J. & A. Strauss (1998), *Basics of Qualitative Research: Techniques and Procedures for Developing Grounded Theory*. London/New York: Sage Publishing.

Dijkstra, G. & J.K. van Donge (2001), What Does the 'Show Case' Show? Evidence of and Lessons from Adjustment in Uganda. *World Development* 29(5): 841–863.

Diop, N. & S. Jaffee (2005), *Fruits and Vegetables: Global Trade and Competition in Fresh and Processed Product Markets*. Washington: World Bank.

Dolan, C. (2001), 'The Good Wife': Struggles over Resources in the Kenyan Horticulture Sector. *Journal of Development Studies* 37(3): 39–70.

Flyvbjerg, B. (2006), Five Misunderstandings about Case-Study Research. *Qualitative Inquiry* 12(2): 219–245.

Fulgencio, K. (2009), Globalisation of the Nile Perch: Assessing the Socio-cultural Implications of the Lake Victoria Fishery in Uganda. *African Journal of Political Science and International Relations* 3(10): 433–442.

Geheb, K., S. Kalloch, M. Medard, A-T. Nyapendi, C. Lwenya & M. Kyangwa (2007), Nile Perch and the Hungry of Lake Victoria: Gender, Status and Food in an East African Fishery. *Food Policy* 33: 85–98.

Guyer, J. (1997), *An Africa Niche Economy: Farming to Feed Ibadan, 1968–88*. Edinburgh: Edinburgh University Press.

Hobson, J. & L. Seabrooke (2007), *Everyday Politics of the World Economy*. Cambridge: Cambridge University Press.

Kolding, J., P. van Zwieten, O. Mkumbo, G. Silsbe & R. Hecky (2008), Are the Lake Victoria Fisheries Threatened by Exploitation of Eutrophication? Towards an Ecosystem-Based Approach to Management. In: G. Bianchi & H.R. Skjolldal, *The Ecosystem Approach to Fisheries*. Rome: FAO.

Long, N. (2001), *Development Sociology: Actor Perspectives*. London/New York: Routledge.

Medard, M. (2012), Relations between People, Relations between Things: Gendered Investment and the Case of the Lake Victoria Fishery, Tanzania. *Signs* 37(3): 555–566.

Msuku, B., H. Mrosso & P. Nsinda (2011), A Critical Look at the Current Gillnet Regulations Meant to Protect the Nile Perch Stocks in Lake Victoria. *Aquatic Ecosystmen Health & Management* 14 (3): 252–259.

Nunan, V. (2010), Mobility and Fisherfolk Livelihoods on Lake Victoria: Implications for Vulnerability and Risk. *Geoforum* 41: 776–785.

Ouma, S. (2012), Creating and Maintaining Global Connections: Agro-business and the Precarious Making of Fresh-Cut Markets. *Journal of Development Studies* 48(3): 322–334.

Ponte, S. & P. Gibbon (2005), *Trading Down. Africa, Value Chains, and the Global Economy*. Philadelphia: Temple University Press.

Pringle, R.M. (2005), The Nile Perch in Lake Victoria: Local Responses and Adaptation. *Africa* 75 (4): 510–537.

Rimmer, D. (2003), Learning about Economic Development from Africa. *African Affairs* 102(408): 469–491.

Swinnen, J. (2007) *Global Supply Chains. Standards and the Poor*. Oxon: CABI.

van Donge, J.K. (1992), Waluguru traders in Dar es Salaam: An Analysis of the Social Construction of Economic Life. *African Affairs* 91(363): 181–205.

van Velsen, J. (1967), The Extended Case Method and Situational Analysis. In: T.S. Epstein (ed.), *The Craft of Social Anthropology*. London: Tavistock.

SECTION 3

Drivers of Food Production

CHAPTER 8

Pressures and Incentives

Urban Growth and Food Production at Tamale's Rural-Urban Interface

Sebastiaan Soeters

This chapter identifies some of the pressures and incentives for smallholder food producers that have resulted from the rapid growth of Africa's cities. The incentives, which derive from an exponential increase in urban demand for key food commodities, are offset by increasing pressure on land as a result of urban expansion. The implications of such pressures and incentives for agricultural communities at the rural-urban interface are considered by comparing two rural communities that are situated along the same road but at different distances from Tamale, the regional capital of northern Ghana. The chapter illustrates how, in the case of Tamale's peri-urban belt, recent pressure on productive resources in communities closer to Tamale due to the urban sprawl outweighs the incentives flowing from an increase in demand. This appears to have resulted in a decline in productivity in communities located closer to Tamale. Communities that are further from the urban market seem to be benefitting from the increased demand resulting from urban growth. The data presented here were collected in 2011 as part of an exploratory survey among farmers.

Introduction

The UN predicts that Sub-Saharan Africa's urban population will account for more than 43% of its total population by 2025, and 56.5% in 2015 (2011), compared with less than 15% in 1950 (Population Division of the Department of Economic and Social Affairs of the United Nations, 2013). In absolute terms, Africa's urban population will have increased from around 34 million in 1950 to 551 million by 2030 (*Ibid.*). And eight of the ten fastest-growing (non-Chinese) cities in the world, according to the 'State of the World's Cities 2012/2013' are currently in Africa (UNHabitat 2012). This trend in much of Africa looks set to continue, especially in West and East Africa (Table 8.1).

Whilst scientists and policymakers alike have been fascinated by the implications of urban growth in Africa, attention has tended to focus on its consequences for economic growth (Barrios & Bertinelli 2006). The implications of rapid urbanization in Africa for specific components of the food economy have not yet received much attention, despite the pervasiveness of the food economy in Africa relative to elsewhere. In general, the explosive expansion of African

 | DOI 10.1163/9789004282698_009

TABLE 8.1 *Urban population, Sub-Saharan African regions, 2011 and 2050, (% of total population)*

	Urban population 2011 (% of total)	Urban population in 2050 (% of total)
Eastern Africa	23.7	44,7
Middle Africa	41,5	61,5
Southern Africa	58.9	74
West Africa	44.9	66,7

SOURCE: UNHABITAT (2012)

cities is good news for Africa's agrarian sector as the ratio of food consumers to food producers is tilting in favour of the producers. The reconfiguration of food suppliers to food 'demanders' is in part being reflected in increasing urban food prices, which in theory provides an incentive for smallholder farmers to increase output. Offsetting the incentives that are emerging from the growing (urban) demand for food are pressures on natural and productive resources as a result of the expansion of these cities. The pressures and incentives are not, however, spread uniformly across space. Indeed, in the context of poorly regulated urban zoning, the conversion of rural land use to urban land use is one of the primary drivers of Africa's rapid urbanization (McGranahan *et al.* 2009). Roadside food producers located close to cities are better positioned to benefit in terms of urban market access but these communities are also most affected by the pressure on land as a result of urban sprawl. This chapter considers the demographic and spatial pressures and incentives for communities in the peri-urban belt around Tamale, the regional capital of northern Ghana, and especially the implications for rural livelihoods. It compares two rural communities – Chanshegu (5 km southwest of Tamale) and Adubiyili (10 km southwest of Tamale) – that are both on the Tamale-Kumasi road.[1]

Pressures and Incentives: Discourses and Dynamics

Since independence, the relationship between cities and the rural areas in Africa has served as a primary focus within shifting development paradigms. During much of the 1950s, 'development' was regarded as the encroachment of the modern sector into the traditional sector (Tacoli 1998). In the dual-economy model, development thus occurred as the African workforce moved from the

1 This is sometimes also referred to as the Tamale-Techiman road.

traditional sector (agriculture) into the modern sector (manufacturing, which was found in cities). By the mid-1960s, it had become clear, however, that the manufacturing sector would not develop quickly enough to absorb the labour coming to the city from the countryside and the emergence of the informal sector served as evidence of this market failure (Hart 1973; Potts 1995). The Urban Bias theory, as it was applied to cities in the global South, added to the increasingly negative perception of cities in Africa (Lipton 1977). However, as Africa's Third Wave set in in the mid-1990s, cities regained developmental significance as 'growth poles' as it was felt that they were required to facilitate the linkages between Africa's agricultural export sectors and global demand. The most recent trend places African cities at the heart of a middle-class consumption revolution. The growth of African cities today serves not only as proof that Africa is developing (since urbanization is seen as driving economic growth) but it also provides potential for an increasingly important private-sector approach to development. Reports, for instance by McKinsey, emphasize how Africa's growing middle class is driving this trend (Leke *et al.* 2010).

The speed of growth of African cities has obvious consequences for the economics of African agriculture. Conceptually, all cities require access to food surpluses produced elsewhere in order to survive and people residing in cities must produce something to exchange directly or indirectly for these food surpluses (O'Sullivan 2009; Steel 2008). The rapid growth of African cities must correspond, therefore, with a rapid increase in the aggregate demand for key food commodities, since the vast majority of urban dwellers do not produce their own food. Furthermore, surpluses that currently feed African cities need to increase to keep pace with the food demands of growing cities. Importantly, the reconfiguration of the rural-urban demographic balance is not simply resulting in a change in the location of food consumption. Since wages in cities are generally higher than in rural areas and most locally produced food commodities are 'normal goods',[2] an increase in the ratio of urban food consumers to food producers causes an exponential increase in the aggregate demand for food. Following on from this, a fundamental premise of basic economic theory is that, in the short term, an increase in aggregate demand drives up prices (as demand outstrips supply). In the long run, producers react, however, to increases in prices and increase production to take

2 'Normal goods' refer to goods that are consumed in greater quantities as people earn more money at about the same rate as the increase in income. In contrast, an 'inferior good' is one where consumption decreases as people earn higher incomes.

advantage of the high prices, which then increases the supply and drives prices back down.

A second important consideration in an attempt to understand the impact of urban growth on local food production relates to the demographics of Africa's urban transition and the impact of demographics on agricultural productivity. Intuitively, the process by which more than half of the total population has come to reside in cities is the result of migration from the rural to the urban areas as people migrate from the countryside to pursue livelihoods in the cities. This frees up land for those who choose to remain in the countryside. But the rise in the number of urban dwellers as a percentage of the total population does not mean an absolute reduction in the number of rural inhabitants *per se*. On the contrary, Figure 8.1 shows that in all of Sub-Saharan Africa's subregions except for Southern Africa (which currently has the smallest rural population), the rural population will continue to grow in absolute terms until at least 2050 and probably beyond, albeit more slowly. Despite Africa's rapid urban growth, an often overlooked fact is that rural population density will generally increase in the foreseeable future. When considered in the context of the increasing amount of land 'grabbed'[3] for the production of biofuels and

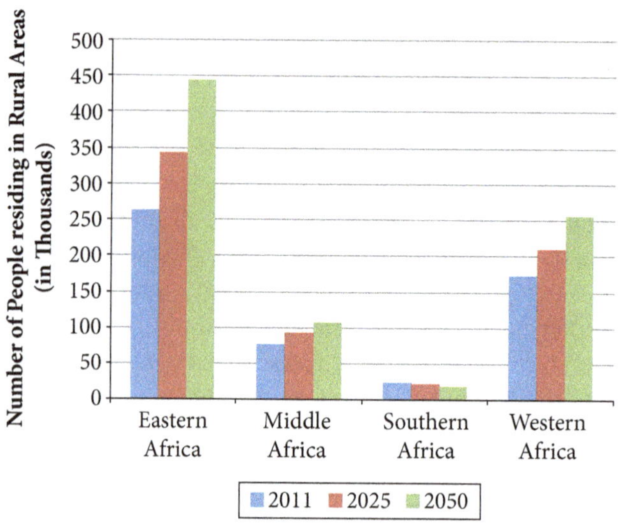

FIGURE 8.1 *Projected rural population growth by African sub-region, 2011, 2025 & 2050*

SOURCE: POPULATION DIVISION OF THE DEPARTMENT OF ECONOMIC AND SOCIAL AFFAIRS OF THE UNITED NATIONS (2013): HTTP://ESA.UN.ORG/UNUP/WALLCHARTS/URBAN-RURAL-AREAS.PDF.

3 'Land grabbing' is currently a popular term but may be misleading as most of the 'grabbed' land in Africa is acquired legally.

other non-food or export uses, pressure on land will increase even though it is being freed up as a result of the population's movement to urban areas.

How population density and the resulting pressure on land impact on agricultural productivity has been widely studied (Boserup, 1966: Hugo 2008; van der Geest 2011; de Brauw *et al.* 2013). Such studies have delivered mixed results, with some analysts arguing that a reduction in population density through out-migration and/or a sustained reduction in fertility rates has a positive impact on agricultural productivity because, as the pressure on land is reduced, the resources available for those who continue to farm increase (de Brauw *et al.* 2013). Others argue, however, that pressure on land is necessary for the sustainable intensification of agricultural production (Tiffen *et al.* 1994). While Malthusian and Boserupian processes (or a combination of the two) have traditionally been applied in the context of studies of rural out-migration, they are equally applicable to cases of increases in population pressure on land as a result of expanding cities. The fundamental dynamic is not dissimilar. It is the impact on agricultural productivity of a changing people-to-land ratio that is interesting and is considered in this chapter.

Study Areas: Tamale, Chanshegu and Adubiyili

Tamale Metropolitan Area (TAMA) comprises one major town that is referred to here as Tamale and 115 villages and covers an area of 750 km^2. Tamale is the regional capital of northern Ghana and was the country's third largest city until recently when it was surpassed by Takoradi, which is the services and hospitality base for Ghana's new oil economy. TAMA is located in Ghana's Guinea Savannah zone and forms the economic node of Ghana's less-developed north (see Map 8.1). TAMA has one rainy season that starts in April/May and runs until September/October, with a mean annual rainfall of 1100 mm. The dry season lasts from November to March. The mean daily temperature ranges from 33°C to 39°C.

Although Tamale is a relatively new city, having developed out of the colonial administrative headquarters set up in 1907, its rate of urbanization has been significant (Table 8.2).

Tamale's rapid demographic growth since 1960 has corresponded with its physical expansion into its hinterland (see Figure 8.2). With weak local government, poor implementation of zoning regulations and unclear land-tenure institutions, Tamale's built-up area has expanded rapidly since independence. Its urban sprawl has spread along the roads heading out of town, especially those that link Tamale to other major markets, notably Kumasi to the south,

MAP 8.1 *Ghana's districts showing Tamale Metropolitan Area (TAMA)*

TABLE 8.2 *Tamale's population growth, 1907–2010*[4]

Year	1911	1921	1931	1948	1960	1984	2000	2010
Population	1435	3901	12,901	16,055	48,055	135,95	293,881	371,351

SOURCES: STANILAND (1977: 1); GOLD COAST CENSUS OFFICE (1911, 1921, 1931, 1948); POPULATION CENSUS OF GHANA (1960, 1970, 1948); POPULATION AND HOUSING CENSUS OF GHANA (2000, 2010)

Ghana's second largest city after Accra. The Tamale-Kumasi road is the busiest and most important of these roads. In the absence of other modes of post-industrial transportation, it serves as the main artery feeding Tamale's local economy, and those farming land along it are regarded as possessing prime

4 This is based on census reports between 1921 and 2010. While the accuracy of these reports is questionable, they are the most reliable sources available. The 1907 figure of 1435 is taken from Staniland (1977) and is based on archival records that are, to my knowledge, no longer available.

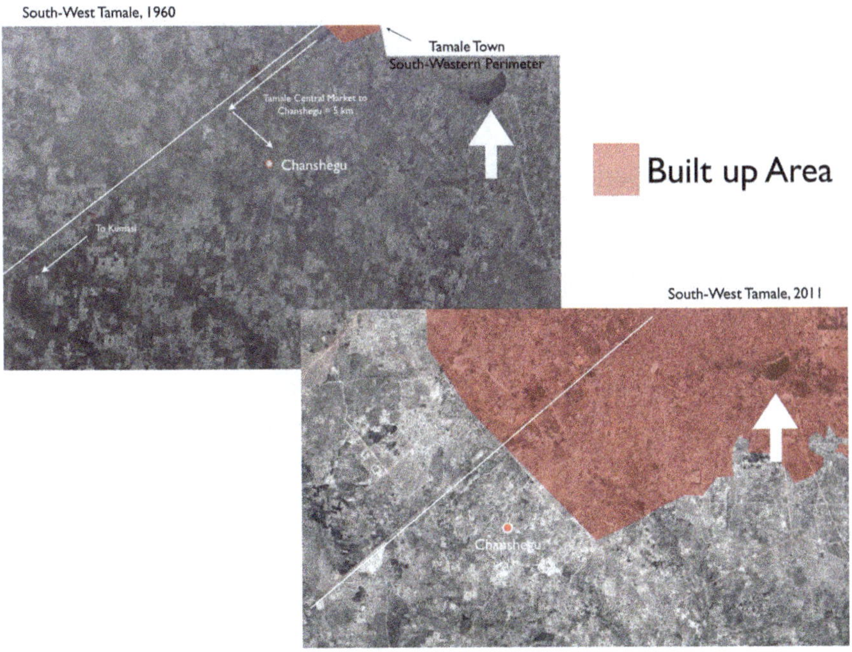

FIGURE 8.2 *Urban sprawl, South-West Tamale, 1960–2011*

property by local farmers. It is telling that land immediately adjacent to the roadside is used by non-agricultural activities, such as cement factories, hotels and churches. Tamale's expansion towards Kumasi has been more aggressive than in other directions.

Chanshegu and Adubiyili are two of the 115 recognized villages in TAMA. In traditional terms, the Dagomba who live in these villages are part of the Kingdom of Dagbon, which has a highly centralized structure with a paramount chief (the *Ya Na*) whose seat is in Yendi, about 90 km east of Tamale. Chanshegu and Adubiyili can both be said to be part of Tamale's peri-urban belt today but their inclusion is relatively recent as they used to be firmly rural according to farmers who indicated that they travelled to Kumasi on the hoof to sell their cattle in the past. Today the two villages are almost exclusively connected to Tamale in economic terms. Its demographic and spatial growth has significantly changed the relationship that the villages have with it as quintessential peri-urban settlements. While this chapter is concerned with peri-urban settlements in general, its focus is more on the agrarian dynamics of becoming peri-urban than purely the agrarian (and livelihood) dynamics of peri-urban settlements as previous studies, such as Prain, Karanja & Lee-Smith (2010), have already explored this aspect.

Agrarian Livelihoods in Chanshegu and Adubiyili

The fieldwork undertaken for the purposes of this chapter was of an exploratory nature and further research is required to draw more conclusive findings. The findings presented here are illustrative and offer plausible explanations. The fieldwork included a series of in-depth interviews and a survey of 20 farmers (household heads) in the settlements of Chanshegu and Adubiyili, at 5 km and 10 km from Tamale's central market respectively. A number of in-depth, semi-structured interviews with farmers, chiefs and other notable persons were also carried out.

Chanshegu and Adubiyili have populations of a similar size. According to Ghana's Population and Housing Census (2000), Chanshegu had 820 residents and 75 houses in 2000. A study conducted eleven years later by students from the University of Development Studies (UDS) concluded that whilst the number of households in Chanshegu had increased to 95, the total number of residents had decreased to 735 (Yussif *et al.* 2011). It is difficult to assess the accuracy of these figures. What is important here is that Chanshegu and Adubiyili are more or less of equal size. In addition, while community-controlled land was poorly defined, interviews with local leaders suggest that the area covered by each of the two settlements (including outlying farmlands) is broadly comparable. Both settlements are ethnically homogenous, with almost all the residents in both communities belonging to the Dagomba ethnic group, the largest in northern Ghana. Livelihoods in the two villages depend primarily on farming, with only limited diversification in off-farm employment. Land is, therefore, the most important productive asset. None of the farmers interviewed used irrigation for farming but relied on the one rainy season per year for water for irrigation purposes.

Maize is regarded as the most important crop culturally and economically, with each of the farmers surveyed in the two communities indicating that their first priority is to produce enough maize to feed their household members. If there are surpluses after subsistence requirements have been met, these are mostly sold to traders in Tamale. While some surpluses are traded onwards from Tamale to the south, farmers indicated that the bulk is sold to retailers and consumers in Tamale itself. Traders from Tamale or elsewhere rarely go to the villages to buy surpluses. Instead, farmers are responsible for the transportation of goods to Tamale. Farmers in Adubiyili make use of and have access to a twice-weekly service truck that, for a small fee, takes their goods into Tamale. Farmers in Chanshegu transport their goods to market by taxi, motorbike or bicycle, depending on the volume. When asked why they thought the service truck stopped in Adubiyili and not in Chanshegu, the farmers in Chanshegu said they thought that their volumes were not large enough to warrant an additional stop. Farmers in both communities indicated that while they made the journey to Kumasi in the past to sell their produce, including livestock, they no longer (or only rarely) go there today. This may be due to the emergence of a specialized service sector (transportation) and/or because Tamale's food economy is now large enough to provide an economic focal point for local farmers. In some instances and wherever possible, farmers employ their own personal networks to market surpluses at Tamale's central market, but it is not regarded as problematic to sell goods to traders outside these networks.

Plot Size

In both Chanshegu and Adubiyili, the settlement-level chief is the custodian of the land and distributes it amongst residents who want to farm there. Although the chief is advised by a council of elders in each village and forms part of the Dagomba chain of command, settlement-level chiefs in rural areas operate fairly autonomously. The farmers interviewed accepted that the land they farmed belonged to the chief and that he could sell land to others outside the community. The Chief of Chanshegu has sold tracts of land to religious organisations to build churches, mosques (see Photo 8.1) and, most notably, 'strangers' living in Tamale who use it primarily to build second homes as an investment. An additional layer of land governance is provided by TAMA, although it appears to have limited influence beyond Tamale's built-up area. Farmers, for instance, indicated that they had no contact with extension services, which TAMA is mandated to provide. Access to other services, such as piped water, is limited. There appeared to be no market for land in Adubiyili, although interviewees indicated that they believed that one would emerge in the future.

The survey indicated that farm sizes in Chanshegu were significantly smaller than those of the farmers interviewed in Adubiyili. The average farm size in Chanshegu was 5.9 acres[5] and in Adubiyili it was 19.5 acres. One farmer in Adubiyili had an exceptionally large farm, which demonstrates the difference in plot sizes between the two settlements. However the median farm size in Chanshegu is 7 acres compared to 15.5 in Adubiyili, which highlights the significant difference in plot sizes between the two settlements (see Table 8.3).

The significantly smaller plot sizes amongst farmers in Chanshegu is likely to be primarily the result of the combination of high fertility rates and low levels of out-migration coupled with a steady increase in the amount of land used for urban-orientated, non-agricultural purposes. Whilst plot sizes may indicate population pressure on land, it is clearly not possible to draw conclusions about productivity based on plot sizes alone. However, since plot size serves as a proxy for population pressure, as illustrated above, population pressure is an important

5 The farmers preferred to use acres rather than hectares as a unit of measurement. One acre = 0.404 hectares.

PHOTO 8.1 *Mosque being built alongside an existing mosque, on land belonging to Chanshegu's traditional authority.*
PHOTO: SEBASTIAAN SOETERS

TABLE 8.3 *Mean and median plot sizes*

	Mean	Median
Chanshegu	5.9	7
Adubiyili	19.5	15.5

N = 20+20
SOURCE: AUTHOR'S SURVEY

determinant of agricultural productivity (in Malthusian and/or Boseropian terms) and there are likely to be differences in farming intensity and productivity between Chanshegu and Adubiyili.

Farmers' Age

The average age of farmers in Chanshegu is 57, compared to 51 in Adubiyili.[6] More qualitative research is required to determine conclusively why such a difference exists or whether it is statistically significant. However, the higher

6 Although many farmers did not know their ages, these were determined by consulting voter identification cards.

average age in Chanshegu coupled with smaller plot sizes, as indicated above, could be explained by the fact that young adult men are no longer able to inherit land from their fathers because the plots are already too small to generate a livelihood. In this case, the average age of farmers would be higher.

Household Size

The mean household size of the farmers interviewed was larger in Chanshegu than in Adubiyili. While the average household size in Chanshegu was 18.1 persons per household, the mean household size in Adubiyili was 14.2 (see Table 8.4).

The fact that households are significantly larger in Chanshegu may indicate that young adults are not leaving their family household to create households of their own. The relative lack of additional divisible land in Chanshegu compared to Adubiyili prevents youth from streaming into farming for themselves. As a result, young adults in Chanshegu are struggling to engage in remunerative opportunities, something that is financially (and culturally) necessary in order to marry and build a house. It is equally plausible that larger household sizes in Chanshegu have resulted from children commuting to school in Tamale on a daily basis. As Adubiyili is further from Tamale, children attending school in Tamale may live with relatives in Tamale or attend boarding school and only return to Adubiyili during school vacations.

Crops Types and Livestock

Maize is regarded as a subsistence crop and all the interviewees in both villages indicated that their first priority in farming is to grow enough maize to feed the household through to the next harvest. Maize is only traded at the market if there is a surplus and/or if cash is required for other purposes. The amount of land dedicated to maize serves not only as an indicator of food availability but also, as a proportion of the total available land, of the amount of remaining land used for crops that are culturally more readily exchanged for money, such as

TABLE 8.4 *Mean and median household size.*

	Mean	Median
Chanshegu	18,1	16,5
Adubiyili	14,2	11

N = 20+20

SOURCE: AUTHOR'S SURVEY.

cassava, yams and sorghum. Farmers in Chanshegu indicated that they used 64% of their land on average for the production of maize, while farmers in Adubiyili employed on average only 45% of their land for maize production. Since plot sizes are bigger in Adubiyili than in Chanshegu, the absolute land available for cash crops after growing sufficient maize for their own consumption in Adubiyili is considerably larger. Farmers in smaller households in Adubiyili therefore produced more maize and had more residual productive resources for the production of cash crops. This was verified by a series of wealth indicators (including the ownership of motorbikes, radios and televisions, the materials from which houses were built and whether interviewees had built their own houses or had inherited them). Farmers interviewed in Adubiyili scored higher than those in Chanshegu. Furthermore, farmers in Chanshegu are more dependent on maize so their livelihoods are more vulnerable in the event of crop losses due to less diversification than is the case in Adubiyili.

The interaction between livestock and arable farmers has been extensively studied, with the general consensus being that the interaction between the livelihoods of livestock keeping and sedentary farming works to increase productivity (Bourn & Wint 1994). Cattle are kept in Chanshegu and Adubiyili and are taken out daily to feed and drink. They return to the village in the evening. In Chanshegu they are kept in a clearing between the compounds at night but no such system appears to be present in Adubiyili. The landowner collects the manure regularly and sells it in Chanshegu. The increasing amount of land used for non-agricultural purposes is said to restrict the paths along which cattle can leave and enter the village in the mornings and evenings. As their movements potentially destroy crops, farming close to the village is regarded as a hazard and not practised. However, it was clear from observations that 'gardening' (the term used to refer to small-scale horticultural production as opposed to 'farming' that is larger in scale and includes maize, millet and cassava) was more prevalent in Chanshegu than in Adubiyili and done on land close to the village. Conducted largely by women, this has the advantage of being done close to the village because garden plots can be easily fenced (farms are too large), thus protecting them from cattle and ruminants as they move in and out of the village. Being near the village also means that women can simultaneously conduct their domestic duties and garden, unlike the men who often remain on the farm further away from the village for the entire day. Women constituted a large proportion of the purchasers of manure, buying from the owner of the land on which the cattle are kept. Whilst the cultivation of horticultural crops represents an agricultural intensification in Chanshegu in the Boserupian sense, not all community members cultivated gardens and the gains made from the increased intensification and extensification of horticultural gardening did not appear to offset the losses incurred by a lack of available land for farming. Furthermore, while there is a clear correlation between livestock keeping and crop farming, this appeared to apply most notably to horticultural gardens cultivated by women and less to larger farms cultivated by men. Finally, the use of manure represents an organic innovation. Although there is undoubtedly a growth in the interaction of livestock and arable farmers, which has led to an intensification of horticulture (Bourn & Wint, 1994), the intensification of agriculture here is limited in scope (and applies primarily to gardeners), with a large number of farmers preferring to use purchased fertilizer.

Technological Uptake

Adopting technologies is fundamental to increasing agricultural productivity (Juma 2011). Technology can refer to the use of fertilizer, ploughing techniques, storage, irrigation, improved seeds, marketing innovations or any other agrarian components that determine productivity. It serves as a good proxy for the intensification process.

Two of the farmers interviewed in Adubiyili had purchased a tractor to plough their own fields but other farmers in the village could hire it (usually paying in bags of maize). Although only two farmers had purchased a tractor, the technology thus spilled over to other farmers in the community who could benefit from the increased ploughing potential that tractors provide. No farmers in Chanshegu owned a tractor. This may have been because the profits accrued from the smaller plot sizes were significantly lower and inhibited such large capital outlays or, alternatively, because the smaller plot sizes meant that the economic gains to be made from the extra ploughing capacity are more than offset by the increased investment made. Tractor use may thus be regarded as an innovation in some agricultural productive systems, but certainly not in all. The farmers in Adubiyili had also installed a rice mill at the entrance to the village. Women were generally tasked with processing rice and bagging it for collection, increasing the value added at source. No such mill (or other such technologies) existed in or immediately around Chanshegu at the time of the survey. In addition, the bagged rice and other surpluses were collected by truck twice a week from Adubiyili and delivered to Tamale's central market. Although the truck stopped at a number of settlements en route to Tamale, it did not stop in Chanshegu and farmers speculated that Chanshegu did not feature on its list of stops because their volumes were smaller. Farmers in Chanshegu took their own produce to market either by bicycle, motorbike or taxi, depending on the quantity. Despite being further from Tamale, Adubiyili's market access appeared to be better than Chanshegu's.

Conclusion

The expansion of cities into their rural hinterlands is one of the major drivers of urbanization in Africa. The pressures and incentives for agrarian livelihoods subjected to urban sprawl are not understood but are a primary determinant of the respective livelihoods. This chapter has drawn some preliminary conclusions about Africa's urban growth and agrarian livelihoods in rural communities in peri-urban belts. It borrowed on notions of land use, population pressure and intensification, themes that have traditionally been linked to migration studies, to create a framework for how rural communities in peri-urban belts are subjected to different pressures and incentives arising out of Africa's urban explosion. While the exploratory nature of the fieldwork prevents conclusive empirical findings being drawn, it has outlined some of the major contours of agrarian livelihoods and urbanization at the farm level and sets a course for much needed further research on linkages between rapid urban growth and agricultural productivity in Africa.

Farmers in Chanshegu are being subjected to immense pressures due to Tamale's expansion but farmers in Adubiyili, as yet unaffected by Tamale's physical expansion, appear to be benefitting from the increased demand for food from its demographic growth. While concrete quantifications cannot be made about productivity, farmers in Adubiyili would seem to be able to better access the incentives from Tamale's urban growth and are less inhibited by the pressure on land from the conversion of tracts of land from rural to urban land uses. This manifests itself in larger farm sizes, smaller households, investments in agricultural capital and increased market access. As a result, whilst farmers in Adubiyili have been incentivized by growing demands for foodstuffs in Tamale, those in Chanshegu have not been able to accrue benefits from the new economic opportunities.

References

Barrios, S. & L. Bertinelli (2006), *Climatic Change and Rural-Urban Migration. The Case of Sub-Saharan Africa*. Louvain-la-Neuve: CORE.

Boserup, E. (1966), *The Conditions of Agricultural Growth: The Economics of Agrarian Change under Population Pressure*. Chicago: Aldine.

Bourn, D. & W. Wint (1994), Livestock, Land Use and Agricultural Intensification in Sub-Saharan Africa. Pastoral Development Network Paper 37a. London: ODI.

de Brauw, A. *et al.* (2013), The Role of Rural-Urban Migration in the Structural Transformation of Sub-Saharan Africa. *World Development* 63; 33–42. http://dx.doi.org/10.1016/j.world dev.2013.10.013. Accessed 8 September 2014.

Ghana Statistical Service (1960), Census of the Population; 1960.

Ghana Statistical Service (2000), Ghana, Population and Housing Census, 2000.

Ghana Statistical Service (2010), Ghana, Population and Housing Census, 2010. Summary of Results. Population by District, Sex and Type of Locality, Northern Region. http://www.statsghana.gov.gh/docfiles/2010phc/Census2010_Summary_report_of_final_results.pdf. Accessed 8 september 2014.

Gold Coast Census Office. Population Census of the Gold Coast and it's Protectorates: 1911, 1921, 1931 & 1948.

Hart, K. (1973), Informal Income Opportunities and Urban Employment in Ghana. *The Journal of Modern African Studies* 11(1): 61–89.

Hugo, G. (2008), *Migration, Development and Environment*. Geneva: International Organization for Migration.

Juma, C. (2011), *The New Harvest: Agricultural Innovation in Africa*. New York: Oxford University Press.

Leke, A., S. Lund, C. Roxburgh & A. van Wamelen (2010), What's Driving Africa's Growth. McKinsey & Company. http://www.mckinsey.com/insights/economic_studies/whats_driving_africas_growth. Accessed 14 April 2014.

Lipton, M. (1977), *Why Poor People Stay Poor: Urban Bias in World Development*. Cambridge: Harvard University Press.

McGranahan, G. *et al.* (2009), Africa's Urban Transition and the Role of Regional Collaboration. IIED Human Settlements Working Paper No. 5.

O'Sullivan, A. (2009), *Urban Economics*. Boston: McGraw-Hill/Irwin.

Population Division of the Department of Economic and Social Affairs of the United Nations (2011), Urban-Rural Areas. http://esa.un.org/unup/wallcharts/urban-rural-areas.pdf. Accessed 16 April 2014.

Population Division of the Department of Economic and Social Affairs of the United Nations (2013), *World Population Prospects, 2012 Revision*.

Potts, D. (1995), 'Shall We Go Home?': Increasing Poverty in African Cities and Migration Processes. *The Geographical Journal* 161(3): 245–264.

Prain, G., N. Karanja & D. Lee-Smith (2010), *African Urban Harvest Agriculture in the Cities of Cameroon, Kenya and Uganda*. New York: Springer.

Staniland, M. (1977), *The Lions of Dagbon*. Canbridge: Cambridge University Press.

Steel, C. (2008), *Hungry City: How Food Shapes Our Lives*. London: Chatto & Windus.

Tacoli, C. (1998), Rural-Urban Interactions: A Guide to the Literature. *Environment and Urbanization* 10(1): 5–22.

Tiffen, M., M. Mortimore & F. Gichuki (1994), *More People, Less Erosion: Environmental Recovery in Kenya*. Chichester: Wiley.

UNHabitat (2012), *State of the World's Cities 2012/2013*. Nairobi: UNHabitat.

van der Geest, K. (2011), *The Dagara Farmer at Home and Away: Migration, Environment and Development in Ghana*. Leiden: Brill.

Yussif, D. *et al.* (2011). Third Trimester Field Programme (TTFP) of Group 169. Tamale Rural, Chanshegu Community. Unpublished Report.

CHAPTER 9

The Dynamics of Urban and Peri-Urban Agriculture

Diana Lee-Smith

This chapter examines the complex dynamics of food production in areas where urban and peri-urban agriculture (UPA) is practised. Urban food demand is the main driver of UPA as it is of agriculture in Africa in general, but the concept has been expanded to encompass hunger and not only the ability to pay for food. UPA provides opportunities for income generation and a pathway out of poverty but the urban poor lack access to the land that could make this possible and have to resort to farming on open spaces, where available, and suffer food insecurity when they are not. UPA contributes to the nutritional needs of urban populations by providing access to fresh vegetables and dairy products and the children of urban livestock keepers are healthier because they consume animal source foods. African urban areas have a nutrient surplus while soil fertility is generally a major constraint in agriculture. UPA institutions need to coordinate with urban waste management experts. African government policies in the region have mostly restricted UPA but new forms of institutions are emerging although farmers' engagement will be required for them to function effectively.

Introduction

While various chapters in this volume address aspects of urban food demand as it affects food production in Africa generally, this chapter examines the complex dynamics of food production within urban areas themselves, namely urban and peri-urban agriculture (UPA). Urban food demand is examined first as the main driver of UPA, as it is of agriculture in general, but it is treated as a concept that encompasses the need for food and not only the ability to pay for it, thus linking it to explanations of urban poverty and food insecurity. The next section explores how UPA functions in relation to the nutritional needs of urban populations, which are often not well addressed in approaches to increased agricultural production. The dynamics of nutrient cycling are then examined, looking at the nutrient surplus in urban waste in relation to soil fertility. Finally, the changing dynamics of government policy on UPA in the African region are discussed in relation to the types of institutions that are emerging and are needed.

An explanation of some of the terms used is required first, along with their grounding in the data available. The production of food in urban areas is the

 | DOI 10.1163/9789004282698_010

main focus in this chapter, although the term 'urban agriculture' also includes food processing and distribution. Mougeot's widely accepted definition of urban agriculture is:

> ... agriculture located within (intra-urban) or on the fringe (peri-urban) of a town, a city or a metropolis, that grows or raises, processes and distributes a diversity of food and non-food products, (re-)uses largely human and material resources, products and services found in and around that urban area, and in turn supplies human and material resources, products and services largely to that urban area.
>
> MOUGEOT (2000: 10).

What distinguishes urban agriculture from other forms of agriculture is that it is an integral part of the urban economic, social and ecological system. About 40% of urban households in Africa are estimated to produce food (Lee-Smith *et al.* forthcoming). Although the definition specifically subsumes urban and peri-urban agriculture (UPA) in the more general term of 'urban agriculture', the two also need to be further differentiated. This is no easy task, however, as UPA is a vulnerable type of land use, and is usually of low priority and frequently illegal.

Peri-urban agricultural production has characteristics similar to any surrounding rural agriculture and merges with it along a rural to urban continuum. At the same time, it tends to verge towards intensification due to urban demands (see Photo 9.1). Intra-urban agriculture is also influenced by urban demand as well as space constraints and other urban conditions such as the availability of waste that can be used as input. Measurements on this continuum have shown that two variables – the proportion of households farming and the size of the enterprise – decrease from peri-urban to urban (David *et al.* 2010; Dongmo *et al.* 2010; Drechsel & Dongus 2010; Lee-Smith *et al.* forthcoming). It has been asserted that the peri-urban zone characterizes an area measuring 15–40 km from an urban centre (Adam 2001; Drechsel *et al.* 2007), although the term is frequently used for areas bigger than this, especially in the case of large urban agglomerations. Some researchers place the peri-urban zone outside urban administrative boundaries, others within it, to distinguish peri-urban from rural areas (Foeken & Owuor 2008).

Limited data suggest that the two variables – the proportion of households farming and the size (land area) of enterprises – also decrease with the increasing size of the urban area, although this has not yet been proven. However, the sets of data on spatial variation and the size of the urban area suggest that urban density and land availability are important variables that influence UPA

PHOTO 9.1 *Intensive urban agriculture in backyard crop-livestock farms that recycle nutrients and provide food and income*
PHOTO: DIANA LEE-SMITH

production and affect the response to urban food demand (Lee-Smith & Memon 1994; Lee-Smith *et al.* forthcoming).

The definition of UPA excludes rural food production by urban households, although multi-locational households straddle the rural–urban divide. Sometimes rural farming by such households contributes more to urban household food security than urban farming does (Foeken & Owuor 2008). This is one of the many aspects of UPA that are hard to measure, although Foeken & Owuor do so effectively for the town of Nakuru, showing that the urban poor are disadvantaged compared to better-off residents but are under-represented among urban farmers (see also Foeken 2006).

The question is whether UPA makes economic sense. Production occurs close to the market and to useful inputs and support services, while these factors combine with scarce land and abundant labour to promote intensification. Data supporting some of these explanations are presented below. UPA may also be economic due to market failures. While many African farmers face low farm-gate prices, food is expensive for consumers because of the high transport costs due to bad infrastructure, the market power of middlemen and a lack of pro-farmer institutions. These factors all create incentives that encourage urbanites to produce their own food rather than to buy it. But UPA can sometimes be economic because not all the costs of production have been internalized, such as the opportunity cost of labour and land.

Urban Demand: The Need for Food with or without the Ability to Pay

Rapid urbanization has been seen in Africa in recent decades and the trend is projected to continue, with an additional 367 million people expected to be living in urban areas between 2000 and 2030. This is more than twice the additional rural population previously anticipated (Prain 2010). Increased urban demands for food affect Africa's agriculture. Food and nutrition dynamics have two aspects, on the one hand there is the purchasing power of the urban population that needs food and, on the other, the lack of purchasing power of the urban hungry. Urban purchasing power and urban hunger are also both factors that are driving people to produce food in urban areas. Two aspects of urban growth, namely household food insecurity and increasing market demand, have been identified as the main drivers of UPA in a forthcoming book on African farming systems on which this chapter is based (Lee-Smith *et al.* forthcoming).

Demand is treated here as a much broader concept than economic demand alone. Urban residents may make demands (or have a need) for food to fill their empty stomachs and/or express their demand for food in the market by buying it. They also engage in UPA to help fill their stomachs or respond to the high demand for food by selling it and generating a living. Some may do both, and many are also saving money by engaging in UPA. These household strategies interact and are inherently hard to measure even though there is no shortage of relevant data.

A survey of poor communities in eleven Southern African cities found that 77% of households were facing food insecurity (Frayne *et al.* 2010). The link between this high incidence of hunger among the urban poor and UPA needs to be adequately understood. The question is addressed in several recent works that suggest access to land for urban food production by the urban poor, the policy environment, spatial geography and the political economy as factors that might account for variations in urban food insecurity among the urban poor (Crush *et al.* 2010; Riley & Legwegoh 2013; Lee-Smith 2013; Lee-Smith *et al.* forthcoming).

As Africa's urban population increases, urban poverty is rising in parallel and has been called the 'urbanization of poverty' (Haddad *et al.* 1999). It is usually assumed that rural-to-urban migration is the predominant factor in urban population increases and likewise in increases in urban poverty. However, studies of urban growth have shown that the natural increase in the urban population and the reclassification of areas from rural to urban as cities expand are at least as important as migration. Migration itself has been shown to be a complex process that is seasonal and cyclical in nature and is not only one-way (Prain 2010).

Multi-locational households, i.e. those that have rural and urban members, defy the capacity of researchers and administrators to classify them. Households or members of households move from the inner urban to peri-urban and rural areas, using both urban and rural resources for agricultural and non-agricultural livelihood strategies to combat poverty and generate a decent living (Prain 2010; Owuor 2006). Labour, agricultural inputs, marketing chains, micro-nutrient flows and social and political ties and obligations, including remittances, flow through rural–urban linkages. They appear to stimulate and intensify both urban and rural agriculture (Satterthwaite & Tacoli 2002; Prain & Lee-Smith 2010).

UPA was assumed to be largely a practice of the urban poor until Foeken & Owuor's work in Nakuru not only confirmed earlier observations that it is a practice engaged in by all income groups but that lower income groups are proportionally under-represented in the activity (Foeken 2006; Foeken & Owuor 2008). So what is the relationship between UPA and poverty? The truth is that there is extensive ignorance surrounding this subject. All the data so far indicate that producing food for household consumption is the main reason that people engage in UPA yet, at the same time, the data also indicate that urban farmers earn good incomes. How does this happen and what do the data mean?

An ODI study has shown that marketing and labour are factors that affect UPA's impact on poverty (Mitchell & Leturque 2010). But surveys show that farming households overwhelmingly produce for their own use and that commercial producers are even consuming part of their production (Prain *et al.* 2010). And yet, urban farming households are generating significant incomes. Farmers in Kampala, Dar es Salaam, Yaoundé and Addis Ababa all had higher than average incomes (Prain & Lee-Smith 2010). Higher incomes among UPA households have been linked to livestock keeping in East Africa and to the fact that owners of livestock are more likely to be commercially active than those who just grow food crops (Foeken 2006; Prain *et al.* 2010). Urban producers of irrigated vegetables may even exceed official minimum salary levels by a factor of 1.6 to 10.

The incomes of farm households are affected by their degree of market orientation as well as a multitude of other variables such as farm size, household labour, the types of crops and animals involved, the availability of cost-saving urban inputs (organic wastes and wastewater), dry-season irrigation and the ability to store, process and preserve products. Households also benefit from processing and marketing, such as dairy products, street foods, sales to kiosks

PHOTO 9.2 *Open spaces provided by Cape Town City Council give the urban poor the chance to feed themselves and generate additional income from intensive urban agriculture.*
PHOTO: DIANA LEE-SMITH

(small local shops) or cleaning and packaging food for sales to supermarkets (van Veenhuizen & Danso 2007). Market research in Kampala shows that the demand for UPA products is high and is not being met in outlets ranging from corner kiosks to supermarkets (Nyapendi *et al.* 2010).

So do the urban poor experience greater levels of poverty and worse food insecurity if they do not engage in UPA? This question was asked in the AFSUN study of food insecurity in eleven Southern African cities (Crush *et al.* 2010). A link between UPA and household food security has been demonstrated from data collected in Kampala (Cole *et al.* 2008) and it could be speculated that the former is also the case. A lack of access to land for urban farming and how this affects residents in high-density urban slums, and female-headed households in particular, has been demonstrated. Women are less able to access land than men in general in Sub-Saharan Africa (Lee-Smith, 2010; 2013).

An association is recognized between urban poverty and living in high-density areas that lack basic services, such as water and sanitation. People residing in such areas are generally only able to engage in food production if they can gain access to unused open space (see Photo 9.2), and these areas are increasingly only to be found in peri-urban areas. Urban and peri-urban farming systems have been classified into backyard and open space sub-systems (FAO 2012; Lee-Smith *et al.* forthcoming). Backyard farmers are generally higher and middle-income earners and also have more secure tenure on the land they use for farming. Such land tends to be more characteristic of intra-urban areas. Micro-gardens, sack gardening and other intensive agricultural technologies have been developed to help slum dwellers and they also facilitate the intensification of food production (Lee-Smith *et al.* forthcoming).

To summarize, the relationship between increasing urban demands for food and UPA is presenting a good opportunity but also a challenge for households to earn a living and escape poverty while at the same time supplementing their food supply. The very poor and female-headed households in African urban areas have fewer chances to use UPA either for their own food security or as a pathway out of poverty. They rely on favourable policies supporting access to land for the urban poor for UPA, or on circumstances of density and geography. Political economy and development history also play a part, the poor being more food-secure in cases where there are diverse actors in urban governance than when food chains are dominated by the corporate sector (Riley & Legwegoh 2013; Lee-Smith *et al.* forthcoming).

UPA's Nutritional Benefits

The high levels of food insecurity among the urban poor in Southern Africa revealed by the AFSUN data were complemented by data on poor dietary diversity among the same population. About 96% of the food intake of the urban poor (both the food-secure and insecure) involved starchy staple foods. However the food-secure had access to eight food groups, while the food-insecure could only access five, two of which were sugar and beverages (Frayne *et al.* 2010).

As early as 1994, a study in a Nairobi slum showed urban-farming households had a higher energy intake and that young children had a better nutritional status in terms of weight-for-height than non-farming households (Mwangi 1995). Later, international UPA studies of the urban poor and food-insecure funded by the FAO and the World Bank found that the practice of urban farming corresponded with greater dietary diversity in ten of the fifteen countries studied (Zezza & Tasciotti 2008; Prain & Dubbeling 2011). Nutrition security (for which food security is necessary but not a sufficient condition) was also linked to aspects of UPA using regression analysis in a detailed study of both urban and peri-urban farming and non-farming households in Kampala in 2003. Although wealth (income and assets) was the main factor affecting household food and nutrition security, various aspects of UPA influenced household food security, namely land size, the keeping of livestock, gender and education. Consuming animal source foods was linked to better child health and nutrition, which indicates that urban livestock keeping can increase both food and nutrition security (Yeudall *et al.* 2008; Sebastian *et al.* 2008).

The amount and types of UPA food vary widely by city and sub-region. Ghanaian backyard farms in the cities of Accra and Kumasi only contribute 10% or less to household food demand (Drechsel *et al.* 2009) and UPA makes a greater contribution to household food security there in terms of fresh produce than calories, as most staples that support the daily calorie intake are derived from rural cereal or tuber production (Drechsel *et al.* 2007). Even in Nakuru, Kenya where extensive UPA production has been recorded, purchased food outweighed both urban and rural food production in importance for urban farming households (Foeken & Owuor 2008). UPA in Nakuru provided 22% of the food intake in farming households and accounted for 8% of the town's overall food and nutritional needs (Foeken 2006). It is in the fresh-produce categories that UPA makes a significant contribution. In Dar-es-Salaam, UPA provided 90% of the city's leafy vegetables and over 60% of its milk (Jacobi *et al.* 2000). And in Cameroon, the traditional leafy vegetables grown in Yaoundé provided 8% of the protein and 40% of the calcium consumed in the city (Bopda *et al.* 2010).

The widespread existence of UPA, as it is documented in some African cities, may ensure better nutrition even in places where it does not provide a high percentage of the farmers' or the cities' food supply. As an example of the role it can play, no severe malnutrition was recorded in Kampala, a city with widespread UPA, during the decade-long civil war in the 1980s and the structural adjustment policies of the 1990s (Cole *et al.* 2008; Atukunda 1998).

A study of nutrition in Yaoundé, where UPA is also abundant, found no major nutritional shortages as people's diets were rich and diversified, and the target of 2500 kcal/person/day was easily met in all income groups. However, a more detailed breakdown of the low-income group found that about 10% were suffering from a chronic deficit of carbohydrates and protein, meaning they were accessing insufficient staples, which are less often supplied by UPA and instead have to be purchased. The very poor obtained their protein and micro-nutrient needs from leafy vegetables and groundnuts, whose nutritional contribution was found to be more significant for the urban poor than for wealthier families. This was because these traditional foods supply a disproportionate share of their protein, minerals and vitamins. The protein of amaranths is particularly noted for its high levels of lysine, an essential amino acid that is often lacking in diets that rely heavily on roots and tubers (which are commonly eaten as staples by the poor). Leafy vegetables are critical to calcium intake as access to dairy products is limited in Yaoundé (Bopda *et al.* 2010).

By contrast, the diet in Kampala is based more on dairy products, as is also the case elsewhere in East African UPA. Both the children and adults in Kampala get much of their protein and many of their micro-nutrients from milk, usually in the form of tea. Dairy cattle, poultry and pigs are raised in the city, whereas the typical livestock in Yaoundé are poultry and pigs (Prain *et al.* 2010).

UPA in backyards and open-space farms complement rural agriculture and can alleviate malnutrition among the urban poor who would otherwise face food insecurity. The efficiency of the national food supply is increased by providing products such as perishables that require rapid delivery (e.g. fresh milk and vegetables) especially where the infrastructure and organization are poor. The contribution of UPA to food and nutrition security (see Photo 9.3) has the potential to be systematically improved through better policy-making (Lee-Smith *et al.* forthcoming).

PHOTO 9.3 *Keeping livestock ensures urban farmers have a good income through value-addition enterprises and it keeps their children healthy.*
PHOTO: DIANA LEE-SMITH

UPA and Urban Waste Reuse

Urban areas produce vast quantities of nutrient surpluses, some of which are found in wastewater while others form part of the urban solid waste that ends up as landfill or on waste dumps. UPA could improve a city's performance in nutrient cycling through soil amendment, which has implications for sustainability. In fact, large urban and peri-urban open spaces in various cities are using wastewater to irrigate UPA crops. The productivity of these irrigated open spaces is high because of the nutrients present and the availability of water, which makes it possible for the same vegetable bed to be used for between three to ten crops annually. This has been extensively documented, especially in West Africa, and has a been a focus of work by local and international institutions, including the International Water Management Institute (IWMI). The potential health risks of using wastewater in agriculture, both rural and urban, have likewise been of concern and new international guidelines on safe wastewater re-usage have been issued as a result (Drechsel *et al.* 2006; WHO 2006).

In addition to wastewater, African urban areas produce millions of tonnes of solid waste per annum, most of which is organic. This suggests a high nutrient-cycling potential, although almost none is being realized at present except by some (unassisted) backyard farmers. An estimated 2223 tonnes of nitrogen (N), the same amount of phosphorus (P) and 3700 tonnes of potassium (K), worth about US$ 2 m, are generated annually in Nairobi alone (Njenga *et al.* 2010). It is not easy to export this waste to farms in rural areas due to marketing constraints. Transporting such waste is not nearly as easy as transporting food and would require organized markets, value addition and subsidies (Drechsel *et al.* 2010).

Data are beginning to emerge about the role played by urban farmers in recycling solid waste on small backyard farms. Quantitative evidence suggests as much as 90% of food waste is being recycled as livestock feed in the backyards of Nakuru, Kenya for example, with the other 10% being used as compost. Crop residues are mostly recycled as feed and about 25% goes to composting, while 48% of livestock farmers used manure for crop cultivation on their own backyard farms (Prain & Lee-Smith 2010). In Yaoundé, Cameroon, 69% of the poultry and pig manure generated in the city and its environs (20,600 tonnes dry weight annually) was utilized in farming, mostly within the same mixed crop-livestock backyard farms that produced it. However 10% was shipped to another city, Bamenda, where it fetched a higher price (Dongmo *et al.* 2010).

The details of the research in Nakuru are revealing. Although virtually all organic domestic waste generated by urban farm households and almost 50%

of the manure generated were reused productively, there were marked differences in the way the different income groups reused manure. Middle-income households with backyard crop-livestock farms achieved the very high manure reuse rate of 88%, while low-income residents in high-density areas only achieved 17%. The logic behind this Is easy to understand as the majority of the recycled manure (61% of 283,000 tonnes wet weight annually) was used directly on the same urban farms (Karanja *et al.* 2010; Lee-Smith *et al.* 2010). Since residents in high-density slums have no backyards where they can easily reuse manure from livestock, much of it (45% of total manure production or 142,000t tonnes fresh weight) is dumped alongside other solid waste (Karanja *et al.* 2010; Lee-Smith *et al.* 2010; Foeken 2006).

The Nakuru findings are consistent with a random household sample survey done in Nairobi in the 1980s. Of the crop growers, 35% used compost and 29% used manure, while 91% and 44% respectively produced these inputs on their own farms. Projected forward, this suggested that roughly 54,500 households in the city would be using compost in 2003, with 41,400 producing it themselves. And 37,700 households would have been using livestock manure to fertilize crops in the city, 16,600 of them producing it on their own mixed-crop and livestock farms. The amount marketed was very low (Njenga *et al.* 2010).

Not enough research is available apart from that cited about what small farmers are now doing in terms of waste and nutrient recycling. Whether the volumes of waste are significant or insignificant in relation to the overall waste burden of any city, small-scale backyard crop-livestock farmers fall outside the framework of policy and/or planning, either by the urban or agriculture sectors (Karanja *et al.* 2010). The organized collection and reuse of livestock waste from low-income farms in high-density areas (as is being initiated in Nakuru) would greatly increase the economic efficiency of food production as well as waste management. Likewise, the planned reuse of urban organic solid waste in urban farming could improve food production and waste management and contribute to improved ecological cycles for sustainability. Open- space irrigated farms using wastewater could likewise achieve higher levels of performance in sustainability and public health through planned management and public-sector support. Policy opposition to urban livestock keeping as well as wastewater farming needs to be rethought in the light of their role in income generation and nutrition as well as nutrient reuse.

In terms of urban sustainability, these aspects of UPA require more research and public-sector policy attention. Urban backyard farms are potential models of good urban ecosystem performance. They could also achieve levels of intensification in food production that might serve as models for rural farming intensification by highlighting the need for soil-quality maintenance. In the same vein, wastewater open-space farms, which are already contributing informally to urban food supply, nutrition and incomes, have the potential to increase their performance and serve as models for urban sustainability.

Government Policy and Institutions on UPA

The key policy challenge for UPA is political recognition. In addition, access to land for UPA by urban poor and female-headed households that are food insecure is a necessary policy goal (Lee-Smith *et al.* forthcoming). As pointed out above, the two policy goals, namely of food security for the urban poor and enterprise development to strengthen the urban food system, need to be clearly separated and both should be addressed in specific policy actions (Lee-Smith 2013).

Addressing food safety and reducing the identified health risks associated with UPA are areas of policy that must also be tackled. There are a multitude of ways that biological (pathogenic) and chemical (toxic) contamination can reach food, i.e. through the air, water and soil. While these different contaminants have been researched, the only comprehensive study of the health risks and benefits from UPA was carried out in Kampala, Uganda in the early 2000s (Cole *et al.* 2008). It established a statistical relationship between consuming animal products and child health and nutrition security, and between urban food security and livestock rearing, suggesting that the benefits of urban livestock keeping balance out the health risks that undoubtedly exist.

The greatest risk to human health from biological contaminants is the widespread absence of sanitation in urban areas. Kampala's farmers use many health-risk mitigation strategies, with over 90% avoiding biological contaminants in milk by consuming it boiled with water and tea leaves. Farmers with access to services (like water) and those who believe that UPA is legal were found to use more and better risk-mitigation strategies (Cole *et al.* 2008). Low-cost options are also available for mitigating the risks of using wastewater while maximizing its benefits (Keraita *et al.* 2008) and many are now permissible under the revised WHO guidelines for wastewater use in irrigation (WHO 2006). In short, the evidence required to make good UPA policy is now available and supportive measures can be put in place to assist farmers in producing safe food given the existing constraints.

In the past, the performance of African governments in institutional and policy development for UPA has been poor. UPA has generally been tolerated, ignored or challenged. In times of emergency, it has tended to be tolerated and even supported and encouraged, for example in the case of Maputo's Green

Zones during the civil war in Mozambique. The same applied in Kampala during Uganda's civil war, although urban food production was not given actual policy support, as was the case in Mozambique. The region's general policy has steadily improved since the 1980s when research findings began to emerge and urban dwellers were suffering the impact of Structural Adjustment Policies (SAPs). In the 1990s, several UN bodies highlighted UPA and the Consultative Group on International Agricultural Research (CGIAR) established the Urban Harvest Initiative (Prain 2010). These and other international partners supported UPA policy development in Africa (Mougeot *et al.* 2010).

Tanzania was one of the earlier examples of a shift in official attitude towards policy and legislation. Its Local Government Act of 1982 and Town and Country Planning Ordinance of 1992 recognized and regulated urban crop and livestock production. The UN-supported Sustainable Dar es Salaam Programme built local government capacity as UPA continued to be incorporated in laws and institutions, such as the Agriculture and Livestock Policy of 1998 and the Urban Vegetable Project organized by the Ministry of Agriculture and Cooperatives that improved urban extension services (van Veenhuizen 2006). Efforts are currently being made to include land for UPA in the Dar es Salaam Master Plan (Halloran & Magid 2013). Negative attitudes towards UPA still persist however, especially in some urban local government bureaucracies (Foeken 2008).

Two international bodies, the Urban Harvest of the CGIAR and the Resource Centres on Urban Agriculture and Food Security (RUAF) headquartered in the Netherlands have helped build stakeholder platforms to integrate research with policy in a number of African countries. So many sectors are potentially involved in or affected by UPA that a negotiated process of new institution-building seems vital. This institutional challenge is not the same as in rural farming and is due to the density of both the population and institutions with a stake not only in UPA but also in urban government. Without clear policy and political leadership, there will be confusion surrounding UPA, with one authority giving awards to the best urban farmers while another prosecutes them (Obuobie *et al.* 2006; Lee-Smith *et al.* forthcoming).

An example of local multi-sectoral institution development was the Health and Urban Agriculture Coordinating Committee in Kampala in the early 2000s. Having begun as a committee that led project-based research linking many institutions and elected and unelected officials, it became a registered organization, the Kampala Urban Food Security, Agriculture and Livestock Coordinating Committee (KUFSALCC) that helped the City Council to develop urban agriculture ordinances in 2006 (Lee-Smith & Prain 2010). However, laws are only as good as the will of those who implement them and subsequent plans to integrate UPA into urban planning failed when there were clashes between those wanting land allocated for low-cost housing with crop cultivation and others with different and competing land claims. Urban land use is deeply contested and access to land and tenure for UPA is the main area where the lack of coordination appears. Nonetheless, authorities in some West African cities have started to offer farmers space with safer water sources. These tend to be in peri-urban locations and the problem of moving farmers from intra-urban areas remains (Lee-Smith *et al.* forthcoming).

Important distinctions emerge when a pro-poor focus is used to analyse legal and institutional UPA frameworks. Institutions acting on behalf of farmers and their associations in active debates with local government have seen beneficial results, for example in Accra, Ghana and Freetown, Sierra Leone (Cabannes 2011). Similarly, the Nairobi and Environs Food Security, Agriculture and Livestock Forum (NEFSALF) was observed to be a stronger institution than the KUFSALCC in neighbouring Uganda because it includes farmers (Prain *et al.* 2010).

Internationally, two different models of local government institution addressing UPA can be identified. In developed countries, especially in North America, Food Policy Councils have become widespread. Typically a civil-society initiative, these bring together the various interests in the urban food chain and relate to (urban) local government in an increasingly formal way. Toronto's Food Policy Council, for example, has always met at City Hall and has been influential in the city government initiating a food strategy in its Health Department. By contrast, the most common way for UPA to become institutionalized in African cities is through the decentralization of local government responsibilities. In 1993, for instance, Uganda's Decentralization Policy led to the creation of Kampala City's Department of Agriculture that developed support programmes for UPA in collaboration with local NGOs. A classification system was established identifying four types of urban and peri-urban agricultural areas within the city, and new ordinances in 2006 governing UPA, urban livestock, meat, fish and dairy handling (Lee-Smith *et al.* forthcoming). Likewise, the devolution of government in Kenya in 2013 led to the creation of Nairobi City County's Agriculture, Livestock and Fisheries Department.

Other African urban local authorities with institutions addressing UPA are Cape Town in South Africa and Addis Ababa in Ethiopia. Such innovation is also ongoing in Ghana, Liberia, Sierra Leone and Bulawayo in Zimbabwe. These participatory multi-stakeholder platforms parallel the urban Food Policy Councils and similar institutional innovations in urban authorities in developed countries. A great deal can be gained by international exchanges on how to make such institutions effective, even though local circumstances may vary. It is equally important to build bottom-up market systems with farmers' networks and organizations, as has been done in Accra, Nairobi, Freetown and Cape Town.

Yet questions remain about the economic externalities of urban farming (public health, nuisance, occupation of public space) that imply an economic cost to non-farmers. Government policies discouraging UPA are usually motivated by such externalities. There is tension, therefore, between some people's hunger and their right to grow food and the needs of others regarding different uses of urban space.

Brazil is the only country that has so far developed a 'Right to Food' policy approach that includes land access for UPA for the food-insecure. African cities have not yet gone for a rights-based approach in their policies. The hangover from the era of prohibitive or *laissez-faire* attitudes to UPA has meant that policies tend to favour market-based support for UPA along with any new-found acceptance of urban farming. This supposition is not yet based on extensive research or policy analysis, although one study of UPA in local government from a political-science perspective suggests that the interests of the poor are generally not addressed (Gore 2008). Evidence from Tanzania shows that poor and female-headed households have not benefitted from UPA policy measures and the UPA ordinances in Kampala have also not provided these groups with any benefits. References to them in early drafts were eliminated as the legislation was reviewed in Council but political interests, and particularly interests in land development, may lead to potential and/or actual neglect of low-income groups' access to UPA (Lee-Smith, 2010; 2013).

Conclusions

The analysis in this chapter suggests that UPA is playing a significant role in Africa's agro-food and nutrition dynamics. The growing urban demand for food there is driving not only food production in the region in general but UPA in particular. An estimated 40% of urban households are producing some food themselves in urban areas, even if it is only a small amount of their own overall consumption. This represents a sizeable amount of food production, especially as urban populations grow.

The nutritional benefits of UPA to low-income urban dwellers may be more significant than the overall volume of food produced, although the policy benefits of allocating land to low-income and female-headed households for food production should not be neglected. With over 75% of low-income urban dwellers being food insecure, UPA is an important policy measure that could address urban hunger and significantly intensify food production due to the efficiencies of space use and nutrient cycling. This potential may also have implications for improvements in agricultural intensification in general, through the proven rural–urban linkages, especially those found in multi-locational households.

Finally, the potential of UPA can best be realized by speeding up the ongoing process of UPA policy and institutional development in Africa. Such institutions work better if urban farmers are involved in their functioning and regional exchange and networking between urban local authorities would help in this respect.

References

Adam, M. (2001), Definition and Boundaries of the Peri-urban Interface: Patterns in the Patchwork. In: P. Drechsel & D. Kunze (eds), *Waste Composting for Urban and Peri-urban Agriculture: Closing the Rural–urban Nutrient Cycle in Sub-Saharan Africa.* Wallingford, UK: CABI Publishing, pp. 193–208.

Atukunda, G. (1998), Urban Agriculture in Kampala, Uganda: Reviewing Research Impacts. Canada: IDRC.

Bopda, A., R. Brummett, S. Dury, P. Elong, S. Foto-Menbohan, J. Gockowski, C. Kana, J. Kengue, R. Ngonthe, C. Nolte, N. Soua, E. Tanawa, Z. Tchouendjeu & L. Temple (2010), Urban Farming Systems in Yaoundé: Building a Mosaic. In: G. Prain, N.K. Karanja & D. Lee-Smith (eds), *African Urban Harvest: Agriculture in the Cities of Cameroon, Kenya and Uganda*, New York/Ottawa: Springer/IDRC, pp. 39–59.

Cabannes, Y. (2011), Pro-poor Legal and Institutional Frameworks for Urban and Peri-urban Agriculture. Draft Study. Rome: FAO.

Cole, D.C., D. Lee-Smith & G.W. Nasinyama (eds) (2008), *Healthy City Harvests: Generating Evidence to Guide Policy on Urban Agriculture.* Kampala/Lima: CIP/ Urban Harvest/Makerere University Press.

Crush, J., A. Hovorka & D. Tevera (2010), *Urban Food Production and Household Food Security in Southern African Cities.* Kingston/Cape Town: Queen's University/AFSUN.

David, S., D. Lee-Smith, J. Kyaligonza, W. Mangeni, S. Kimeze, L. Aliguma, A. Lubowa & G. Nasinyama (2010), Changing Trends in Urban Agriculture in Kampala. In: G. Prain, N.K. Karanja & D. Lee-Smith (eds), *African Urban Harvest: Agriculture in the Cities of Cameroon, Kenya and Uganda.* New York/Ottawa: Springer/IDRC, pp. 97–122.

Dongmo, T., F. Meffeja, J.M. Fotso & C. Nolte (2010), Crop-Livestock Integration in the Urban Farming Systems of Yaoundé. In: G. Prain, N.K. Karanja & D. Lee-Smith (eds), *African Urban Harvest: Agriculture in the Cities of Cameroon, Kenya and Uganda.* New York/Ottawa: Springer/IDRC, pp. 61–70.

Drechsel, P., O. Cofie & D. Danso (2010), Closing the Rural–urban Food and Nutrient Loops in West Africa: A Reality Check. *Urban Agriculture Magazine* 23: 8–10.

Drechsel, P. & S. Dongus (2010), Dynamics and Sustainability of Urban Agriculture: Examples from Sub-Saharan Africa. *Sustainability Science* 5(1): 69–78.

Drechsel, P., S. Graefe & M. Fink (2007), *Rural–urban Food, Nutrient and Virtual Water Flows in Selected West African Cities.* Colombo: International Water Management Institute.

Drechsel, P., S. Graefe, M. Sonou & O. Cofie (2006), *Informal Irrigation in Urban West Africa: An Overview.* Colombo: International Water Management Institute.

Drechsel, P., E.O. Sarpong & L. Hope (2009), Urban Farming and Household Food Security. *Urban Agriculture Magazine* 22: 6.

FAO (2012), *Growing Greener Cities in Africa: First Status Report on Urban and Peri-urban Horticulture in Africa.* Rome: FAO.

Foeken, D. (2006), *'To Subsidize My Income': Urban Farming in an East African Town.* Leiden: Brill.

Foeken, D. (2008), Urban Agriculture and the Urban Poor in East Africa: Does Policy Matter? In: M. Rutten, A. Leliveld & D. Foeken (eds), *Inside Poverty and Development in Africa. Critical Reflections on Pro-poor Policies.* Leiden: Brill, pp. 225–254.

Foeken, D. & S.O. Owuor (2008), Farming as a Livelihood Source for the Urban Poor of Nakuru, Kenya. *Geoforum* 39(6): 1978–1990.

Frayne, B., W. Pendleton, J. Crush, B. Acquah, J. Battersby-Lennard, E. Bras, A. Chiweza, T. Dlamini, R. Fincham, F. Kroll, C. Leduka, A. Mosha, C. Mulenga, P. Mvula, A. Pomuti, I. Raimundo, M. Rudolph, S. Ruysenaa, N. Simelane, D. Tevera, M. Tsoka, G. Tawodzera & L. Zanamwe (2010), *The State of Urban Food Insecurity in Southern Africa.* Kingston/Cape Town: Queen's University/AFSUN.

Gore, C. (2008), Healthy Urban Food Production and Local Government. In: D.C. Cole, D. Lee-Smith & G.W. Nasinyama (eds), *Healthy City Harvests: Generating Evidence to Guide Policy on Urban Agriculture.* Kampala/Lima: CIP/Urban Harvest/Makerere University Press, pp. 49–65.

Haddad, L., M. Ruel & J. Garrett (1999), Are Urban Poverty and Undernutrition Growing? Some Newly Assembled Evidence. *World Development* 27(11): 1891–1904.

Halloran, A. & J. Magid (2013), Planning the Unplanned: Incorporating Agriculture as an Urban Land Use into the Dar es Salaam Master Plan and Beyond. *Environment and Urbanization* 25: 541.

Jacobi, P., J. Amend & S. Kiango (2000), Urban Agriculture in Dar es Salaam: Providing an Indispensable Part of the Diet. In: N. Bakker, M. Dubbeling, S. Gündel, U. Sabel-Koschela & H. de Zeeuw (eds), *Growing Cities, Growing Food: Urban Agriculture on the Policy Agenda. A Reader on Urban Agriculture.* Feldafing: Deutsche Stiftung für Internationale Entwicklung.

Karanja, N., M. Njenga, P. Kuria Gathuru, A. Karanja & P.M. Munyao (2010), Crop-Livestock-Waste Interactions in Nakuru's Urban Agriculture. In: G. Prain, N.K. Karanja

& D. Lee-Smith (eds), *African Urban Harvest: Agriculture in the Cities of Cameroon, Kenya and Uganda*. New York/Ottawa: Springer/IDRC, pp. 213–228.

Keraita, B., B. Jimenez & P. Drechsel (2008), Extent and Implications of Agricultural Reuse of Untreated, Partly Treated and Diluted Wastewater in Developing Countries. *CABI Reviews: Perspectives in Agriculture, Veterinary Science, Nutrition and Natural Resources* 3(58).

Lee-Smith, D. (2010), Cities Feeding People: An Update on Urban Agriculture in Equatorial Africa. *Environment and Urbanization* 22(2): 483–499.

Lee-Smith, D. (2013), Which Way for UPA in Africa? *CITY* 17(1): 69–84.

Lee-Smith, D., N. Karanja, M. Njenga, T. Dongmo & G. Prain (2010), Nutrient Cycles in Three African Cities. *Urban Agriculture Magazine* 23: 17–19.

Lee-Smith, D. & P.A. Memon (1994), Urban Agriculture in Kenya. In: A.G. Egziabher, D. Lee-Smith, D.G. Maxwell, P.A. Memon, L.J.A. Mougeot & C.J. Sawio (eds), *Cities Feeding People: An Examination of Urban Agriculture in East Africa*. Ottawa: IDRC, pp. 67–84.

Lee-Smith, D. & G. Prain (2010), The Contribution of Research-Development Partnerships to Building Urban Agriculture Policy. In: G. Prain, N.K. Karanja & D. Lee-Smith (eds), *African Urban Harvest: Agriculture in the Cities of Cameroon, Kenya and Uganda*. New York/Ottawa: Springer/IDRC, pp. 287–308.

Lee-Smith, D., G. Prain, O. Cofie, R. van Veenhuizen & N. Karanja (forthcoming), Urban and Peri-urban Farming Systems (UPUFS). In: J. Dixon, D. Garrity & J-M. Boffa (eds), *Farming Systems and Food Security in Sub-Saharan Africa: Priorities for Science and Policy*. London: Earthscan.

Mitchell, J. & H. Leturque (2010), Using Value Chain Analysis to Increase the Impact of Urban Farming. *Urban Agriculture Magazine* 24: 21–23.

Mougeot, L.J.A. (2000), Urban Agriculture: Definition, Presence, Potentials and Risks. In: N. Bakker, M. Dubbeling, S. Gündel, U. Sabel-Koschela & H. de Zeeuw (eds), *Growing Cities, Growing Food: Urban Agriculture on the Policy Agenda. A Reader on Urban Agriculture*. Feldafing: Deutsche Stiftung für Internationale Entwicklung.

Mougeot, L., F. Gasengayire, D. Lee-Smith, G. Prain & H. de Zeeuw (2010), IDRC and Its Partners in Sub-Saharan Africa 2000–2008. In: G. Prain, N.K. Karanja & D. Lee-Smith (eds), *African Urban Harvest: Agriculture in the Cities of Cameroon, Kenya and Uganda*. New York/Ottawa: Springer/IDRC, pp. 267–285.

Mwangi, A.M., (1995), *The Role of Urban Agriculture for Food Security in Low-Income Areas in Nairobi*. Leiden: African Studies Centre.

Njenga, M., D. Romney, N. Karanja, P. Kuria Gathuru, S. Kimani, S. Carsan & W. Frost (2010), Recycling Nutrients from Organic Wastes in Kenya's Capital City. In: G. Prain, N.K. Karanja & D. Lee-Smith (eds), *African Urban Harvest: Agriculture in the Cities of Cameroon, Kenya and Uganda*. New York/Ottawa: Springer/IDRC, pp. 193–212.

Nyapendi, R., R. Best, S. Ferris & J. Jagwe (2010), Identifying Market Opportunities for Urban and Peri-urban Farmers in Kampala. In: G. Prain, N.K. Karanja & D. Lee-Smith (eds), *African Urban Harvest: Agriculture in the Cities of Cameroon, Kenya and Uganda*. New York/Ottawa: Springer/IDRC, pp. 139–165.

Obuobie, E., B. Keraita, G. Danso, P. Amoah, O.O. Cofie, L. Raschid-Sally & P. Drechsel (2006), *Irrigated Urban Vegetable Production in Ghana: Characteristics, Benefits and Risks*. Accra: IWMI/RUAF Foundation/IDRC/CPWF.

Owuor, S.O. (2006), *Bridging the Urban–rural Divide. Multi-spatial Livelihoods in Nakuru Town, Kenya*. Leiden: African Studies Centre.

Prain, G. (2010), The Institutional and Regional Context. In: G. Prain, N.K. Karanja & D. Lee-Smith (eds), *African Urban Harvest: Agriculture in the Cities of Cameroon, Kenya and Uganda*. New York/Ottawa: Springer/IDRC, pp. 1–11.

Prain, G. & M. Dubbeling (2011), *Urban Agriculture: A Sustainable Solution to Alleviating Urban Poverty, Addressing the Food Crisis, and Adapting to Climate Change. Case Studies of the Cities of Accra, Nairobi, Lima, and Bangalore*. Leusden: RUAF Foundation.

Prain, G., N.K. Karanja & D. Lee-Smith (eds) (2010), *African Urban Harvest: Agriculture in the Cities of Cameroon, Kenya and Uganda*. New York/Ottawa: Springer/IDRC.

Prain, G. & D. Lee-Smith (2010), Urban Agriculture in Africa: What Has Been Learnt? In: G. Prain, N.K. Karanja & D. Lee-Smith (eds), *African Urban Harvest: Agriculture in the Cities of Cameroon, Kenya and Uganda*. New York/Ottawa: Springer/IDRC, pp. 13–35.

Riley, L. & A. Legwegoh (2013), Comparative Urban Food Geographies in Blantyre and Gaborone. *African Geographical Review*. http://dx.doi.org/10.1080/19376812.2013.805148.

Satterthwaite, D. & C. Tacoli (2002), Seeking an Understanding of Poverty that Recognizes Rural–urban Differences and Rural–urban Linkages. In: C. Rakodi & T. Lloyd-Jones (eds), *Urban Livelihoods: A People-Centred Approach to Reducing Poverty*. London: Earthscan, pp. 52–70.

Sebastian, R., A. Lubowa, F. Yeudall, D. Cole & S. Ibrahim (2008), The Association between Household Food Security and Urban Farming in Kampala. In: D.C. Cole, D. Lee-Smith & G.W. Nasinyama (eds), *Healthy City Harvests: Generating Evidence to Guide Policy on Urban Agriculture*. Kampala/Lima: CIP/Urban Harvest/Makerere University Press, pp. 69–87.

van Veenhuizen, R. (ed.) (2006), *Cities Farming for the Future: Urban Agriculture for Green and Productive Cities*. Leusden: RUAF Foundation, IDRC & IIRR.

van Veenhuizen, R. & G. Danso (2007), Profitability and Sustainability of Urban Agriculture. FAO Agricultural Management, Marketing and Finance Occasional Paper No. 19. Rome: FAO.

WHO (2006), *Guidelines for the Safe Use of Wastewater, Excreta and Grey Water: Wastewater Use in Agriculture. Vol. 2*. Geneva: World Health Organization.

Yeudall, Y., R. Sebastian, A. Lubowa, J. Kikafunda, D. Cole & S. Ibrahim (2008), Nutritional Security of Children of Urban Farmers. In: D.C. Cole, D. Lee-Smith & G.W. Nasinyama (eds), *Healthy City Harvests: Generating Evidence to Guide Policy on Urban Agriculture*. Kampala/Lima: CIP/Urban Harvest/Makerere University Press, pp. 89–103.

Zezza, A. & L. Tasciotti (2008), *Does Urban Agriculture Enhance Dietary Diversity? Empirical Evidence from a Sample of Developing Countries*. Rome: FAO.

CHAPTER 10

From Suitcase Farmers to Telephone Farmers

Agriculture and Diversified Livelihoods among Urban Professionals

Melle Leenstra

The world food crisis in 2008 refocused global attention on Africa's underutilized agricultural potential and calls for a Green Revolution for Africa were reignited. A dominant paradigm suggests that if African smallholders had access to efficient markets for inputs and outputs, they would adopt new technologies and increase productivity. This would not only increase access to food but would also create an avenue out of poverty for the smallholders involved. This chapter however argues that for Africa to unlock its full agricultural potential, agriculture needs to move from being a semi-subsistence livelihood activity to being a professional commercial venture. It picks up the thread of the literature on rural livelihood diversification and deagrarianization. In the turbulent decades since decolonization, descendants of the former 'peasant class' have built livelihoods in which agriculture fits into a range of income-generating activities. This is not only the case for peasant farmers but also for those with clearly urban-professional identities whose portfolios of livelihood activities include agriculture. In Kenya, this Africa-wide phenomenon is called 'telephone farming'. Compared to the average small farmer, telephone farmers have better access to the resources needed to invest in commercial agriculture, such as land and capital. They appear more interested in adopting non-traditional agricultural technologies. At the same time, they are faced with serious constraints concerning the creation of profitable ventures as they lack the social and human capital needed to compensate for their time constraints. Further research is needed to experiment with agricultural business models embedded in contemporary socioeconomic realities.

Introduction

After having risen steadily since 2002, global food prices shot up unprecedentedly in 2007/2008 (Mitchell 2008). This was caused by structural factors in combination with incidental harvest losses due to drought and floods. These structural factors included a growing population and increasing consumption, alternative uses for food (animal feed and biofuels), an increase in input prices associated with high oil prices and a decline in agricultural productivity in certain parts of the world (World Bank 2012). The increase in food prices soon caught international public attention and was termed the World Food Crisis

 | DOI 10.1163/9789004282698_011

(*New York Times* 2008). Though prices rapidly decreased throughout 2009 and 2010, they soared again in 2011, which prompted an international policy response. African agricultural development was seen as a policy priority following nearly a decade of disinvestment in the sector. This energized calls for a Green Revolution for Africa, a technology and market-led transformation of African agriculture to allow Africa to feed itself and help feed the rest of the world too.

Calls for a green revolution were not new to Africa and can be traced back to the agricultural development projects in Mexico of the 1940s before spreading elsewhere in Latin America and Asia in the 1950s and beyond (Rockefeller Foundation 2006). The aim of the green revolution was to generate a 'large increase in crop production in developing countries achieved by the use of artificial fertilizers, pesticides, and high-yield crop varieties'.[1] To support this, various international agricultural research institutes were established under CGIAR (Consultative Group on International Agricultural Research), as were numerous national research institutes.[2] Since the 1970s such institutes have invested extensively in research on Africa. Although the green revolution is considered to have been generally successful in Asia and parts of Latin America, it is widely considered to have failed in Africa (Holt-Gimenez 2006).

Some of the arguments for this failure are technical. It is, for instance, argued that the wrong cultivars and crops have been improved. The focus has been on developments in growing wheat, rice and yellow maize rather than crops such as sweet potato, cassava and tropical white maize.[3] In addition, it is argued that recommendations for fertilizers are often not well suited to the prevailing soil conditions. Such technical critiques argue for a better adjustment of the green revolution to local circumstances (Voortman 2013). Much of the critique against the green revolution is ideological and driven by a mistrust of the corporate interests of agri-technological companies. The green revolution is also equated with large-scale monocultures that are related to environmental damage. The alternative paradigm promoted is food sovereignty, which implies less dependence on markets but a focus on ecological farming

1 Definition from Oxford Dictionaries: http://www.oxforddictionaries.com. Accessed 22 February 2014.

2 Such as the Kenyan Agricultural Research Institute (KARI) in Kenya, the Zambian Agricultural Research Institute (ZARI) in Zambia and the National Agriculture Research Organisation (NARO) in Uganda.

3 See: http://foodsecuritysm.wordpress.com/2011/01/13/why-did-the-original-green-revolution-not-reach-africa/. Accessed 22 February 2014.

methods, local seeds, support for small holders and improved access to land for the poor (Holt-Gimenez 2006).

This chapter argues on the basis of a literature review, prior research and personal experience that neither the classic green revolution paradigm nor the food sovereignty paradigm alone can provide realistic approaches to unlocking Africa's agricultural potential. Both paradigms are blind to the socioeconomic and cultural realities of African farming systems. Instead, a deeper understanding is needed of agriculture in the context of diversified livelihood systems. African farmers indisputably need supportive institutional and policy frameworks, markets for inputs, for finance and for selling their produce as well as the freedom to make their own decisions about which technological packages best suit their needs. However, it is argued here that African farmers need the space to apply innovative farming systems and business models that are attuned to their social economic realities but also treat farming as primarily a business rather than just as a part of a broader livelihood strategy.

Small Is Beautiful?

Most policies aimed at agricultural development in Africa focus on smallholder farmers. For instance, the World Bank (2007) states in its 2008 World Development Report that 'using agriculture as the basis for economic growth… requires a productivity revolution in smallholder farming'. This is understandable as smallholders are a defining feature of African agriculture and many poor Africans are engaged in small-scale semi-substance agriculture (FAO 2010).

In its appeal for a new Green Revolution for Africa, the Rockefeller Foundation describes the situation of just one of the 180 million small-scale, semi-subsistence farms in Sub-Saharan Africa. This farmer and her family have no access to irrigation or chemical fertilizers and so the farm is dependent on nature for its water. The household markets a small proportion of its produce in good years and members go hungry in bad years. This traps the farmer and her children in poverty, malnutrition and disease and leaves them bereft of education and opportunities to find non-farm employment. It is, however, argued that if the farmer could gain access to better seeds, fertilizer, tools and advice on agricultural practices through the market (agro dealers), she would be able to escape the poverty trap she is currently in (Rockefeller Foundation 2006).

Empirical evidence on the economics of technology adoption by smallholders and the nature of rural livelihoods casts doubt on this paradigm. On the basis of an analysis of data on crop improvement interventions, Harris & Orr

(2014) conclude that returns for smallholders on improved crop production technologies alone are too low to lift them above the poverty line. This is due, in large part, to the small (and reducing) size of farms available to smallholders practising rain-fed agriculture. Collier & Dercon (2009) also question the evidence for a policy focus on smallholders as engines for growth and poverty reduction, arguing that labour productivity is needed to increase agricultural production but that smallholders and the institutions that sustain them are weak agents in this. In their view, African smallholders are active in agriculture by default rather than because they have explicitly chosen to be entrepreneurs. They see instead scope for large-scale farmers as commercial enterprises interacting with smallholders as a means of achieving economies of scale for processing, marketing and technology adoption.

Scepticism about whether small African farms can provide the basis to replicate Asia's Green Revolution is shared by Ellis (2005), who draws on an extensive body of literature on African rural livelihoods. He points out significant differences between the institutional contexts in which African and Asian farmers operated as being the reasons for the failure of the Green Revolution in Africa in the past. African farmers are faced with volatile prices for outputs and a fragmented provision of agricultural support policies, following a history of inefficient parastatal marketing boards and subsequent liberalization. At the same time, farm sizes have gradually decreased due to rigid inheritance norms in societies in Sub-Saharan Africa in a context of sustained population growth. As a way of coping with the price and weather risks associated with farming, African smallholders have diversified their livelihoods. In fact in the mid-1990s, researchers noted such rural livelihood diversification that there was even talk of 'depeasantification' or 'deagrarianization', as illustrated by a publication with the apt title *Farewell to Farms* (Bryceson & Jamal 1997).

Livelihood Diversification

Farming is generally seen as an occupation rather than merely an activity.[4] This notion is fed by the historical experiences of European and Asian societies that were shaped by feudalism and, subsequently, industrialization. They are to a certain degree marked by occupational specialization, and class or cast identities were partly shaped by the economic role one played in society. In much of Africa, historical experiences have been different. Since the colonial

4 Particularly by outsiders looking at Africa. including policy makers, scientists and business people from abroad or those formally educated in 'western' paradigms.

period, Africa has undergone tremendous rural-urban migration and many rural folk identify themselves socially as farmers. At the same time, closer analysis shows that rural-urban migration is not a one-way process (Ferguson 1999) and many rural dwellers depend on non-farm incomes for a part of their livelihood (Bryceson & Jamal 1997; Ellis & Freeman 2005).

There is a sense of this livelihood diversification in informal contacts with farmers. Let me here share a personal encounter with a man in his thirties who I met at a farmers' fair in Central Kenya. I initially framed him as a vegetable farmer since he had been introduced to me as a beneficiary of a Dutch horticultural project. He told me that he was renting an acre of land to grow French beans for export. As we walked past the displays at the fair, he stopped to admire a silage chopper. While I was confused about how this could be of use to a bean grower, he explained that it would process fodder for his dairy cows. He already owned one and also had a heifer in calf and was using the proceeds from growing beans to invest in dairy farming. This provided him with liquidity: with this steady income he would be more attractive to banks so he could access loans for bigger investments. I then asked what other business he was involved in. He mentioned that he had a small *duka*, a kiosk from which he sold household sundries. When asked what he would do if his businesses were profitable, he explained that he wanted to get into real estate and build houses both for himself and to rent out.[5]

Rural livelihood diversification is a well-studied phenomenon and goes far beyond anecdotal experiences. Ontita (2012), for example, details how rural dwellers in Nyamira, a tea-growing area in the west of Kenya, ventured in and out of agriculture. He describes how young men start out working on their parents' farms before finding opportunities to work off-farm as migrant plantation workers or in self-employment or (for those able to get formal qualifications) in salaried positions, such as teaching. Many of them will return at some time to agriculture, investing some of the resources they have accumulated from off-farm work into agriculture (e.g. procuring tea or pyrethrum seedlings or buying dairy cows). While still maintaining an agrarian identity, off-farm activities (e.g. trade, transport, producing and selling local alcohol, providing spiritual services) for some still take priority over farming at certain times in their lives, though many may revert to farming when they get older.

The processes of deagrarianization and livelihood diversification are logical and rational. Historically, African peasants were only partially integrated into markets and in the late-colonial period marketing structures were set up to incorporate smallholders in the production of cash crops – such as tea, coffee, pyrethrum, coffee and maize – parallel to the plantation systems. This was not

5 Encounter with a farmer at Karatina Agribusiness and Finance Fair, 4 October 2012.

necessarily the result of a conscious colonial policy and was in large part due to the agency of African farmers discovering and propagating opportunities.[6] Especially in the period of state-led economic development following independence, state marketing boards (crop parastatals) dominated farmers' marketing opportunities. Agricultural policy in this era provided extension, subsidized inputs, transportation and guaranteed output prices. At the same time, the system extracted resources from the rural economy by fixing prices and exchange rates to the detriment of farmers. In addition, many parastatals were taken over by political interests and were subsequently mismanaged. This, in turn, sparked the Structural Adjustment Programmes that disbanded systems that had provided inputs, extension and price guarantees. The private sector only partially managed to fill the gaps left by the crop parastatals and smallholders in many parts of Africa have since faced increased input prices, a lack of extension services and volatile prices for their produce. This made farming for the market a risky and often unrewarding occupation and spurred the process of deagrarianization (Bryceson & Jamal 1997; Ellis & Freeman 2005).

Livelihood diversification is a rational response to the risks African farmers face. These not only include risks from the market but also those related to climate (floods, droughts, crop disease) and social factors (disease, theft, fraud). To protect themselves from risk but also to extend their consumption across cropping cycles, smallholder farmers thus manage a varied portfolio of on-farm and off-farm income-generating activities (Ellis 2005). A popular way of managing the risks involved in farming is to abandon farming (at least temporarily) by either migrating in pursuit of paid employment or by pursuing formal education with a view of obtaining a salaried job. Such strategies are not only applied by individuals but are also used between generations as a way of investing in the possibility that one's dependents will support one later in life. At the same time, for those who leave farming temporarily, a return to agriculture is a tested social safety net for the time when they can no longer depend on formal employment (Ontita 2012; Leenstra 2012).

Looking at farming from the perspective of African villagers supports Collier & Dercon's (2009) assertion that agriculture is a default option and part of a lifestyle for most small-scale farmers rather than a decision to engage in entrepreneurship. Peasant farmers engage in agriculture to sustain themselves and ensure access to income that will be consumed and invested off-farm. By investing in

6 Ontita (2012) describes, on the one hand, how maize was promoted as a cash crop due to the wartime demand for food and, on the other, tea spread to smallholder systems by plantation workers bringing seedlings back to their home villages.

education, religion, housing and off-farm businesses, such as kiosks and taxis,[7] there is no reason to be derisive of this strategy that has shaped contemporary Sub-Saharan African societies. It is the engine that has shaped human capital and provided social security and financial assets in the form of housing. It has fed burgeoning populations and produced much of the primary produce that non-oil-producing or mineral-exporting African countries have depended on for their foreign exchange.

This farming system, which is based on part-time, amateur farming, is not sustainable as African populations are growing fast. In much of Africa, especially in densely populated areas with good conditions for rain-fed agriculture, such as Western and Central Kenya, plots are constantly declining in size as land is subdivided between multiple heirs under prevailing inheritance norms (Ontita 2012; Ellis 2005). Theoretically, this can be compensated for by agricultural intensification but the reality of farming as a part of diversified livelihood strategies means that farmers are loath to invest in the inputs and labour needed for intensification.[8] Moreover, decreasing plot sizes are further reducing the economies of scale needed to adopt the technologies required to intensify land use and raise incomes. In addition, as farmers do not want to invest in fertilizer or the labour-intensive practices of conservation farming, this leads to a gradual depletion of soil fertility. Reliance on low-input, low-output, part-time smallholder farming will leave African countries with an increased demand for food and income and a degraded agricultural base from which to provide for their requirements. Such a scenario is neither socially nor economically or ecologically sustainable.

Telephone Farmers

Livelihood diversification is not only part of the reality of poor people in rural areas but also of better-off town or city dwellers with professional identities that may appear to be unrelated to farming. For my PhD research on the Zambian health sector, I had not initially expected to touch on agriculture but in the course of interviewing health workers about their lives and careers, it turned out that most were working 'beyond the payslip', with many of them actively investing in agriculture. In fact, some of these health workers ran

7 For individuals, this may mean *bodaboda* motorcycle taxis, though collectively self-help groups or saving cooperatives are known to invest in fleets of *matatus* or minibus taxis.

8 Not only capital but also labour is scarce due to competition with other off-farm opportunities.

larger-scale and more technologically advanced farming ventures than rural villagers who had a primarily agrarian identity.

Let us look at three examples. A nurse tutor from Lusaka, whose husband was working for the Ministry of Finance, had 20 ha of land in a resettlement scheme where she grew maize using privately procured fertilizer. This resulted in a turnover of nearly €10,000 in a planting season. A driver from the Ministry of Health had 13 ha of land just outside Lusaka where he grew tomatoes and other vegetables under irrigation. And finally, there was an environmental health technician posted in a rural village in Luapula Province who had planted 200 oil palms in an area where this was a newly identified cash crop (Leenstra 2012: 137–138).

The engagement of these members of the professional elite in agriculture or other businesses reflects a strategy of livelihood diversification similar to that of villagers. Despite earning well above the national average, these professionals' salaries were far outstripped by their expenditures and aspirations. These included the desire to educate their children, claims made of them by dependent relatives, the wish to invest in housing and the need to prepare for retirement in a country where pension schemes do not offer a secure income (Leenstra 2012, Ch. 5).

While this phenomenon has seen little explicit academic or policy attention, it is not unique to Zambia. Social engagement with African friends, colleagues and work contacts across the continent has confirmed that this is widespread and has a special relevance in this era of mobile money and IT applications. Such people are called 'telephone farmers' but the term is not new, despite the connotations of mobile phones and *M-pesa*.

There is a fleeting reference to this concept in the work of Christopher Leo, a Canadian political scientist who conducted research in Kenya in the late 1970s. In his study of the Million Acre Scheme that was set up to resettle Africans on previously white-owned estates, Leo (1984) described how the programme benefited 'progressive farmers' rather than smallholders and former squatters whose resettlement was grudgingly accepted by policymakers. Due to the bias of World Bank funders, the 'progressive farmers' gained better access to development funds, extension services and better terms for purchasing land than their 'non-progressive' colleagues. However, as the former squatters were more dependent on agriculture for their livelihoods than the progressive farmers who were often engaged in other business or in salaried employment in town, small farmers were more successful in using the land productively. Leo (*Ibid.*: 192–193) wrote about this latter group as follows: 'They – and many others like them elsewhere in Kenya – became what some administrative officials referred to as telephone farmers, people who concentrate on more lucrative activities while farming "over the weekend and by telephone"'. A later reference appears in a study on labour conditions on large farms where some of the farm owners were found to be absentee landowners and were referred to as weekend farmers, suitcase farmers or telephone farmers (Foeken & Verstrate 1992: 18).

Despite its prevalence, telephone farming is a somewhat hidden phenomenon, partly due to the moral and ideological connotations associated with the phenomenon in the few studies that touch on this type of farming. Leo (1984) reacted to the World Bank's modernist favouring of 'progressive farmers', asserting that it was 'a piece of bourgeois dogma'. In a study of rice growing in Northern Ghana published in the early 1980s, Konings (1984) frames the activities of telephone farmers in the discourse of the Cold War class struggle and talks about 'capitalist farmers' when referring to military officers, politicians, managers and professionals who were engaging in rice farming to the detriment of the autochthonous peasant population. As in the case of the Million Acre Scheme, these 'urban-based ... petty-bourgeois fractions in society' (*Ibid.*: 97) benefitted from unfair access to extension services, land and public financing. But even beyond the ideological polarization of the Cold War era, there are reasons why telephone farming is not part of the general discourse. While 'working beyond the pay check' is common, it clashes with official concepts of professionalism and bureaucracy and is pejoratively referred to as 'moonlighting' (Anders 2005). It is part of the everyday reality of urban professionals in Africa to straddle the urban-rural dichotomy, i.e. to live in the urban reality while maintaining a foothold in the countryside. However, they tend to stress their modern, urban, professional identity rather than their rural identity as a part-time villager.

Attempts have been made since the 1990s to deconstruct this urban-rural dichotomy that influenced the study of deagrarianization referred to above. Interest in the dynamics between the rural and the urban, Foeken & Owuor (2006) conducted an in-depth study into urban agriculture that also touched on the phenomenon of telephone farmers and the rural farming activities of urban dwellers. As a part of this study, Owuor (2003) conducted a survey of rural livelihood sources among urban households and found that 56% of respondents were engaging in rural farming to supplement their food and/or incomes. Strikingly, these farmers used high levels of farm inputs: 83% used chemical fertilizer and 70% planted improved seeds or seedlings. These findings showed urban dwellers relying on rural farming to subsidize their urban livelihoods due to conditions of urban poverty and took a moral stance regarding rural farming by urban households. As the activity is framed in the context of livelihood coping strategies related to urban poverty, it is not presented as an elite pastime as one might have expected it would be.

Many donor-funded, NGO-implemented development projects traditionally overlook the farming activities of telephone farmers as they can easily be construed as being elitist and not, therefore, belonging to the target group of development assistance. Instead, the focus is on making agricultural markets work for the poor and providing access to appropriate technology. However, in view of the doubts about the effectiveness of targeting a green revolution at the poorest of the poor, one could ask whether a change of paradigm is needed if the ultimate aim is to increase Africa's ability to feed itself and trade agricultural commodities sustainably with the rest of the world.[9] This would require agricultural entrepreneurship based on viable local business models that might mean fostering agricultural business development of the more entrepreneurial smallholders. Building on the rural enterprises of those Africans embedded in urban life could also be an option. Inclusive growth means building bridges and linkages between urban and rural spaces, between modernity and tradition and technology and natural resources.

Unlocking the Potential of Telephone Farmers

The Netherlands Embassy in Nairobi commissioned a study in 2013 to explore the phenomenon of telephone farmers and opportunities for unlocking the entrepreneurial potential of these farmers. Projects were formulated several years earlier to link smallholders to markets but doubts have been expressed about whether providing market opportunities alone would be sufficient incentive for farmers to produce enough to meet market demand. The embassy felt that investment in agricultural entrepreneurship was needed to allow Kenyan farmers to benefit fully from market opportunities (Makumire *et al.* 2013). An additional motivation was that, due to the integration of development policy and the promotion of Dutch expertise, investments and trade, investment in agricultural entrepreneurship would be an investment in a market for the Dutch agribusiness sector (Embassy of the Kingdom of the Netherlands Nairobi 2014).

The quick-scan study by Makumire *et al.* (2013) confirmed the prevalence of telephone farming as a socioeconomic reality on the basis of interviews with bankers, officials, agricultural development practitioners and telephone farmers themselves. Specific figures are however lacking regarding the number

9 One could argue that targeting technology adoption and market linkages to help the most vulnerable in society confuses the issues of social and economic policy.

of absentee land owners there are, how much land telephone farmers own and how much lies idle or is underutilized. Kenya has not held a land census in recent years[10] although a Kenyan government official guesstimated that between 15% and 20% of Kenyan farmers fell into this group.

Makumire *et al.* (*Ibid.*) confirmed that telephone farmers had access to land and modest investment capital but that they faced serious management challenges. Interviews with telephone farmers gave the impression that many of them had sunk serious amounts of money into their farms without commensurate returns, i.e. they were effectively subsidizing the venture from other income sources. Respondents felt there were opportunities to provide business development support to assist telephone farmers establish profitable ventures. However, they doubted the feasibility of the Embassy's initial hypothesis that this could be done by fostering joint ventures between telephone farmers and farm managers. The reason was a lack of social capital, namely trust.

For analytical clarity, Makumire *et al.* distinguished two types of telephone farmers[11]: (i) people that inherited land and relied on relatives or casual caretakers to manage their farm in their absence; and (ii) people that purchased land as an investment and had an interest in recouping their costs. Both types of telephone farmers struggled with the quality of the managers they relied on. This is best illustrated by conversations I had with two telephone farmers. One complained that his caretaker had no business sense and had, for example, sold a pick-up truck full of French beans for less than the cost of transporting them. When the land owner expressed his exasperation, the caretaker claimed that the sum of money was better than nothing, not fully realizing the implications of what he had done. However, the telephone farmer admitted that the caretaker had one redeeming quality: he was honest. Another telephone farmer had the opposite experience with his farm manager who he had taken on because of his positive CV as he had worked on the Vice President's farm. After harvesting, the yield was so low that the telephone farmer was convinced the manager must have pilfered his inputs.

The study by Makumire *et al.* identified lack of communication and trust and poor technical skills and farm management as key bottlenecks affecting

10 Cabinet Secretary of Agriculture, Livestock and Fisheries, personal communication.

11 This is just a snapshot of a diversity of farming operations. Other anecdotes include urban professionals leasing land for wheat farming; business clubs collectively purchasing land for farming; and multi-local households where the man worked in town and his wife farmed upcountry, which is a common farming model approaching the idea of the telephone farmer.

further growth for telephone farmers. As they spend most of their time away from their farms, they have problems monitoring its day-to-day activities. Farm managers take advantage of prevailing information asymmetries, underreport yields and pilfer farm inputs. At the same time, they see themselves as underpaid (which judging from some accounts they are) and feel little incentive to be productive and trustworthy. Trust is one of the reasons that some telephone farmers prefer to employ relatives as caretakers but this can aggravate the skills problem. Kenya has a general shortage of practical farm management skills because educational institutions tend to focus on the theoretical aspects rather than offering opportunities for practical training. In addition, farm managers are usually generalists who lack technical production and also business and marketing skills (*Ibid.*: 9–10).

The bottlenecks faced by telephone farmers are, however, not only attributable to their caretakers but also to themselves. Bankers interviewed by Makumire *et al.* suggested that telephone farmers have a poor mindset when it comes to farming and lack dedication due to their other sources of income. This echoes Leo's (1985) conclusions on telephone farmers in the Million Acre Scheme. Rather than having a strong incentive to invest in the development of their farms, they appear to view them more as retirement homes. Telephone farmers have relatively easy access to finance as banks consider them important clients due to their other income and assets. To access loans, some make use of business plans prepared for them by consultants but they often fail to understand them. This can result in failure to make a profit due to bad farm and business management and loans granted for telephone farming are sometimes not even spent on the farm but are diverted to other business activities. Another shortcoming of telephone farmers is their lack of a business approach to farming. Instead of basing crop-related decisions on market analysis, the farmers base it on personal interest or on what others are doing. Like smallholders, telephone farmers tend to overdiversify, which results in a patchwork of unrelated income-generating activities (*Ibid.*: 8–9).

Analysis of the shortcomings faced by telephone farmers suggests that they could benefit from business coaching and access to relevant technical and marketing advice. Considering the challenges faced by businesses and the pursuit of food security, there may be a case for investing public funds in this area. However, it is also good to study private initiatives related to telephone farming. One such initiative is a venture set up in a town several kilometres outside Nairobi called the Distant Farmer Dairy Farm. This encourages those interested in investing in dairy farming but with little time or interest to be actively involved in the day-to-day care of animals to purchase a dairy cow. The farm will then care for the animal for a fee and transfer the proceeds of any marketed milk to the 'distant farmer'.[12] There are other similar initiatives in the tea and coffee sectors. Early in 2013, the Kenya Tea Development Agency announced a plan to manage tea farms owned by absentee farmers on their behalf for a fee (*Business Daily* 2013) and another company also manages coffee estates for their owners.[13] These initiatives still have to be studied in detail to assess their feasibility.

Conclusion

This chapter has argued that the pursuit of food security by increasing Africa's agricultural productive capacity cannot be based on technical solutions alone. It must also involve socio-culturally embedded entrepreneurship. When exploring agricultural business models, it is vital to look beyond the well-trodden paths of either peasants or plantations as a diversity of commercial farming models with unexplored potential lies between these two extremes. This is the 'missing middle' that is seemingly being ignored by policy makers who tend to focus on either poor and vulnerable smallholders or the large commercial farming schemes. To increase Africa's agricultural production, it is important to invest in focused agricultural entrepreneurship. And to find entry points to do so, it is vital to align investments with the initiatives and investments of enterprising *wananchi*.[14]

A prevalent but understudied agricultural business model is the rural agricultural business engagement of urban professionals. This chapter has by no means provided a definitive analysis of the phenomenon of telephone farming but is an appeal for further in-depth research into the practice of agricultural entrepreneurship in contemporary Africa. A clearer perspective of the diverse character of this group is required as well as a better sense of its prevalence. Additional analysis should focus on the opportunities and challenges faced by the group as well as supportive policies that could play a role in unlocking their productive capacity. This calls for experimental research involving researchers, practitioners and policy makers.

12 See: http://dairysuccess.blogspot.nl/2013/06/a-dairy-farming-opportunity-in-kenya.html.

13 JKUAT scientist, personal communication.

14 The citizens in respective African countries.

References

Anders, G. (2005), Civil Servants in Malawi: Cultural Dualism, Moonlighting and Corruption in the Shadow of Good Governance. PhD Thesis. Erasmus University School of Law.

Bryceson, D.F. & V. Jamal (eds) (1997), *Farewell to Farms: Deagrarianisation and Employment in Africa*. Aldershot: Ashgate.

Business Daily (2013), KTDA Offers To Manage Absentee Landlords' Tea Farms at a Fee. 12 January.

Collier, P. & S. Dercon (2009), African Agriculture in 50 Years: Smallholders in a Rapidly Changing World?. Paper for the FAO's How to Feed the World in 2050? Conference. Rome: 12–13 October.

Ellis, F. (2005), Small Farms, Livelihood Diversification and Rural-urban Transitions: Strategic Issues in Sub-Saharan Africa. Paper presented at the workshop on 'The Future of Small Farms'. Wye, UK: 26–29 June.

Ellis, F. & H.A. Freeman (eds) (2005), *Rural Livelihoods and Poverty Reduction Policies*. London: Routledge.

Embassy of the Kingdom of the Netherlands Nairobi (2014), *Multi Annual Strategic Plan 2014–2017*. Nairobi.

FAO (2010), *Policies and Institutions to Support Smallholder Agriculture. Committee on Agriculture*. Rome: FAO.

Ferguson, J. (1999), *Expectations of Modernity: Myths and Meanings of Urban Life on the Zambian Copperbelt*. Berkeley, Ca.: University of California Press.

Foeken, D. & L. Verstrate (1992), *Labour Conditions on Large Farms in Trans Nzoia District, Kenya*. Leiden: African Studies Centre.

Foeken, D. & S.O. Owuor (2006), Farming as a Livelihood Source for Urban Dwellers: Results from a Research Project in Nakuru, Kenya. ASC Infosheet 1. Leiden: African Studies Centre.

Harris, D. & A. Orr (2014), Is Rainfed Agriculture Really a Pathway from Poverty? *Agricultural Systems* 123: 84–96.

Holt-Gimenez, E. (2006), *Ten Reasons Why the Rockefeller and the Bill and Melinda Gates Foundations' Alliance for Another Green Revolution Will Not Solve the Problems of Poverty and Hunger in Sub-Saharan Africa*. Oakland: Institute for Food and Development Policy (Food First).

Konings, P. (1984), Capitalist Rice farming and Land Allocation in Northern Ghana. *Journal of Legal Pluralism and Unofficial Law* 22: 89–119.

Leenstra, M. (2012), *Beyond the Façade: Instrumentalisation of the Zambian Health Sector*. Leiden: African Studies Centre.

Leo, C. (1984), *Land and Class in Kenya*. Toronto: University of Toronto Press.

Makumire, T., H. Nijhoff & K. de Vries (2013), *Unlocking the Potential of Telephone Farmers*. Wageningen: WUR/CDI.

Mitchell, D. (2008), A Note on Rising Food Prices. Policy Research Working Paper 4682. Washington DC: World Bank.

New York Times (2008), The World Food Crisis, 10 April.

Ontita, E. (2012), *Resources and Opportunity: The Architecture of Livelihoods in Rural Kenya*. Eugene: Wipf & Stock.

Owuor, S.O. (2003), Rural Livelihood Sources for Urban Households. A Study of Nakuru Town, Kenya. ASC Working Paper No. 51. Leiden: African Studies Centre.

Rockefeller Foundation (2006), *Africa's Turn: A New Green Revolution for the 21st Century*. New York: Rockefeller Foundation.

Voortman, R.L. (2013), Why the Green Revolution Failed in Sub-Saharan Africa. *Rural 21, International Journal for Rural Development* 47(3): 32–33.

World Bank (2007), World Development Report 2008: Agriculture for Development. Washington DC: World Bank.

World Bank (2012), Responding to Higher and More Volatile World Food Prices. Report 68420-GLB. Washington DC: World Bank.

SECTION 4

Institutional Issues

CHAPTER 11

National Agricultural Research Systems in Africa

Olubunmi Abayomi Omotesho and Abraham Falola

The quest for a solution to the challenges facing agricultural development in Africa calls for sound, systematic research to increase the stock of knowledge for agricultural policy formulation. This can, however, only be achieved if there is a proactive agricultural research framework. This chapter explores the way agricultural research is currently organized in Africa, reviews the activities and research endeavours of some notable research-based organizations and identifies the challenges facing the establishment of an efficient agricultural research system. It is argued that even though a great deal of effort has been made by relevant research organizations, the agricultural research system in Africa is still inefficient. Strengthening research institutions, linking the agricultural research agenda to national development priorities and having a sustainable agricultural research policy are recommended. It is also suggested that the coordination, interaction, interlinkages, partnerships and networks among system agents are improved, sustainable sources of funding are sourced and a mechanism to support the various components of the system is established.

Introduction

About 65% of Africa's labour force is employed in agriculture, with the sector accounting for 32% of the continent's Gross Domestic Product (GDP) (Alliance for a Green Revolution in Africa [AGRA] 2013) and almost 60% of its total export earnings (IFPRI 2009). In addition, agriculture is the main source of income for 90% of Africa's rural population (ECA 2005; Seck *et al.* 2013). Against this backdrop, the sector contributes to a reduction in poverty and hunger, creates employment and helps prevent rural-urban migration. African governments and development agencies have established research systems through which systematic investigations into how the agricultural sector can be improved are carried out and their findings are disseminated for agricultural policy formulation. This chapter explores the need for an efficient agricultural research system in Africa and reviews the activities of some of the notable agricultural research-based organizations in an attempt to identify the challenges facing the establishment of an efficient agricultural research system.

 | DOI 10.1163/9789004282698_012

An agricultural research system could be seen as encompassing all public and private institutions that devote their activities full time or partially to agricultural research and that are committed to the national, regional or global research agenda. Such institutions may be formal or informal. The formal ones include: those whose mandate is to carry out research only; higher educational institutions that are involved in teaching and research; technical departments of ministries or development agencies that engage in adaptive research programmes; non-governmental organizations (NGOs); and the private sector (FAO 2002). The informal institutions are: farmers, artisans and other local people engaged in farmers' own innovations and experiments. These research systems are analogous but differ in the fact that the formal ones are centralized, powerful and visible, while the informal ones are dispersed, weak and hard for professionals to easily recognize.

Though each country or sub-region in Africa has its own institutional and organizational history, most agricultural research systems, especially in the Sub-Saharan region, have colonial antecedents. These took root in the late-nineteenth century when Europeans first colonized the continent. The desire to secure adequate, low-cost supplies of raw materials from indigenous or introduced tropical crops, such as oil palm, cocoa, coffee, tea and cotton, for which no ready substitutes were available in Europe, resulted in the colonial administrators introducing botanical gardens, model farms and experimental stations to screen and propagate these tropical export crops. Some of these institutions have been transformed into today's research systems (Roseboom *et al.* 2006).

It is argued here that national agricultural research systems in Africa are characterized by some typical attributes. First, the control and direction of the activities in agricultural research systems are usually top down, with little interference from farmers and commodity groups who are the end users of any research innovations. In addition, the systems are subject to intensive political control and so changes in government can result in changes in policies within the systems. The research systems are also characterized by limited professional development activities, training, motivation and support funds. In the same vein, linkages among these agricultural research entities are often weak, even though they are critical for the interface between technology generation, education and technology transfer.

Research organizations in Africa are famous for acting independently of each other, with minimal coordination. This undermines the feedback loop that is necessary for developing responsive research agendas and compromises access to knowledge (Anandajayasekeram 2011). Another distinguishing feature of the agricultural research system in Africa is the regional distribution of research capacity in terms of facilities and resources. Most public research in Africa is still dominated by government research agencies that represent 81% of the total research capacity, universities contribute 18%, while the private and non-profit sectors contribute the remaining 1% (IAC 2004; Anandajayasekeram 2011). The research capacity at universities is small and fragmented in most countries but is growing slightly faster than that of non-academic agencies (Michelsen & Hartwich 2004; Anandajayasekeram 2011).

The Need for an Efficient Agricultural Research System

An efficient research system in African countries is needed to achieve food security. A recent report by the FAO (FAO 2013) revealed that 223 m people were undernourished in Sub-Saharan Africa between 2011 and 2013, which is 26.5% of the global figure and puts the region in second place behind Southern Asia. An effective agricultural research system is thus needed to develop, implement and disseminate technologies and innovations that can increase agricultural productivity and provide Africans with unhindered physical, social and economic access to sufficient, safe and nutritious food.

An efficient system would enable researchers to base their activities on the needs and interests of their client groups and the outcome of any research would be directly related to the priorities of the users and not based on preconceived ideas. This is vital if one wants research to have an impact. In addition, the systematic organization of an agricultural research system promotes linkages among the different players in agricultural research, such as research institutes/agencies, NGOs and commodity groups. This provides the basis for cooperation and complementarity and encourages support from the different research agents and relevant stakeholders in terms of financial, material and human resources.

An efficient agricultural research system is also desirable to enhance protection of new research knowledge, inventions and discoveries from analogues, unauthorized copies/reproductions, counterfeits and other unfair practices and forms of competition. This is important to align value captured from new technologies and inventions and thus maximize the incentives and motivation for creative activities and inventiveness in the agricultural research system. Invariably, the generation, diffusion, exploitation and application of research output is effective. This can also facilitate creativity, investment and the transfer of knowledge in agricultural research.

Efficient agricultural research systems are needed to develop, sustain, strengthen or promote an agro-food revolution in Africa. In the 1960s, the Green Revolution was introduced as a series of research, development and technological transfer initiatives aimed at saving millions of people from hunger and starvation. This was done by developing high-yielding varieties of cereal grains, the expansion of irrigation infrastructure, the modernization of management techniques and the distribution of hybrid seeds, synthetic fertilizers and pesticides to farmers (Hazell 2011). In Africa, this revolution is being promoted by organizations like the Alliance for a Green Revolution in Africa (AGRA) and the Africa Green Revolution Forum (AGRF). And recently, attention has also been given to the Rainbow Revolution, involving issues related to the environment, biodiversity, rights, diversification, value addition, capturing markets, enhancing incomes and addressing gender parity in agricultural production as well as increased agricultural production (Agricultural Sector Coordinational Unit Kenya, 2012). Improving the quality of products, food safety and equity in economic growth are also important, as is limiting the negative effects on the environment. Strong agricultural research systems are needed to achieve these goals.

Components of Agricultural Research Systems

The major components of agricultural research systems in Africa are the international research centres, regional research organizations, sub-regional institutions, national agricultural research systems (NARSs) and public and private research institutes under the umbrella of the national institutes. Some of these organizations are examined below. Their activities tend to overlap but they are all looking for solutions to the problems facing agricultural development (Figure 11.1)

The UN's Food and Agricultural Organization (FAO)

The UN's Food and Agricultural Organization (FAO) is an intergovernmental organization that was set up to promote efforts aimed at achieving global food security. The organization has 194 states as members, two associate members and one member organization, namely the European Union.[1] Its main goals are the eradication of hunger, food insecurity and malnutrition; the elimination of poverty; the enhancement of economic and social progress for all; and the sustainable management and utilization of natural resources.

1 http://www.fao.org/about/who-we-are/en/.

FIGURE 11.1 *Levels of Agricultural Research Systems*
SOURCE: AUTHORS' OWN DESIGN

Regarding research, the FAO promotes sustainable agriculture through the collection, analysis and dissemination of agricultural research data that aid development. It organizes a forum where policymakers and experts from around the world can convene to discuss how to forge agreements on major food and agricultural issues. And it also mobilizes and manages funds to ensure that its field offices achieve their goals.

The International Fund for Agricultural Development (IFAD)

The International Fund for Agricultural Development (IFAD) is an international agency that was established in 1977 following the 1974 World Food Conference. The conference was organized in an attempt to address the food crisis of the early 1970s that primarily affected Sahelian countries in Africa. IFAD focuses on country-specific solutions for the eradication of rural poverty in developing countries by empowering the poor rural (both men and women) to achieve higher incomes and improved food security. The agency collaborates with governments, donors and non-governmental organizations in its efforts to increase access for the rural poor to financial services, markets, technology, land and other natural resources.

As part of its contribution to research efforts in Africa, IFAD provides grant support to agricultural research centres working to improve the well-being of the rural poor through sustainable increases in agricultural productivity based on the wider adoption of improved cropping systems. In this regard, IFAD creates or enhances the capacity of rural farmers at field level for farmer-participatory research and extension (FPRE) programmes. It also focuses on generating motivation among farmers and participation in local institutions – through practical demonstrations of the utility of FPRE and by showing tangible results – so that the approach can be replicated and sustained using local and national resources.

Forum for Agricultural Research in Africa (FARA)

The Forum for Agricultural Research in Africa (FARA) is an apex organization that brings together the major stakeholders in agricultural research and development in Africa. The forum was set up in 2001 with a view to fostering continental and global networking in an attempt to reinforce the capacities of Africa's national agricultural research systems and sub-regional organizations. FARA aims to contribute to the sustainable achievement of significant broad-based agricultural growth in Africa, and specifically to sustainable improvements in agricultural productivity, competitiveness and markets in Africa.

FARA works with the four primary sub-regional research organizations (SROs) in Africa: the North Africa Research Organization; the West and Central African Council for Agricultural Development (CORAF/WECARD); the Centre for Coordination of Agricultural Resources and Development for Southern Africa (CCARDESA); and the Association for Strengthening Agricultural Research in East and Central Africa (ASARECA). It makes its contribution to agricultural research and development through four mutually reinforcing Networking Support Functions (NSF). They are (i) advocacy and policy; (ii) access to knowledge and technologies; (iii) capacity strengthening; and (iv) partnership and strategic alliances.

CGIAR-IITA

The International Institute of Tropical Agriculture (IITA) is one of fifteen research centres in the Consultative Group on International Agricultural Research (CGIAR). It was founded in 1967 and has its headquarters in Ibadan, Oyo State, Nigeria. The institute is an independent, non-profit research organization that is innovative in its solutions to hunger, malnutrition and poverty and meets the development needs of tropical countries. It does this by implementing strategies to enhance crop quality and productivity, reduce producer and consumer risks, and generate wealth from agriculture.

The group conducts research on biotechnology and genetic improvements, natural-resource management, the social sciences and agribusiness, and plant production and plant health. It is also involved in nine CGIAR research programmes on: integrated systems for the humid tropics; policies, institutions, and markets; maize; roots, tubers and bananas; grain legumes; agriculture for improved nutrition and health; water, land and ecosystems; climate change, agriculture and food security; and a gene bank. To achieve its project goals, IITA sets up partnerships with other CGIAR centres, international and national research institutions, universities, extension specialists, seed producers, farmers and their associations, and non-government organizations.

IITA engages in training activities as part of its strategies to improve food security in Sub-Saharan Africa. These activities are used to facilitate research collaboration between IITA and national agricultural research and extension systems (NARES). The Institute also assists in upgrading the skills of national scientists by actively engaging partners in its Research for Development (R4D) projects and through its non-degree training courses that are available to research associates and assistants (RTAs).

West Africa Agriculture Productivity Programme (WAAPP)

The West Africa Agriculture Productivity Program (WAAPP) is sponsored by the World Bank and the Japanese Policy and Human Resource Development (PHRD) and aims to increase agricultural productivity and growth in the Economic Community of West African States (ECOWAS) sub-region. The programme's goal is to contribute to sustained agricultural productivity and growth in the region's top priority commodity subsectors: yam, cassava, rice, maize, sorghum, poultry, goats, sheep and aquaculture. WAAPP's development objective is to accelerate the adoption of improved technologies in participating countries' top agricultural commodity priority areas, which are aligned with the sub-region's agricultural commodity priorities. The WAAPP project was first implemented in three countries – Senegal, Mali and Ghana – in 2007 and has since been extended to include others including Nigeria, Burkina Faso, Côte d'Ivoire, Benin, Guinea, Liberia, Sierra Leone, Gambia, Niger and Togo.

Since its inception, WAAPP has provided funds for agricultural research into the national agricultural research systems of the participating countries in the context of regional research coordination and monitoring through the West and Central African Council for Agricultural Research and Development (WECARD). This is with a view to encouraging the integrated development of agricultural research into the Technology Generation and Dissemination (TGD) continuum across the West African sub-region. At a continent-wide level, WAAPP works in line with the Comprehensive Africa Agriculture Development Program (CAADP) on the strategy for agricultural growth of the African Union/New Partnership for Africa's Development (NEPAD). And at a sub-regional level, WAAPP provides a vehicle for implementing the agricultural productivity agenda in participating countries by specifying project components in line with their agricultural transformation agenda.

Sasakawa Institutions (SAA)

Sasakawa Africa Association (SAA) is an international agricultural development NGO registered in Geneva, Switzerland. It was co-founded in 1986 by Nobel Laureate Dr Norman Borlaug (who was awarded the 1970 Nobel Peace Prize), Japanese philanthropist Ryoichi Sasakawa (Nippon Foundation) and former US President Jimmy Carter. SAA's vision is to ensure a more food-secure rural Africa and increasing numbers of entrepreneurial smallholder farmers. Its mission is to transform African extension advisory services in partner countries to ensure greater family food security and a more profitable participation in commercial activities along the value chain, while respecting natural resources. Since it was founded, SAA has worked with the Carter Center's Global 2000 Program to establish the Sasakawa Global 2000 (SG2000) agricultural programmes in 14 Sub-Saharan countries: Ethiopia, Mali, Nigeria, Uganda, Ghana, Sudan, Tanzania, Benin, Togo, Mozambique, Eritrea, Guinea, Burkina Faso and Malawi.

The SG2000 programme is tasked with improving farmers' welfare by focusing on five thematic areas and enhancing crop productivity in the participating countries. This is done by using demonstration plots to train farmers in cultivation skills and agricultural technologies, including enhanced seeds and fertilizers. In addition, SG2000 focuses on post-harvest and agro-processing with the goal of producing processed goods of sufficient quality to market and promoting mechanized processing and improved storage techniques and technologies. It also promotes public-private partnerships and market access, which entails a multi-faceted approach to marketing and branding. It assists resource-poor farmers to access agricultural inputs and helps producers by organizing farmers' associations to bundle the produce of multiple farms together for sale to high-volume buyers through collective marketing.

SG2000 is active in the area of human-resource development too. This involves improving the skills and expanding the influence of agricultural extension workers, which is carried out in collaboration with the Sasakawa Africa Fund for Extension Education (SAFE), SAA's sister organization. In this respect, SAA works with African universities and colleges to build effective curricula to expand and strengthen the knowledge and skills of frontline agricultural and rural development advisory service providers so they can serve the needs of smallholder farm families more effectively.

SG2000's final thematic focus is monitoring, evaluating, learning and sharing to ensure that its resources are optimally allocated through an objective quantitative analysis across the diverse areas where its programmes operate. Evaluating and monitoring SAA activities helps the association to understand and address producers' needs and priorities, and facilitates the sharing of knowledge both within SAA and with outside organizations and governmental bodies.

National Agricultural Research Systems (NARS) in Africa

NARSs form the cornerstone of the global agricultural research system as they work to improve agricultural productivity, profitability and farmers' welfare in different countries. These systems are made up of various national agricultural research institutes (NARI), agricultural universities, centres for agricultural research in non-agricultural universities, private-sector firms, local non-governmental organizations and agricultural producers/commodity groups. Although some have long histories (such as that in Tanzania), many NARSs in Africa are composed of relatively young institutions.

In Africa, NARSs are dominated by public-sector NARIs that are saddled with the responsibility of carrying out applied and adaptive research, work with farming communities to identify future research needs and provide support to other institutions within the NARS. They also provide support and feedback on national development plans and policies. The agricultural universities and centres for agricultural research in non-agricultural universities help to improve people's technical skills when it comes to agricultural research. These institutions facilitate the development and transfer of technology and improve a nation's ability to absorb improved technologies from elsewhere. Universities also provide technical training and develop the researchers of tomorrow.

The contribution of private-sector research systems in NARSs in African countries has improved recently in terms of expenditure and personnel numbers (Kirsten *et al.* 2010a; 2010b). This varies, however, depending on the level of the country's economic development, policy environment, patents and legislation on plant-breeder rights and various market factors. In some cases, the private research organizations form partnerships with foreign multinationals and operate as subsidiaries of these large foreign companies. As such, the majority of innovations are either imported or adapted and distributed under licence from international firms or parent companies. Most of the agricultural research done by these private sector organizations is thus largely adaptive or done for testing purposes and to comply with the regulatory authorities (Kirsten *et al.* 2010b).

There are also local NGOs that work mainly at the farmer and community levels and serve as intermediaries between resource-poor rural households and research, donor, financial and other organizations. Their flexibility, informality, social commitment and participatory style are especially suited to the complex task of rural development aimed at alleviating poverty, particularly in countries where special interests or political instability may render governments and bureaucracies unresponsive to the needs of the poor (FAO 1996).

Agricultural producers and commodity groups constitute formidable agents of research for development in the region but their place in the research system is limited in Africa. These groups include both small- and large-scale farmers, transporters, traders, processors, retailers and consumers who are potentially important in setting research priorities relevant to their needs and directly or indirectly contributing funds for research. Examples of such organizations involved in articulating their demands on research systems can be found in Senegal and Kenya.

NARSs vary greatly in the role they play, the number of researchers they have, the level of their research efforts and the funding and resources available to them. This is consistent with the diverse nature of agro-ecological conditions and the location-specificity of small-farm production and natural-resource management opportunities in different countries.

Agricultural Research: The Nigerian Case

Nigeria currently has the largest and most elaborate national agricultural research system in Sub-Saharan Africa.[2] It is coordinated by the Agricultural Research Council of Nigeria (ARCN) that was established in 2006 to address the challenges faced by the Nigerian agricultural research system. Its key components include 15 agricultural research institutes, 11 federal colleges of agriculture, 3 agricultural universities and 50 faculties of agriculture and veterinary medicine in federal universities. Their research output is used by farmers, commodity groups and other users to improve agricultural productivity and users' welfare (Figure 11.2).

2 www.efina.org.ng.

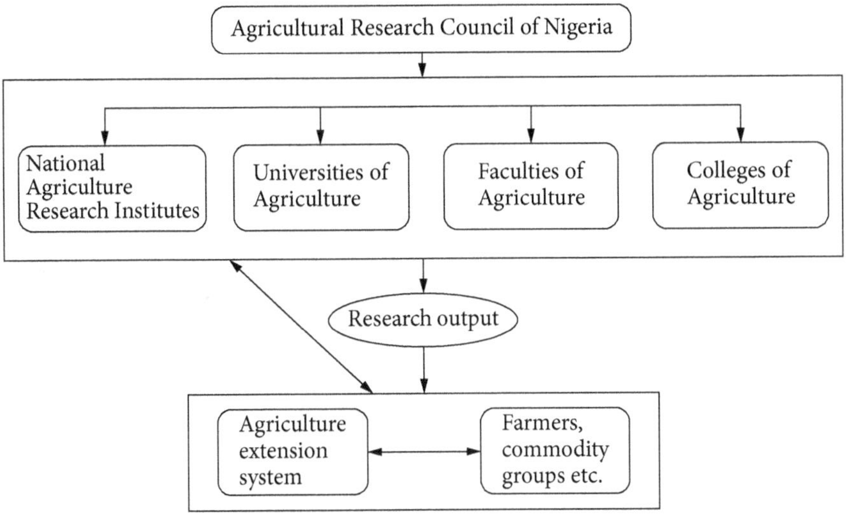

FIGURE 11.2 *The organization of Nigeria's agricultural research*
SOURCE: AUTHORS' OWN DESIGN

ARCN's mission is to achieve significant improvements in agricultural productivity, marketing and competitiveness through the generation of appropriate technologies, policy options and knowledge management of the agricultural research system. The ARCN does this by working with several international and regional organizations on agricultural and research policy issues. These include CGIAR (IITA, IFPRI, International Livestock Research Institute [ILRI], International Center of Tropical Agriculture or Centro Internacional de Agricultura Tropical [CIAT], Africa Rice Center [WARDA] and International Crops Research Institute for the Semi-arid Tropics [ICRISAT]), FARA, the FAO, the World Bank, the African Development Bank (AfDB), the Indian Council of Agricultural Research (ICAR), the Brazilian Agricultural Research Corporation (EMBRAPA) and the Kenya Agricultural Research Institute (KARI). Despite this drive, however, reports by Ragasa *et al.* (2011) and Uganneya *et al.* (2013) show that the Nigerian agricultural research system still has a long way to go in achieving its objectives. First, the collaboration between the ARCN, its institutional components and its external partners is still limited in terms of formal linkages. This is due to inadequate funding for interaction, the lack of a forum for interaction, poor leadership and coordination within the research organizations, the weak implementation of some parts of the Memorandum of Understanding and a lack of mobilization and capacity building (Ragasa *et al.* 2010). Likewise, formal linkages among ARCN's component research institutes are hampered by a communication gap, a lack of sustainability of collaborative efforts, programme overlap and bureaucracy (Ragasa *et al.* 2010).

Constraints to an Efficient Agricultural Research System in Africa

A myriad of problems are known to plague efficient agricultural research systems in Africa. These vary between countries and/or research entities because each country/research organization has its own peculiar structure and framework. However, a lot of these constraints are common to the different countries/research entities because of similar socioeconomic circumstances and/or their stage of development. The problems are highlighted below in broad categories. Although they are not exhaustive and sometimes not mutually exclusive, they touch on the major bottlenecks to efficient agricultural research systems in Africa.

The first issue is the problem of a poor environment for research, which manifests itself in different forms such as an inadequate national commitment to research; policy instability; a lack of strategic planning for agricultural research in member countries; limited societal appreciation for the importance of agricultural research in development; the poor remuneration of researchers; the wrong research intent; fatigue among respondents; a lukewarm attitude to research findings among users; a lack of interactive academic and professional societies; a lack of access to scientific information that is already available on the region and globally; poorly developed end users of research output (i.e. industries, private sector, civil society, etc.); researchers failing to disseminate their research findings beyond journals and donors; and poor administrative support for research both at governmental and institutional level (Ministry of Agriculture and Land Affairs, Republic of South Africa 2008).

There is also limited manpower for agricultural research. This bottleneck is related to the lack of a stable agricultural manpower policy to guide training in research methodology for agricultural professionals; rapid staff turnover, inadequate technical support (i.e. technologists, statisticians, etc.); an inability to retain trained manpower due to poor levels of remuneration and the lack of an enabling environment; poor career perspectives; gender discrimination; a shortage of qualified agricultural researchers; and a lack of specialists in agriculture's core sub-disciplines (FARA 2006; ADBG 2010; Oladele *et al.* 2010).

Inadequate research manpower has serious consequences for the effectiveness of the African agricultural research system. It leads to poorly completed, half-completed or abandoned research projects and also makes any lag in the knowledge production process unduly long (Idachaba 1997; Batte & Wanzala, 2009; Oladele *et al.* 2010). A rapid staff turnover, for instance, prevents a national agricultural research organization from building up a continuous body of knowledge or creating a niche position for itself that is uniquely recognized by the scientific community.

There is also the problem of inadequate infrastructure for research. This is typified by a lack of appropriate laboratory space, equipment and supplies; deteriorating library facilities; erratic electricity supplies; poor access to current (published) agriculture-related research; and inadequate transportation for research purposes. Others include poor and uncertain potable water supplies and the physical insecurity of life and property in many institutes, including the safety of field crops, animals and research plots (Flaherty *et al.* 2010).

There are weak or sometimes non-existent linkages among research institutes in Africa (Dimelu & Anyanwu, 2008; Oladele 2010). This is especially true in the case of linkages between research institutes and universities. The reasons for this include: the different 'parent' ministries of research institutes on one hand and the universities on the other hand; the fear among the leadership of research institutes of losing power to or being 'swallowed' up by a university or being forced into the performance criteria career system of universities, with particular reference to the quantity and quality of scientific publications; poor political will; and/or a lack of appreciation of the value of functional linkages between research institutes and universities. The problem of poor linkages also manifests itself in relations between public and private research institutes, and national and regional institutes. As a result, NARSs in Africa are not able to explore possible synergies and this limits their overall effectiveness.

Another major problem is lack of access to modern technology. In many African countries, there is poor access to new scientific and technological tools that could facilitate research due to the initial costs involved in setting up a system. This constraint is also clear in the organizations' inadequate telecommunication facilities: low computer penetration; poor computer skills on the part of researchers; and intermittent and unpredictable email and Internet connectivity in many regions (Salau & Saingbe, 2008; Ogunkunle & Fomsi 2010; Ogunjobi & Fagbami 2012).

A supply-driven approach to research innovations also constitutes a major obstacle (Mbabu & Ochieng 2006). Most agricultural research systems in Africa still adopt a first-generation approach to innovation: the push for technology, which is largely supply driven This involves developing new agricultural techniques and disseminating them to farmers as end users without a strong focus on farmers' real needs. Though some research organizations have 'graduated' to the second-generation approach to innovation – the need or demand-pull approach that focuses on consumer or market-determined needs – only a few research agencies have adopted other improved methods. These include the third, fourth and fifth generations that are common in other parts of the world (Rothwell 1994; Mbabu & Ochieng 2006;). The third generation of innovation combines the first and second generations in a push-pull relationship.

The fourth generation, in turn, integrates marketing, R&D activities, suppliers and leading customers, whilst the fifth generation involves broader systems of integration and networking models, including strategic partnerships with suppliers and customers, and collaborative marketing and research arrangements (Terziovski *et al.* 2001).

Funding constraints also exist due to inadequate investment in agricultural research, which can be seen in the poor funding by governments and other deployment agencies. Added to this is a lack of or inadequate government grants to initiate and sustain research in member countries; donor-driven interest towards selected research programmes that may not be in line with national priorities; a lack of local competences when competing for international research funding; and insufficient local private-sector funding for agricultural research (The World Bank Group, n.d.; FARA 2006).

Inadequate funds result in uncertainties and risks in the financial planning and management of African agricultural research systems (Beintema & Stads 2011). And the non-availability of funds in the amounts and at the times they are most needed also reduces the productivity of research and researchers. Insufficient research materials may be purchased and/or not at the time when they are most needed and research projects can be abruptly scaled down or even abandoned. A lack of funds also results in the depreciation and obsolescence of human capital due to an inability to retrain research staff, to update their skills and to allow them to acquire new technical competences. Inefficacies in resource allocation thus arise as a result of improvising and substituting research inputs that may not be efficient when it comes to achieving certain research objectives.

Finally, there is little attention given to safeguarding intellectual property. Many national agricultural research systems in Africa are not well positioned to commercialize any technologies they generate. This limits the level of originality in agricultural research efforts across the continent. The problem results from a low level of awareness among people about intellectual property protection (IPP) or intellectual property rights (IPR); an absence of clear and enforceable IPRs for agricultural research; and the lack of a clear institutional framework (policies, regulation and administration) for IPRs in agricultural research; and few formally domesticated and operationalized international conventions for the protection of IPRs in African agricultural research systems. Added to this is the lack of essential orientation, structure and capabilities for service delivery on agricultural research products on the part of existing IPR-administering agencies in African countries and weak intellectual property management policies at national agricultural institutes and faculties/colleges of agriculture in Africa (Kameri-Mbote 2005; Ministry of Agriculture and Land Affairs, Republic of South Africa 2008; Ngwediagi 2009; Shadrack 2011; Adediji 2012).

Conclusion

Significant progress has been made in improving the state of agricultural research systems in Africa but there are still challenges facing the effectiveness of the system. There is a need, therefore, for governments at all levels in Africa to align their national research priority areas with the needs of the end users (the agricultural producers and commodity groups). It is recommended that African governments draw up effective national research policies and strategies and build national agricultural research capacities, particularly through resource allocation, the training of researchers, the strengthening of research institutions and the establishment of coordination mechanisms. Appropriate political support for agricultural research systems should be established, implemented and/or strengthened.

As a matter of urgency, research agencies should be improved at regional, sub-regional and national levels. Closely related to this is the need for adequate funding for research and human resources for agricultural systems. Appropriate funding diversification strategies are required to ensure adequate, predictable and sustainable agricultural research systems at the national level. Research agencies should also be encouraged to expand their collaboration with other partners in agricultural research to facilitate needs-based research within their agricultural systems.

Research organizations, development agencies and relevant stakeholders in Africa have to develop modern technological facilities that will facilitate the dissemination and communication of research results. This could be done by developing a system-wide knowledge management and ICT policy, strategy and infrastructure to allow increased sharing of agricultural research results and information.

At a global level, the African Union and the United Nations could develop an umbrella organizational structure that would facilitate linkages and interactions among complementary research institutions. In the same vein, public-private research partnerships, regional research networks and research-farmer-extension linkages should be strengthened and overhauled. Efforts ought to be geared at functionally integrating research institutes and universities. This would enable agricultural research systems in Africa to explore possible complementarities between their institutes and universities in research, training and extension programmes.

Strategies aimed at improving the dissemination and communication of research need strengthening too. Better and stronger linkages between researchers and extension agents have to be facilitated as the latter are closer to farmers and other users of research outcomes than the former. The dissemination of results should go beyond journals and donors, and a more comprehensive system of communicating research findings could involve, for example, seminars for target groups such as agricultural producers, commodity groups, farmers, retailers and processors.

References

Adediji, D. (2012), Enforcement of Intellectual Property Rights in Nigeria: A Critique. *TradeMarks Journal* 2(31).

African Development Bank Group (ABDG) (2010), Agriculture Sector Strategy 2010–2014. http://www.afdb.org/fileadmin/uploads/afdb/Documents/Policy-Documents/Agriculture%20Sector%20Strategy%2010-14.pdf. Accessed 23 February 2014.

Agricultural Sector Coordination Unit (ASCU) Kenya (2012), National Agricultural Research System Policy. http://www.ascu.go.ke/DOCS/NARS_policy_final (2).pdf. Accessed 2 January 2014.

Alliance for a Green Revolution in Africa (AGRA) (2013), Africa Agriculture Status Report: Focus on Staple Crops. Nairobi, Kenya. http://agra-alliance.org/download/53397 7a50dbc7/. Accessed 8 September 2014.

Anandajayasekeram, P. (2011), The Role of Agricultural R&D within the Agricultural Innovation Systems Framework. Conference Working Paper No. 6. http://www.ifpri.org/sites/default/files/publications/ananda.pdf. Accessed 2 January 2014.

Batte, R. & S. Wanzala (2009), Profiling Higher Education Institutions: Case Studies from Eastern and Southern Africa: A Report to RUFORUM and CTA. http://knowledge.cta.int/Dossiers/CTA-and-S-T/Selected-publications/Profiling-Higher-Education-Institutions. Accessed 27 February 2014.

Beintema, N.M. & G.J. Stads (2011), African Agricultural R&D in the New Millennium: Progress for Some, Challenges for Many. Washington, DC: Food Policy Report. International Food Policy Research Institute.

Dimelu, M.U. & A.C. Anyanwu (2008), Linkage Behavior and Practices of Agencies in the Agricultural Innovation Transfer Sub-System in Southeastern Nigeria: Issues for Agricultural Extension Policy. *Journal of Agricultural Extension* 12(2): 9–23.

ECA (2005), Economic Report on Africa. Meeting the Challenges of Unemployment and Poverty in Africa. Addis Ababa: Economic Commission for Africa.

FAO (1996), Role of Research in Global Food Security and Agricultural Development. Technical Background Document.

FAO (2002), Impact of Foreign Assistance on Institutional Development of National Agricultural Research Systems in Sub-Saharan Africa. FAO Research and Technology Paper 10. http://ftp.fao.org/docrep/fao/005/y4349e/y4349e00.pdf. Accessed 3 January 2014.

FAO (2013), The State of Food Insecurity in the World. http://www.fao.org/docrep/018/i3434e/i3434e.pdf. Accessed 2 January 2014.

Flaherty, K., G. Ayoola, J. Ogbodo & N. Beintema (2010), Nigeria: Recent Development in Agricultural Research. http://www.asti.cgiar.org/pdf/Nigeria-Note.pdf. Accessed 1 March 2014.

Forum for Agricultural Research in Africa (FARA) (2006), Agricultural Research Delivery in Africa: An Assessment of the Requirements for Efficient, Effective and Productive National Agricultural Research Systems in Africa. http://www.fara-africa.org/media/uploads/library/docs/fara_publications/archived_publications/NARS_Assessment_exec_sum.pdf.pdf. Accessed 22 February 2014.

Hazell, P.B.R. (2011), The Asian Green Revolution. IFPRI Discussion Paper 00911. http://www.ifpri.org/sites/default/files/publications/ifpridp00911.pdf. Accessed 3 January 2014.

IAC (Inter-Academy Council) (2004), Realizing the Promise and Potential of African Agriculture in Science and Technology Strategies for Improving Agricultural Productivity and Food Security in Africa. Amsterdam: InterAcademy Council.

Idachaba, F.S. (1997), Instability of National Agricultural Research Systems in Sub-Saharan Africa: Lessons from Nigeria. ISNAR Research Report No. 13. The Hague: International Service for National Agricultural Research.

International Food Policy Research Institute (IFPRI) (2009), Agriculture's Critical Role in Africa's Development. Media briefing on GM crops for African farmers, 19 May 2009. http://www.ifpri.org/publication/agriculture-s-critical-role-africa-s-development. Accessed 3 January 2014.

Kameri-Mbote, P. (2005), Intellectual Property Protection in Africa: An Assessment of the Status of Laws, Research and Policy Analysis on Intellectual Property Rights in Kenya. IELRC Working Paper 2005–2.

Kirsten, J., C. Haankuku & R. Stander (2010), Assessing Private-sector Agriculture Research and Innovation in South Africa. Paper presented at the joint 3rd African Association of Agricultural Economists (AAAE) and 48th Agricultural Economists Association of South Africa (AEASA) Conference, Cape Town, 19–23 September.

Kirsten, J., R. Stander & C. Haankuku (2010), Measuring Private Research and Innovation in South Asia and Sub-Saharan Africa: A South Africa Country Report. http://www.asti.cgiar.org/pdf/private-sector/SouthAfrica-PS-Report.pdf. Accessed 21 March 2014.

Mbabu, A.N. & C. Ochieng (2006), Building an Agricultural Research for Development System in Africa. International Food Policy Research Institute (ISNAR Division) Discussion Paper 8.

Michelsen, H. & F. Hartwich (2004), University-based Agricultural Research: A Comparative Study in Sub-Saharan Africa. ISNAR Research Report 26. The Hague: International Service for National Agricultural Research.

Ministry of Agriculture and Land Affairs, Republic of South Africa (2008), National Agricultural Research and Development Strategy. http://www.daff.gov.za/docs/researchP/RD_Strat.pdf. Accessed 21 February 2014.

Ngwediagi, P. (2009), Establishment of a Plant Breeders' Rights System in Tanzania: Achievements and Challenges. In: CAS-IP NPI. 2009. Institutionalization of Intellectual Property Management: Case Studies from Four Agricultural Research Institutions in Developing Countries. Rome: CAS-IP.

Ogunjobi, T.E. & O.O. Fagbami (2012), Use of the Internet by Researchers in Agricultural Research Institutes in Ibadan, Oyo State. *International Journal of Library and Information Science* 4(4): 52–56.

Ogunkunle, R.A. & E.F. Fomsi (2010), Internet Literacy for Research Development among Lecturers in Tertiary Institutions in Rivers State, Nigeria. *Journal of Emerging Trends in Educational Research and Policy Studies* 1(2): 55–60.

Oladele, O.I. (2010), Research-Extension-Farmers Linkage System on Banana and Plantain (Musa spp.) in Nigeria: The Diffusion of Innovations. In: T. Dubois *et al.* (eds), *Proceedings of IC on Banana & Plantain in Africa*. Leuven: ISHS, pp. 837–842.

Oladele, O.I., S.K. Subair & N.V. Sebina (2010), Knowledge and Utilization of Job Enrichment Techniques among District Agricultural Officers in Botswana. *African Journal of Agricultural Research* 5(21): 2918–2924.

Ragasa, C., S. Babu, A.S. Abdullahi & B.Y. Abubakar (2010), Strengthening Innovation Capacity of Nigerian Agricultural Research Organizations. IFPRI Discussion Paper 01050.

Ragasa, C.R., A.S. Abdullahi & G.O. Essegbey (2011), Measuring R&D Performance from an Innovation Systems Perspective: An Illustration from the Nigeria and Ghana Agricultural Research Systems. ASTI/IFPRI-FARA Conference Working Paper 14. http://www.ifpri.org/sites/default/files/publications/ragasa.pdf. Accessed 28 February 2014.

Roseboom, J., P.G. Pardey & N.M. Beintema (2006), The Changing Organizational Basis of African Agricultural Research. EPTD Discussion Paper No. 37.

Rothwell, R. (1994), Towards Fifth-generation Process Innovation. *International Marketing Review* 11(1): 7–31.

Salau, E.S. & N.D. Saingbe (2008), Access and Utilization of Information and Communication Technologies (ICTs) among Agricultural Researchers and Extension Workers in Selected Institutions in Nasarawa State of Nigeria. *Production Agriculture and Technology Journal* 4(2): 1–11.

Seck, P.A., A. Agboh-Noameshie, A. Diagne & I. Bamba (2013), Repackaging Agricultural Research for Greater Impact on Agricultural Growth in Africa. *Journal of Food Security* 1(2): 30–41.

Shadrack, R.M. (2011), Agricultural Research's Role in Enhancing Agricultural Productivity in South Africa. In: 'How the Private and the Public Sectors Use Intellectual Property to Enhance Agricultural Productivity', Proceedings of a seminar at the World Intellectual Property Organization (WIPO) Geneva, 14 June 2011, pp. 36–59.

Terziovski, M., D. Samson & L. Glassop.(2001), Creating Core Competence through the Management of Organizational Innovation. Research Paper, Foundation for Sustainable Economic Development, Melbourne, Australia.

Uganneya, S., R. Ape & J. Udensi (2013), Agricultural Research Information System of Nigeria: An Assessment of Institutional Effectiveness. *Library Philosophy and Practice* (*e-journal*), Paper No. 998.

World Bank Group (n.d.), Achievements and Problems in Development of National Agricultural Research Systems. http://lnweb90.worldbank.org/oed/oeddoclib.nsf/DocUNIDViewForJavaSearch/258EF4BB538B4A4B852567F5005D3705. Accessed 22 February 2014.

Web References

http://www.ifad.org/ The International Fund for Agricultural Development (IFAD). Accessed 23 February 2014.

http://www.efina.org.ng/media-centre/events/workshops/agriculture-the-new-frontier-access-to-information-on-nigeria-s-agricultural-sector/force download/226. The Role of Research in Agricultural Development. Accessed 27 February 2014.

http://www.fao.org/about/en/ Food and Agriculture Organization of the United Nations. Accessed 23 February 2014.

http://www.saa-safe.org. Sasakawa Africa Association (SAA). Accessed 2 January 2014.

http://www.iita.org/search/-/journal_content/56/25357/3587627. Collaboration in the framework of the World Bank's West Africa Agricultural Productivity Program (WAAPP). Accessed 8 September 2014.

http://fara.infosysplus.org/fara/fara_index.php/tpl/fara_ainfo/URL_NAME/fara. Forum for Agricultural Research in Africa. Accessed 2 January 2014.

CHAPTER 12

Contributions of Small- and Large-Scale Farms and Foreign and Local Investments to Agricultural Growth

The Nigerian Example

Sheu-Usman Akanbi and Akinyinka Akinyoade

Nigeria's agricultural sector has made a substantial contribution to the country's GDP *since 1960. It currently employs about 65% of the population and provides employment for about 90% of the rural population. This chapter offers a synopsis of agricultural growth performance in Nigeria, taking cognizance of the driving factors vis-à-vis small- and large-scale farms, and local and foreign participation in the sector. Analysis shows that, as the major engine of rural growth and livelihood improvement in Nigeria, small-scale agriculture is highly constrained due to a number of factors such as limited technical and financial support, indifference among the youth to farming, inadequate government policies and reliance on other livelihood sources. This leaves the country with few alternatives as large-scale farming is still in an emergent stage. Sustained small-scale, local participation, with aspects foreign investment as a necessary complement, could go a long way to promoting agricultural growth in Nigeria where there is a looming fear of food insecurity in the current oil-driven economy. The implications are relevant for other African countries with an agricultural sector structured in a similar way to Nigeria's.*

Introduction

The agricultural sector has contributed significantly to Nigeria's GDP since the 1960s and currently provides employment for about 90% of the rural population. However, the boom-and-bust cycles of crude-oil receipts, with the attendant negative impact on agricultural potential and increasing demands for food to meet the nutritional needs of Nigeria's ever-increasing population, call for a revision of production strategies and capital for agricultural investment (NPC 2013). More specifically, there have been annual calls for economic diversification and renewed attention for the agriculture sector including the need for rural transformation, redistribution, poverty alleviation and socioeconomic development (Stewart 2000). The present approach to boosting agricultural-sector performance revolves around the debate as to whether to promote large-scale farming or to continue with small-scale farming and whether or not to

 | DOI 10.1163/9789004282698_013

encourage foreign or local investment in agriculture. This chapter examines the present state of Nigerian agriculture and the factors driving its growth, with special reference to differentials of catalysts, such as small-scale and large-scale farms, and foreign investment inflows in current agricultural performance.

Debates on small-scale versus large-scale farming, as well as the proportion of foreign or local investment that is desirable for the growth of the agricultural sector, have been well documented. Some have argued that changes in farming practices can have negative ecological effects due to increases in fertilizer and pesticide use (Firbank *et al.* 2003), that the trend towards larger farm sizes is bad for the environment[1] and may encourage land grabbing (de Schutter 2011), and that 'small-farm efficiency' allows a cross-sectoral and multi-occupational character for contemporary rural livelihoods and can lead to a reduction in rural poverty in low-income countries (Ellis & Biggs 2001). The opposing view is that small-scale agricultural production equates to 'local eating' (Deusing 1995). However results from empirical research by Gorton & Davidova (2004) on farm efficiency in six Central and East European (CEE) countries show no clear-cut evidence of corporate farms being inherently less efficient in all farming activities than family farms, and 'where significant differences have been found in favour of family farms against the average corporate farm, the best corporate farms still tend to perform as well as the best family farms'. The study also indicated that in countries 'where small family farms are well established and managed continuously by the present farm household, they appear to be less inefficient compared to larger cohorts as against countries where small family farms are a relatively new phenomenon'. The research presented here borrows from this perspective, especially as nearly 90% of the country's rural population are still employed in the agricultural sector. The analysis is based on Kwara State of Nigeria where Shonga farms provide ready case-study sites of small-scale farm holders using local capital side by side and, in certain cases, interwoven with large-scale farm operators. The latter are essentially 'white farmers' who were evicted by the Zimbabwean government but who were invited by Kwara State Government to set up farming operations in Nigeria to boost the region's food production. Field surveys and the existing literature have been used to collect pertinent information to address the small-scale versus large-scale farm holdings and foreign versus local capital conundrum in the quest to boost Nigeria's agricultural performance.

This chapter is structured as follows. The first section considers the current state of agriculture in Nigeria and also discusses small-scale farms. with special

1 Data from ESRC Adoption of Sustainable Agricultural Technologies project at Manchester University in 1996 reported by Rigby & Cáceres (2001).

attention being devoted to the Fadama Project. This is followed by an examination of large-scale farms that focuses on the Shonga Zimbabwean farmers. The subsequent sections discuss the drivers of agricultural growth performance in Nigeria and international perspectives of small-scale versus large-scale farming before presenting some conclusions and policy implications.

The Current State of Nigerian Agriculture

Nigeria is the largest geographical area in the West African sub-region, covering 923,768 km^2. The country lies entirely within the tropics and has two main vegetation zones: the rain forest and Savannah zones. The agricultural sector is an admixture of small- and large-scale farming and foreign and local participation (Apata *et al.* 2011). In the decade after independence (i.e. the 1960s), the Nigerian economy could reasonably be described as an agricultural economy because agriculture served as its overall engine of growth (Ogen 2003; Alkali 1997). The agricultural sector contributed up to 60% of GDP in the 1960s and, despite Nigerian peasant farmers' reliance on traditional tools and indigenous farming methods, these same farmers produced 70% of the country's exports and 95% of its food needs (Lawal 1997). From the standpoint of occupational distribution and contribution to GDP, agriculture was the leading sector but was neglected during the heyday of the oil boom in the 1970s (Akinyoade *et al.* 2013). Ever since, Nigeria has been suffering from an insufficiency of basic food items that has necessitated resorting to food imports to augment shortfalls in production at home (Olagbaju & Falola 1996).

Nigeria's economic growth has largely been accounted for by resilient agricultural growth. According to the Nigeria Vision 2020 First Implementation Plan for 2010–2013, the agricultural sector contributed 73% to the growth of GDP between 1999 and 2009. Real growth averaged about 7% per annum from 2004 to 2008 but declined to 5% and 4% in 2011 and 2013 respectively. The agricultural sector added value to the tune of 42% to GDP within the same period.

Distribution of Farm Holdings by Size in Nigeria

The size distribution of farm holdings in Nigeria is categorized as follows: small-scale farms (from 0.10 ha to 5.99 ha); medium-scale farms (from 6.0 ha to 9.99 ha); and large-scale farms (bigger than 10 ha) (Apata *et al.* 2011). These

2 Leadership (2013), Editorial, 'Nigeria's Outrageous Food Import Bill', Leadership (Abuja) http://allafrica.com/stories/201311240262.html 23 November 2013, accessed June 23, 2014.

classes constituted the following percentages of farm sizes in 2004: 84.5%, 11.3% and 4.2% respectively (Oksana 2005; Antman & Mckenzie 2005; Dorward *et al.* 2005). When judged by international standards, where all farms less than 10 ha are classified as small, it is clear that 95.8% of all farm holdings in Nigeria in 2003 (or a total of 46.1 m holdings) were small-scale farms, while the remaining 2.033 m holdings were medium-scale (Apata *et al.* 2011). The farm-size data presented in Table 12.1 show that marginal and small farms constitute about 80% of all the total farm holdings in Nigeria in 2006.

Small-Scale Farms

Most African countries have a rural-driven agricultural system that has left production in the hands of small-scale farmers. These farms are operated under complex systems of cultivation, such as bush fallowing or shifting cultivation, that vary depending on the types of crops and the area's prevailing socio-cultural norms (Ayoola 2001). The above-mentioned system of farming is characterized by the use of basic farm tools, dependency on rain, low levels of inputs and low outputs and incomes, and the farmers often bear the physical and financial brunt. Most of the farm labour is performed by the individual farmer and members of his family. The farming system is therefore not capable of attaining optimization and/or maximization as well as productivities characteristic of modern and scientific systems of farming (Olayide & Heady 1982). According to Brown (2012), smaller rural farmers do not have the capacity to manipulate the supply-and-demand scenario with any degree of sophistication because their

TABLE 12.1 *Farm-size demographics, Nigeria, 2006*

Category	Size (ha)	Average size (ha)	Total holdings (%)	Area (%)	Irrigated area (%)
Marginal farms	< 1.0	23	56	23	0.3
Small farms	1–2	1.42	24	36	2.2
Semi-medium	2–4	2.69	11	21	21.8
Medium	4–10	4.87	06	11	33.7
Large	> 10	13.51	03	9	42.2
All farms		2.25	**100**	**100**	**100.0**

SOURCE: NATIONAL BUREAU OF STATISTICS, ABUJA, NIGERIA (2006): APATA *ET AL.* (2011)

production methods are primarily of a subsistence nature. This evidently prompted the Nigerian government to initiate the World Bank/African Development-assisted Fadama Development Project, a community-based cooperative project that is supported by a subsidized agricultural credit system (World Bank 2003; 2008) and, more recently, the Federal Government of Nigeria's Agricultural Transformation Agenda: The Growth Enhancement Scheme. These two schemes, which were primarily initiated to assist in not only scaling up the farm holdings of the mainly subsistence and small-scale rural farmers in Nigeria but also to enhance their productive capacity, are examined below.

The National Fadama Development Project (Fadama I, II & III)

Fadama is the Hausa word for irrigable land. This is usually a low-lying plain with shallow aquifers along Nigeria's major river systems. Such land is particularly suitable for irrigated production and fishing, and traditionally provided feed and water for livestock (Ingawa *et al.* 2004). *Fadama* also refers to a seasonally flooded area used for farming during the dry season. It is defined as an alluvial lowland area that has been formed by erosion and the depositional actions of nearby rivers and streams (Qureshi 1989). These areas encompass land and water resources that could easily be developed for irrigation agriculture (World Bank 1994). If they spread over a large area, they are often referred to as 'wetlands' (Nkonya *et al.* 2008; Blench & Ingawa 2004) and are considered to have high potential for economic development with appropriate investments in infrastructure, household assets and technical assistance. The realization of this led to the Fadama I and II projects that successfully refined approaches for the improved utilization of such land (World Bank 2003, 2008).

Fadama 1 was implemented as a pilot between 1993 and 1999 and focused on the cultivation of rice and vegetable crops that were well suited to the country's *fadama* areas. The infrastructural emphasis was on providing tube wells, wash bores, boreholes and pumps for farmers through simple credit arrangements aimed at boosting aggregate crop output (Nkonya *et al.* 2008). Though the project was judged as a success overall, it had a number of shortcomings. It helped producers increase their output but did not provide farmers with processing, storage and marketing facilities (downstream agricultural sector). As a result, much of their output was either not sold at all or sold at low prices due to a supply glut (World Bank 2003). Fadama 1 also witnessed continuous conflicts between the joint users of the communities' land and water resources. From the lessons learnt, the new all-inclusive Fadama II was set up and subsequently implemented between 2004 and 2009. The target beneficiaries this time included not only the traditionally favoured crop farmers but also pastoralists, livestock rearers, fish culturists and other fisher folk. The project, which was declared disbursement effective on 27 May 2004, was funded by the World Bank and the African Development Bank to the tune of US$100 m and US$30 m respectively (World Bank 2003). Eighteen Nigerian states participated in this phase: twelve were funded by the World Bank and the other six (including Kwara State) were funded by the African Development Bank. Fadama II was also judged a success and even won a World Bank award. This positive outcome subsequently prompted other states to show an interest in the third phase.

Fadama III was initiated in October 2009 to tackle low productivity among Nigeria's smallholder farmers. At US$250 m, it was the biggest World Bank-funded agricultural project to date, although the Nigerian side was expected to provide counterpart funding to the tune of US$200 million (World Bank 2008). Like Fadama II, the new project adopted a community-driven development (CDD) approach that included:

- supporting the provision of marketing infrastructure
- empowering stakeholders
- improving mechanisms for conflict resolution
- supporting the establishment of rural non-farm enterprises to avoid food losses
- focusing on the contribution of women; and
- supporting improved management, increased food production and socially inclusive management of resources in the *fadama* areas

The project also employed a culture of participatory local development planning, while laying the basis for private-sector-led community-driven smallholder irrigation and socially inclusive management of the natural-resource base of the *fadama* areas. This thus increased the per capita incomes of *fadama* users, which is consistent with the Millennium Development Goals (World Bank 2008). Fadama III, which is expected to end by 31 March 2014, was implemented in 36 states and also in the Federal Capital Territory (FCT). A maximum of 20 Local Government Areas (LGAs) in each state were approved as project location sites by the respective State Ministries of Agriculture and Natural Resources. From the World Bank-assisted IDA credit of US$250 m, US$4,658,600 was loaned to Kwara State. This was the average amount received by each of the participating states.

The beneficiaries of Fadama III were individuals living on less than US$1 a day and were drawn from: (i) the rural poor engaged in economic activities (farmers, pastoralists, fishermen, nomads, traders, processors, hunters and gatherers as well as other economic interest groups) and (ii) relatively disadvantaged groups (women, including widows, the handicapped, people living with HIV/AIDS and youth). Project beneficiaries were encouraged to organize themselves into economic interest groups, called Fadama User Groups (FUGs), each having a minimum of 20 members and with benefits accruing to roughly 15 additional household members for each FUG member. The FUGs were also organized into Fadama Community Associations (FCAs) as apex organizations. Fifteen FUGs, on average, constituted an FCA at community level.

At the end of December 2013, a total of 115 FCAs with 1323 FUGs were registered in Kwara State, the equivalent of US$2,250,209.2 being disbursed to 106 of the FCAs. To date, the project has implemented 1302 subprojects made up of small-scale community-owned infrastructure and productive assets with accompanying capacity building and/or advisory services, including input support. Of these, 93 are on-going while 1209 have been fully completed and 92.9% of the completed subprojects (598 units) are now fully functioning. A total of 997 units of productive assets, with an acquisition value of US$577,071.42, have been acquired by 1069 FUGs and are been utilized effectively. This has resulted in savings of US$135,792.18 (23.5%), which is expected to be used for the necessary repairs and/or replacement of productive assets from time to time (Kwara State Fadama Development Project 2013) .

Fadama III was implemented by employing an innovative Local Development Planning (LDP) tool and the success of the project was predicated on community-driven development mechanisms. An impact assessment was last done at the project's mid-term stage and the results indicated that the development objectives had been met, with over 75% of the targeted beneficiaries having been reached. This achievement surpassed the project baseline study's expectation of increasing yield per hectare of crop cultivation by at least 20%. A percentage increase of 34% was achieved by the mid-term project evaluation. However, the cumulative impact of this bank-assisted project can better be attested to after the on-going beneficiaries' impact assessment of the project has been released.

Nigeria's Agricultural Transformation Agenda: The Growth Enhancement Scheme (GES)

The Nigerian Minister for Agriculture and Rural Development, Dr Akinwumi Adesina, reached out to Nigerian small-scale farmers with subsidized inputs through the Electronic Wallet System (e-wallet) initiative in 2012 (Adesina 2013). This was set up to allow farmers to obtain seeds and fertilizer using electronic vouchers sent directly to their mobile phones, thus eliminating the need for transactions with middle men. In the first year, the Minister reported that 1.5 m farmers had received their subsidized seeds and fertilizer based on requests made via their mobile phone. The system appeared to generate a market revolution as it enabled seeds and fertilizer companies to develop their supply chains to reach farmers across the country. To assess the programme, a survey was conducted in Kwara State by the authors of this chapter, with fifty interviews being held with registered farmers in ten of the sixteen local government areas in the state (i.e. five interviews per local government area). The Nigerian government's Growth Enhancement Scheme had some shortcomings as it was revealed that:

- farmers only received two of the six bags of fertilizer required per ha, in comparison with 2010 when they accessed fertilizer directly from the state government and each 'fortunate' farmer had access to six bags of fertilizer per ha
- 35% of the farmers claimed that not all the promised allocations of other inputs were provided
- administrative and logistic issues discouraged 26% of respondents from redeeming the voucher that allowed them to then collect agro-inputs
- communication problems existed between Cellulant (the project's IT consultants), government officials and registered farmers

However, 72% of the farmers interviewed confirmed that the new system was far more efficient and effective than the previous one, which was constrained by corrupt government officials and unscrupulous middlemen. As the GES programme is only in its second year of operation, it may be too early to assess its true contribution to agricultural growth in Nigeria effectively.

Large-Scale Farms

In a bid to address the drift towards a mono-cultural crude-oil economy, the Nigerian government began to encourage the large-scale commercial production of food crops in 1975, having earlier enjoyed a boom in plantation farming. Large-scale cocoa, rubber and oil-palm plantations dominated in southern Nigeria and the Numan, Lafiagi and Sunti sugar factories, driven by the small- and large-scale sugarcane out-grower schemes, were set up in the north. The out-grower scheme was equally boosted by the then newly established Nigerian River Basins Development Authorities and the World Bank-assisted State Agricultural Development Projects in the 1970s and 1980s respectively (Omotesho 1981; Lawal 1997). Between 1979 and 1999, local participation in farming was bolstered by the emergence of large-scale private/corporate farm holdings. Among these were the corporate Guinness farms at Mokwa, the Leventis farm at Ilesha, Northcott at Jalingo, and the Dana farms in Zaria. Folawiyo integrated farms at Ilora, Abiola farms (in various locations), Obasanjo integrated farms at

Abeokuta, Maizube farms at Minna, Nyako and Nasko farms in Adamawa and Niger States respectively were among the privately based, local participation in large-scale farming in Nigeria (FMANR 2010).

In addition, the Nigerian Agricultural and Cooperative Bank (NACB) was established in 1973 to inject oil wealth into the agricultural sector through the provision of credit facilities for agriculture and agro-related businesses (Olagunju 2000). Between 1995 and 1998, the government embarked on reforming the lending policies of the Agricultural Credit Guarantee Scheme (ACGS) to allow easier access to agricultural credit schemes. It also established the Calabar Export Processing Zone (EPZ) and initiated the Enugu, Kaduna, Jos and Lagos EPZs that each specialized in specific food and export crops. In fact, the National Rolling Plan for 1996–1998 assumed that, by the year 2000, Nigeria would be able to feed its own population, have the capacity to process agricultural raw materials for local industries and for export, and have significantly increased the contribution of the agricultural sector to the country's GDP (Lawal 1997). In spite of these efforts, it is worth noting that Nigeria became a net importer of various agricultural products from the mid-1970s onwards (Table 12.2).

Nigeria imported 153,000 metric tons of palm oil at a cost of US$92 m in 1982 and 55,000 metric tons of cotton valued at US$92 million (Alkali 1997). Between 1973 and 1980, a total of 7.07 m tons of wheat, 1.62 m tons of rice and 431,000 tons of maize were also imported. The cost of food imports in Nigeria thus rose from approx. US$80 million in the 1960s to US$147 million in the 1970s and to about US$190 million in 1988 (*Ibid.*). Nigeria spent an average of US$60 m on rice imports annually in the 1990s. The country is still a large food importer and

TABLE 12.2 *Composition of agricultural imports (2009–2011) (in million)*

	2009	2010	2011
Live animals, animal products	156,978.5	190,034.3	293,995.5
Vegetable products	268,641.9	270,774.5	892,422.7
Prepared foodstuffs, beverages, spirits and wine, tobacco	176,927.9	207,717.9	479,901.2

SOURCE: NBS/NPC (2012)

Note: 160 = US$1.

PHOTO 12.1 *Hunter Coetzee's banana and pineapple plantation, Shonga, Kwara State, Nigeria*
PHOTO: SHEU-USMAN AKANBI

spent as much as N 356 billion (US$2.2 bn) on importing parboiled rice, N 217 billion (US$1.4 bn) on sugar and N 97 billion (US$0.6 bn) on fish in 2012.[2]

Shonga 'Zimbabwe' Farmers in Kwara State, Nigeria

Investment in large-scale farming would seem to be gaining importance in Nigeria. With an estimated population of over 170 million inhabitants in mid-2013, which represents 16% of Africa's total population, there is overarching concern about the country's ability to achieve self-sufficiency in food production and source adequate raw materials for its agro-related industries on a sustainable basis (PRB 2013; Oni *et al.* 2009; Nkonya *et al.* 2010; Eboh 2011). It was perhaps the realization of this fact that spurred the government of Kwara State in Nigeria's middle belt to initiate a number of programmes geared at improving the state's level of agricultural production. Notable among these interventions in the agricultural sector was the large-scale commercial Shonga farm venture (Photo 12.1).

In 2003, the State Governor, Dr Bukola Saraki, encouraged some of the 'white' farmers who were being evicted from Zimbabwe to relocate to Shonga in Edu Local Government Area based on a public-private partnership construction with the state government. To ensure effective take-up, the government provided

PHOTO 12.2 *Reid Irvint's dairy farm in Shonga, Kwara State, Nigeria*
PHOTO: SHEU-USMAN AKANBI

incentives such as land, infrastructural facilities and guarantees for bank loans. Each of the initial farmers received a 25-year lease of 1000 ha of land. Although the acquisitioning of land to establish these large-scale farms met with fierce local resistance at first, particularly from two of the communities (Dumagi and Faigi), the government succeeded in effectively repressing any protest (Mustapha 2011). A total of 1687 ha was divided between the first fifteen Zimbabwe farmers in Nigeria in 2005 for the cultivation of crops ranging from grains like maize, cowpea, soya beans and groundnut to root tubers like cassava.

Of the initial fifteen farmers, seven are still in business today: four in poultry, two in dairy and one in crop farming. Only two indigenous poultry farms (Zarm farm at Erinle and Rear Admiral Afolayan's farm at Osi), alongside the Coga and Thai crop farms at Bode-Saadu and Malete respectively, are among the few farms that could compete with the so-called Kwa-Zimbo farms on scale of production in Kwara State. Recently however, on a visit by one of the authors to the Shonga farms, a government extension agent revealed that six of the crop farmers had left as a result of not being able to secure funding to embark on an effective irrigation system, which had thus limited their capacity to produce year-round and increase the farm's profits.

The local farmers who were evicted from their ancestral land were relocated to new farming areas. They received grants of N 20,000 (US$130) per ha for land clearing, N 11,000 (US$70) for land preparation and N 3000 (US$20) for the procurement of hybrid seeds. In addition, they were given up to 720 bicycles as their communities were a considerable distance from their new farm locations (Mustapha 2011). But one important aspect of the Memorandum of Understanding between the Zimbabwean farmers and the Kwara Government has still not been effected, namely the establishment of the farm that was to be managed by the commercial farmers on behalf of the state. This farm was expected to serve as a training, skills and knowledge transfer centre for local farmers and government extension agents but the state government has never released the funds required for the venture.

The Shonga commercial farms can hardly be described as a model for foreign investment. Based on evidence of expertise on the ground, government spending on the new commercial farmers reveals more of an outflow of local resources. Mustapha (2011) concluded that (Photo 12.2):

> ... despite the occasional rhetorical reference to local farmers, the central plank of Governor Saraki's ultimate strategy was not the transformation of peasantry, but the creation of a new agrarian technology-savvy class by-passing old peasantry.

It has been difficult to conduct a comparative cost-revenue or cost-benefit analysis between the Zimbabwean farmers and local farmers since 2004. Production statistics have not been made available either by the Zimbabwean farmers or their host, namely Kwara State Government. However, pioneering research work comparing the farming activities of the Shonga farmers and local farmers by Musa *et al.* (2008) offers an insight into farm efficiencies in the first year of crop production. How close a farm is to achieving its maximum potential (output) as defined by the size and a given set of inputs is the measure of its technical efficiency. The study involved 50 local and 10 foreign farms and the foreign farmers were found to have a larger farm size than the domestic farmers. Farm sizes ranged from 0.40–3.60 ha for local farms to between 120 ha and 272 ha for foreign farms in the first year. The small-scale farm holdings of local farmers were the result of a low capital base and a lack of land for larger production areas.

Musa *et al.* (2008) compared the efficiency of crop production in the Shonga farm area and concluded that local farmers competed favourably with the commercial farmers regarding production. This is similar to Gorton & Davidova's (2004) findings in six Central and Eastern European countries. The technical efficiency estimates for the Shonga farms varied greatly between local farmers, ranging from 18% to 99.7% with a mean technical efficiency of 71%, while those for the commercial farmers ranged between 91% and 100%. The implication is that if local farmers were given the same leverage, they would be able to achieve similar results regarding technical efficiency.

Other Nigerian states. like Lagos, Ebonyi, Rivers, Osun, Kano and Kogi states, have invested considerably in large-scale farming since 2003. Nasarawa State Government also made an effort to host Zimbabwean farmers but the initiative failed because, unlike Kwara State where the government gave priority attention to the foreign farmers, Nasarawa State Government did not facilitate the counterpart financing of their farms and the farmers left (Mustapha 2011) (Photo 12.3).

Drivers of Agricultural Growth Performance

Small-scale units used to account for a substantial share of Nigeria's agricultural production (Apata *et al.* 2011) but since the oil boom in the 1970s, economic growth and the increasing commercialization of agriculture have tipped the balance in favour of productive efficiency and effectiveness from small- to

PHOTO 12.3 *Government officials interacting with Paul Rezholm on his dairy farm in Shonga, Kwara State*

PHOTO: SHEU-USMAN AKANBI

large-scale farms (Olubiyo & Adewumi 1997). However, since 95.8% of all farm holdings in Nigeria in 2004 were classified as small-scale farms, it can be inferred that these small-scale farms must have contributed significantly to the nation's economic growth (Apata *et al.* 2011). This therefore underpins the need to examine the growth factors and the drivers of growth performance of these small-scale subsistence farms that dominate the agricultural sector of the Nigerian economy.

Growth refers to an increase in the amount of output over time. Agricultural growth is associated with performance in four constituent subsectors: crops, livestock, fisheries and forestry. The crop subsector had, on average, the largest share of growth, followed by the livestock, forestry and fisheries sectors that grew substantially in the oil-boom period post-1974. Currently, the crop subsector contributes up to 85% of the country's agricultural GDP, while livestock contributes about 10%, fisheries around 4% and forestry about 1%. On a comparative basis, the crop sector's share declined up to the early1980s while the performance of the fisheries sector has steadily improved over the past 50 years. Nevertheless, the crop subsector remains overall the dominant agricultural economy in Nigeria (CBN 2008; 2009).

While the agricultural sector may have contributed significantly to improved growth performance in Nigeria in recent years, its actual contribution appears to fall way short of its potential. The quality of agricultural growth remains questionable considering the low productivity, poor economic competitiveness and weak linkages to other sectors seen (Oni *et al.* 2009; Nkonya *et al.* 2010; Eboh 2011). Moreover, evidence shows that the increases in agricultural output have been due to an expansion in the amount of land under cultivation rather than an increase in productivity (Eboh 2011; UNDP 2009). This growth pattern, which has been mainly driven by land expansion, is not transformative but nominal in nature and merely additive. In contrast, transformative growth occurs when increases in agricultural output are driven by structural and 'real' improvements in production functions and by the economic competitiveness of agricultural value chains. Such a (desirable) growth pattern can be brought about by the improved use of more resource-efficient production inputs (improved seeds, fertilizers, agrochemicals, water) coupled with the right institutional conditions (product market, agricultural extension, agricultural credit) and efficient infrastructure. Growth in agricultural productivity and farm incomes are prerequisites for structural transformation. Increased farm incomes push up the demand for non-farm products and, in turn, stimulate the growth of small and medium enterprises (UNECA 2005; Timmer & Akkus 2008). In Nigeria, where agricultural growth is mainly the result of expansion of farm land area, doubts are being expressed about the longer-term sustainability of current growth records.

A key empirical study that examined agricultural growth in relation to the contributions of small-scale local subsistence farmers is that by Apata *et al.* (2011). It simulated two periods: one before and one after the Structural Adjustment Programme (SAP) era. Their findings reveal that, in the post-SAP period, Nigeria's subsistence agriculture was an important shock absorber against further agricultural output declines. They suggested that subsistence farmers enjoy a greater potential for Total Factor Productivity (TFP) growth through technology transfer. TFP is a way to measure productivity taking into account all factors of production. The linear effect of human capital on TFP is known to provide support to the role of educational attainment in enhancing domestic innovation in agriculture and international trade in promoting technology transfer. It points to the importance of education in facilitating the assimilation of foreign improvements in technology. Increased competitiveness in subsistence agriculture and domestic production do not invariably result in sufficient restructuring that would induce overall growth in Nigeria's agricultural sector. There are now critical constraints to agricultural policy effectiveness that include policy instability and inconsistencies, a narrow base of policy formulation, poor policy implementation and a weak institutional framework for policy coordination.

The Computable General Equilibrium (CGE) model developed along the lines of models as described in Lofgren *et al.* (2002) was adopted by Apata *et al.* (2011) in their study. CGE models are a class of economic models that use current economic data to estimate how an economy might react to changes in policy, technology or other external factors. The Social Accounting Matrix (SAM) was also used to disaggregate primary agricultural production not by product but by farm type, which enabled the study to distinguish institutional and economic characteristics.

Productivity of Small-Scale and Large-Scale Farms

The issue of scale of production and investment in agricultural development is closely connected to the issue of productivity (Rigby & Cáceres 2001). Sustainable intensification for the production of food from the same land area while reducing the environmental impact is important for smallholders and large-scale farming (Godfray *et al.* 2010) and can be achieved by stimulating local agriculture and providing smallholders with increased certainty about demands for their products. Gorton & Davidova (2004) also highlight the need to avoid simplistic comparisons in terms of yields per ha but to account for the differential use of inputs. They note that scale economies are temporary disequilibrium phenomena that persist only under specific circumstances. Moreover, caution should be taken when using large and small farms' dichotomies in empirical studies: they are arbitrary divisions of an essentially continuous variable (Doran 1985) and are sometimes unclear (Guthman 2004). In other regions of the world too, plot sizes are more important than the size of the farm (Nguyen *et al.* 1996).

In the World Bank Development Report (2008), Collier sets out three priorities that could help overcome the world food crisis: (i) moving to large-scale commercial farms instead of peasant or smallholder farming, (ii) promoting genetically modified organisms; and (iii) reducing (distorting) subsidies for biofuels in the US. However Derek & Alain differ with Collier in the same report on the point of modernizing agriculture and assert that smallholder farming is a proven strategy for accelerating growth and reducing poverty and hunger. They contend that smallholders have shown themselves to be efficient commercial farmers, when given a chance, as can be seen in the Asian Green Revolution experience that has been led by smallholders in countries like India and China since the 1960s. While recognizing that large-scale agriculture has a place in some land-abundant areas of Africa if it is driven by markets rather than subsidies and the rights of current land users are adequately protected, Derek & Alain feel that it would be mistaken to forsake the proven power of smallholders to jumpstart growth and reduce poverty and hunger in Africa and elsewhere (World Bank 2007).

This assertion is true for Nigeria, where the literature has identified the positive role of subsistence agriculture in driving national economic growth. For instance, Apata *et al.* (2011) noted that during the downturn in the economy in the 1980s, Nigeria's subsistence agriculture was an important shock absorber. Although Olubiyo & Adewumi (1997) posited that large-scale farms in Nigeria were more efficient when compared to small-scale farms, Apata *et al.* (2011) proved that whereas a simulated financial crisis should have increased the relative competitiveness of large-scale crop farms versus small-scale farms, this was not the case based on Nigeria's experiences in the 1980s. It was also observed that the agricultural sector responded in a mixed fashion when small-scale and large farms were compared regarding their contributions to growth. And successive government reforms showed that efficiency-enhancing institutional change would benefit both large-scale and small-scale farms.

Local farmers could thus be given the chance to prove their worth compared to large-scale farmers. Promoting smallholder farming is, as Derek & Alain noted, not 'romantic populism' but sound economic and social policy that reflects empirical evidence. While acknowledging the efforts of the Kwara State Government to develop large-scale commercial farms and the relative success of the Shonga farms, Mustapha (2011) expressed the need for caution based on the selective support given to 'Nigeria-Zimbabwe' farmers and averred that:

> However, for both political and livelihood reasons, agricultural policy cannot proceed simply by sweeping aside the peasantry that continues to produce the major bulk of the food supply with little or no official support.

This type of assessment might have served as the premise on which the current Nigerian Minister of Agriculture, Akinwumi Adeshina, based the Growth Enhancing Scheme that is primarily focused on small-scale subsistence Nigerian farmers.

Conclusion and Policy Implications

The majority of Africa's population still live in rural areas and agriculture is their main form of economic livelihood. The analysis in this chapter has highlighted the fundamental role of agriculture in Africa's quest for food security and economic development and pointed to the range and interaction of constraining influences on agricultural growth. Large-scale farming is desirable but is fraught with challenges that have implications for Africa's agricultural growth and performance. Transformative growth in the agricultural sector calls for pragmatic interrelated policy measures. These include the following.

Capacity Building at All Levels

Capacity building at the farm level is crucial for improving crop, soil and water management, enhancing the demand for and use of better and more efficient production inputs, and increasing the financial absorptive capacity of farmers. Tied to farm-level capacity building is the need to reinvent public agricultural-sector institutions for the effective and efficient delivery of financial services, agricultural extension services and farmers' education, provision of agro-inputs, expansion of irrigation, and market development and support.

Providing Subsidized Agro-inputs to Farmers

The provision of agricultural subsidies for fertilizer, farm implements and equipment would boost agricultural production. In addition, the introduction of unique schemes, like the e-wallet system, could facilitate better access to farm inputs that would enable both small-scale and large-scale farmers to expand their scales of production. In addition, all the existing credit facilities and schemes that have been put in place to assist farmers should be strengthened so that farmers can have better access to soft loans. These should have attractive single-digit interest rates with no stringent conditions attached.

Accessing Foreign Direct Investment: A Necessary Complement

Although sustained small-scale, local participation is advocated, foreign direct investment needs to complement local investments in agricultural development in Sub-Saharan Africa, particularly in Nigeria, where food insecurity may occur in the future in spite of the country's current oil-driven economy. Another important consideration in relation to planning is land reform. Land policies have important implications for the rural poor in that their incomes depend on the extent to which they control land and its output. Given the relatively small size of the large-scale commercial agriculture sector in Nigeria, and indeed in Africa in general, and the country's potential economic growth, investments in land development and a nationally contextual land reform policy are crucial for the sector's growth.

References

Adesina, A.A. (2013), *Nigeria's Agricultural Transformation Agenda*. Keynote Address by Minister of Agriculture and Rural Development, Federal Republic of Nigeria at the annual congress of the Southern African Confederation of Agricultural Unions at the Federal Ministry of Agriculture and Rural Development, Dar es Salaam, Tanzania.

Akinyoade, A., T. Dietz, D. Foeken & W. Klaver (2013), Agricultural Dynamics and Food Security Trends in Nigeria. Developmental Regimes in Africa (DRA) Project ASC-AFCA Collaborative World Research Group: Agro-Food Clusters in Africa (AFCA) Research Report 2013-ASC-1. London/Leiden.

Alkali, R.A. (1997), *The Bank and Nigeria: Cornucopia or Pandora's Box?*. Kaduna: Baraka Press, pp. 15–16.

Antman, F. & D. Mckenzie (2005), Poverty Traps and Non-linear Dynamics with Measurement Error and Individual Heterogeneity. Mimeo. Stanford University.

Apata, T.G., A. Folayan, O.M. Apata & J. Akinlua (2011), The Economic Role of Nigerian's Subsistence Agriculture in Transition: Implication for Rural Development. 85th annual conference of the Agricultural Economics Society, Warwick University.

Ayoola, G.B. (2001), *Essays on the Agricultural Economy: Book of Readings on Agricultural Development Policy and Administration in Nigeria*. Ibadan: T.M.A. Publishers.

Blench, R.M. & S.A. Ingawa (2004), *A Practical Guide for National Fadama Development Project II on Conflict and Management.* Fadama: World Bank PCF/Government of Nigeria, pp. 1–19.

Brown, G. (2012), The Agricultural and Rural Development Transformation Agenda of Nigeria. 15th ARMTI lecture series. 16 February.

CBN (2008), Annual Report and Financial Statements. Abuja: Central Bank of Nigeria (CBN).

CBN (2009), Macroeconometric Model of the Nigerian Economy. Abuja: Central Bank of Nigeria (CBN).

de Schutter, O. (2011), How Not To Think of Land-grabbing: Three Critiques of Large-scale Investments in Farmland. *The Journal of Peasant Studies* 38(2): 249–279.

Deusing, W. (1995), Is Organic Enough? *The Natural Farmer* 2(27).

Doran, H.E. (1985), 'Small' or 'Large' Farms: Some Methodological Considerations. *American Journal of Agric Economics* (1): 130–132.

Dorward, A., J. Kydd & J. Poulton (2005), Coordination Risk and Cost Impacts on Economic Development in Poor Rural Areas. Agricultural Economic Society Conference, Nottingham.

Eboh, E.C. (2011), Agricultural Economy of Nigeria: Paradoxes and Crossroads of Multimodal Nature. 56th Inaugural lecture, University of Nigeria.

Ellis, F. & S. Biggs (2001), Evolving Themes in Rural Development 1950s–2000s. *Development Policy Review* 19(4): 437–448.

Federal Ministry of Agriculture and Natural Resources (FMANR), (2010), National Agriculture Investment Plan 2011–2014, ECDWAP/CAADP/Process, Federal Ministry of Agriculture and Natural Resources, Nigeria.

Firbank, L.G., M.S. Heard, I.P. Woiwod, C. Hawes, A.J. Haughton, G.T. Champion, R.J. Scott, M.O. Hill, A.M. Dewar, G.R. Squire, M.J. May, D.R. Brooks, D.A. Bohan, R.E. Daniels, J.L. Osborne, D.B. Roy, H.I.J. Black, P. Rothery & J.N. Perry (2003), Methodological Insights. An Introduction to the Farm-scale Evaluations of Genetically Modified Herbicide-tolerant Crops, *Journal of Applied Ecology* (40): 2–16.

Godfray, H.C.J., J.R. Beddington, I.R. Crute, L. Haddad, D. Lawrence, J.F. Muir, J. Pretty, S. Robinson. S.M. Thomas & C. Toulmin (2010), Food Security: The Challenge of Feeding Nine Billion People, *Science* 327(812); doi: 10.1126/science.1185383.

Gorton, M. & S. Davidova (2004), Farm Productivity and Efficiency in the CEE Applicant Countries: A Synthesis of Results. *Agricultural Economics* 30(2004): 1–16.

Guthman, J. (2004), The Trouble with 'Organic Lite' in California: A Rejoinder to the 'Conventionalisation' Debate. *SociologiaRuralis* 44(3): 1–16.

Ingawa, S.A., A.A. Oredipe, K. Idefor & C. Okafor (eds) (2004), Facilitators Project Implementation Manual. Second National Fadama Development Project (Fadama 11). ABuja: Federal Ministry of Agriculture and Rural Development.

Kwara State Fadama Development Project (2013), 7th World Bank/Federal Government of Nigeria Mission Report on Kwara State Fadama Development Project.

Lawal, A.A. (1997), The Economy and the State from the Pre-colonial Times to the Present. In: A. Osuntokun & A. Olukoju (eds), *Nigerian Peoples and Cultures*. Ibadan: Davidson, pp. 195–198.

Lofgren, H., R.L. Harris & S. Robinson (2002), *A Standard Computable General Equilibrium* (CGE) *Model in* GAM, Microcomputers in Policy Research 5. Washington DC: International Food Policy Research Institute.

Musa, J.J., O. Chukwu & S. Bello (2008), Foreign Direct Investment in Nigeria: Efficiency of Foreign and Local Farms in Edu Local Government Area of Kwara State. *International Business Management* 2(5): 195–200.

Mustapha, A. (2011), Zimbabwe Farmers in Nigeria: Exceptional Farmers or Spectacular Support? *African Affairs* 110(4): 1–26.

National Planning Commission (NPC) (2013), National Planning Commission. Source: NPC's Compilation and Computation from NBS's Trade Statistics: Table on Composition of Agricultural Imports (2009–2011) (in N million). http://www.nigerianstat.gov.ng/pages/download/61.

Nguyen, T., E. Cheng & C. Findlay (1996), Land Fragmentation and Farm Productivity in China in the 1990s. *China Economic Review* (2): 169–180.

Nkonya, E., J. Pender, E. Kato, O. Oni, D. Phillips & S. Ehui (2010), *Options for Enhancing Agricultural Productivity in Nigeria.* NSSP 11. Abuja: International Food Policy Research Institute (IFPRI).

Nkonya, E., D. Philip, T. Mogues, J. Pander, M.K. Yahaya, G. Adebowale, T. Arokoyo & O. Ogen (2008), Impacts of a Pro-Poor Community Driven Development Project in Nigeria. IFPRI Discussion Paper 00756. IFPRI.

Ogen, O. (2003), *Patterns of Economic Growth and Development in Nigeria since 1960*. In: S.O. Arifalo & G. Ajayi (eds), *Essays in Nigerian Contemporary History*. Lagos: First Academic Publishers, pp. 231–234.

Oksana, N., (2005), Small Farms: Current Status and Key Trends. In: Proceedings of Research Workshop on the Future of Small Farms. Organized by International Food Policy Research Institute (IFPRI)/2020 initiative and Overseas Development Institute (ODI), Imperial College, London.

Ogunremi, G.O. & E.K. Faluyi (eds) (1996), *An Economic History of West Africa since 1750*. Ibadan: Rex Charles. 263.

Olagbaju, J. & T. Falola (1996), *Post-independence Economic Changes and Development in West Africa*. In: G.O. Ogunremi & E.K. Faluyi (eds), *An Economic History of West Africa since 1750*. Ibadan: Rex Charles.

Olagunju, M.A. (2000), *Economic Issues in Nigeria's Development*. In: J.O. Akinbi (ed.), *Towards a Better Nigeria*. Ibadan: Ben Quality Press.

Olayide, S.O. & E.O. Heady (1982), *Agriculture Production Economics*. Ibadan: Ibadan University Press. Pg 38.

Olubiyo, S.O. & M.O. Adewumi (1997), A Comparative Study of Resource Allocation in Small and Large Scale Farms in Kwara State. *Counterpoint Science Edition* 7(1): 66–74.

Omotesho, O.A. (1981), Irrigation and Crop Production in Nigeria: A Case Study of the Kano River Project, Kadawa, Kano. Unpublished MSc Thesis. University of Ibadan.

Oni, O., E. Nkonya, J. Pender, D. Phillips & E. Kato (2009), Trends and Drivers of Agricultural Productivity in Nigeria. NSSP Report 1. Abuja: International Food Policy Research Institute (IFPRI).

Population Reference Bureau (2013), World Population Data Sheet 2013, September 2013. Population Reference Bureau: www.prb.org/pdf13/2013-population-data-sheet_eng.pdf. Accessed 11 March 2014.

Qureshi, M. (1989), Water Management in Fadama Crops. Paper presented at the National Workshop on Fadama and Irrigation Development in Nigeria, Bauchi, Nigeria.

Rigby, D. & D. Cáceres (2001), Organic Farming and the Sustainability of Agricultural Systems. *Agricultural Systems* 68(2001): 21–40.

Stewart, R. (2000), Welcome Address. In: Proceedings of the 7th World Sugar Farmers' Conference. Durban.

Timmer, C.P. & S. Akkus (2008), The Structural Transformation as a Pathway Out of Poverty: Analytics, Empirics and Politics. Working, Paper No. 150. Washington DC: Center for Global Development.

United Nations Development Programme (UNDP) (2009), Human Development Report Nigeria. 2008–2009: Achieving Growth with Equity. Abuja: UNDP.

United Nations Economic Commission for Africa (UNECA) (2005), Structural Transformation to Break Away from Rural Poverty, Chapter 4 in Economic Report on Africa 2005. http://www.uneca.org/sites/default/files/publications/era2005full.pdf. Accessed 8 September 2014.

World Bank (2007), World Development Report 2008, *Agriculture for Development*. http://siteresources.worldbank.org/INTWDR2008/Resources/WDR_00_book.pdf. Accessed 9 September 2014.

World Bank (1994). *A Review of World Bank Experience in Irrigation*. Washington DC: World Bank.

World Bank (2003), Nigeria, Project Appraisal Document, World Bank-assisted Nigeria, Second National Fadama Development Project. http://www.worldbank.org/en/country/nigeria/projects/operational-documents?qterm=+Second+National+Fada%2520ma+Development+Project+&lang_exact=English. Accessed 19 February 2014.

World Bank. (2008), Project Appraisal Document, World Bank-assisted Nigeria. Third National Fadama Development Project. http://www.worldbank.org/en/country/nigeria/projects/operational-documents?qterm=+third-National+Fadama+Develo%2520pment+Project+&lang_exact=English. Accessed 19 February 2014.

CHAPTER 13

Loss and Damage from Droughts and Floods in Rural Africa[1]

Kees van der Geest and Koko Warner

'Loss and Damage' refers to the adverse effects of climate variability and climate change that occur despite efforts in global mitigation and local adaptation. This chapter reports on five case studies on loss and damage as a result of drought and floods in rural areas in Burkina Faso, Ethiopia, Kenya, Mozambique and The Gambia using a questionnaire survey (N = 1,973) and participatory research tools. The research goes beyond existing knowledge on adaptation and coping mechanisms by examining the limits, constraints and residual impact of household measures to cope and adapt. Only 28% of the households surveyed successfully avoided loss and damage from drought or floods but those that introduced agricultural adaptations, such as planting drought-resistant crop varieties, or diversified their livelihoods with non-farm activities were significantly more successful than others. Migration, on the other hand, was associated with the failure to avoid loss and damage. The chapter uses examples from five case studies and is structured around three 'loss and damage pathways' to illustrate the consequences of not being able to cope and adapt adequately. The findings presented here indicate that climate-related losses and damage are already a reality in many rural African communities.

Introduction

African agriculture is extremely vulnerable to climatic stressors because most of the continent's crop cultivation is rain-fed and natural climate variability is

1 The data presented in this chapter were gathered in the context of the Loss and Damage in Vulnerable Countries Initiative. As part of this project, the United Nations University conducted nine case studies in collaboration with national research institutes. In addition to the five African case studies presented here, fieldwork was also done in Bangladesh, Bhutan, Micronesia and Nepal. Funding came from the Climate and Development Knowledge Network (CDKN). The Africa Climate Policy Centre (ACPC) also funded three case studies (in Ethiopia, Burkina Faso and Mozambique). The authors wish to thank Fatima Denton and Tom Owiyo of ACPC for coordinating the work. A special thanks goes to the national researchers – Denis Opiyo Opondo, Sidat Yaffa, Alemseged Tamiru Haile, Ange-Benjamin Brida and Seydou Traore – and their teams for gathering the case study data. For more information on the project, see www.lossanddamage.net

 | DOI 10.1163/9789004282698_014

high (Boko *et al.* 2007; Roudier *et al.* 2011). Institutional support to farmers and access to crop insurance are also limited (Warner *et al.* 2012a). Drought and floods regularly cause havoc for African farmers and impact on farming systems, as do pests and plagues, volatile commodity prices, poor access to markets and storage facilities, and high levels of poverty that all act as constraints on adaptive capacity. Climate change and increasing climate variability are compounding these challenges (Thornton *et al.* 2011; Hertel *et al.* 2010). With current trajectories for greenhouse gas emissions, it is unlikely that global temperature increases can be kept within the aimed 2°C range and this could have serious consequences in the future (Meinshausen *et al.* 2009; van Vliet *et al.* 2012). The impact could severely affect water availability, agriculture and food security in Africa (Boko *et al.* 2007; Fung *et al.* 2011; Thornton *et al.* 2011).

Regional predictions of precipitation changes in Africa are still uncertain and vary between regions (Boko *et al.* 2007; Foresight 2011). Moreover, correlations between rainfall and crop yields are not straightforward, especially in sub-humid areas (Zaal *et al.* 2004). Recent studies using statistical, econometric and process-based predictions have found that some areas are at risk of severe declines in crop yields, while others will benefit from climatic changes that may improve conditions for agriculture (Müller *et al.* 2011). Still, of all the continents, the largest drought-induced yield decreases are expected in Africa (Li *et al.* 2009) and the same applies to second-order impacts on food prices and poverty (Hertel *et al.* 2010). While the focus of these studies is mostly on increased drought risk, higher flood risks can be equally devastating (Tschakert *et al.* 2010; Di Baldassarre *et al.* 2010) and temperature rises are also of concern due to crop sensitivities to heat and increased evaporation (Müller *et al.* 2011; Roudier *et al.* 2011). The most serious impact of climate change can be expected in regions that are currently situated at the limits of crop cultivation. In an area currently populated by 35 million people in the West African Sahel and parts of East and Southeast Africa, crop cultivation is projected to no longer be viable in the future (Jones & Thornton 2009). If these projections became reality, such areas would become increasingly uninhabitable and anyone living there would face the risk of crossing interacting boundaries of biophysical impacts of climate change and the social dimensions of adaptation limits (Adger *et al.* 2009; O'Brien & Wolf 2010; Warner *et al.* 2013).

While natural and human systems will continually change and adjust to stressors (Adger *et al.* 2003), adaptation deficits, barriers and limits exist and can undermine livelihood sustainability, impede sustainable development and compromise human well-being (Burton 2009; Preston *et al.* 2013; Kates *et al.* 2012). A key question, and also the focus of this chapter, is what patterns of loss and damage emerge in human systems, such as African agriculture and rural livelihoods, when efforts to adjust are insufficient?

The following definition of loss and damage was used for the community-based field research discussed in this article: loss and damage refers to the negative effects of climate variability and climate change that people have not been able to cope with or adapt to.

Households incur loss and damage when they are not able to respond adequately to climate-related stressors and when adaptation and coping measures have costs or adverse side-effects. Costs and effects can be both economic and non-economic. The underlying causes of loss and damage are social vulnerability – exposure to climatic stressors and lack of adaptive capacity – but are also related to inadequate efforts to reduce global emissions of greenhouse gases. Losses and damage from climate change depend to a large extent on the intensity of climatic disruption that, in turn, depends on global mitigation efforts.

The empirical research reported in this chapter aims to enhance an understanding of how the interaction of climate-related stressors with agricultural livelihoods and social vulnerability is creating particular patterns of loss and damage today in rural African communities. The case studies gathered data in four research domains:

(i) ***Climate stressors:*** Manifestations of climate variability and climate change that may involve either extreme weather-related events or more gradual changes. The focus in these case studies was on drought, floods and longer-term climatic changes, particularly in rainfall patterns.

(ii) ***Societal impact:*** The adverse effects of climate-related stressors on people's food production, income sources, health, physical assets and other aspects of well-being depend to a large extent on vulnerability, which is determined by exposure to climatic stressors and coping capacity. Special attention is devoted to the impact on agriculture in this chapter.

(iii) ***Responses:*** What is being done to cope with the impact of drought and floods? And what is being done to adapt to changing rainfall patterns and changes in drought and flood risk? The study explicitly distinguishes 'coping' from 'adaptation', as they are different types of responses to different types of stressors (van der Geest 2004; Birkmann 2011). Coping strategies are short-term responses to the impact of specific events, such as drought and floods. Adaptation involves longer-term responses to changing conditions. Adaptation measures are often adopted in response to a mix of climatic and non-climatic changes (Moser & Ekstrom 2010) and can be proactive and reactive (Berrang-Ford *et al.* 2011).

(iv) ***(Residual) loss and damage:*** These are the negative effects of climatic stressors that people have not been able to cope with or adapt to. Loss and damage occurs when people cannot respond to climate stressors at all; if their coping and adaptation measures are not efficient enough; if these measures have costs; and if they have adverse long-term effects on livelihood sustainability (Warner & van der Geest 2013). Costs and adverse effects include non-economic losses, which often elude quantification but can cause severe harm to human well-being and undermine sustainable development (Morrissey & Oliver-Smith 2013).

Though data were gathered on specific climatic events (drought and floods) as well as slow-onset changes (in rainfall patterns), this chapter focuses primarily on the impact of drought and floods on livelihoods and the coping strategies adopted in response. In line with the overarching theme of this volume, additional attention is given to agricultural impacts and adaptations.

Research Methods

The research presented here generated primary data providing a people's perspective on loss and damage. Research was conducted in five African countries – Burkina Faso, Ethiopia, the Gambia, Kenya and Mozambique – in districts that are vulnerable to the impact of climate-related stressors. National research teams from each country gathered quantitative and qualitative data in household surveys (n = 1973) and during more than 100 focus-group discussions, expert consultations and in-depth interviews with some of the people affected. Researchers also compared meteorological data with local perceptions of climatic stressors. Table 13.1 provides an overview of the study areas and climate stressors on which each of the case studies focused.

The survey instrument (sample size: 304 to 465 households per case study) assessed the impact of drought and floods on crops, livestock, other livelihood activities and assets; household strategies to deal with these; and the extent to which the measures adopted were successful in avoiding loss and damage. A template questionnaire was designed and later adapted for each case study to suit the characteristics of local livelihood systems and environments. The first section of the survey instrument enquired into household socioeconomic and demographic characteristics and sources of food and income. The information gathered in this section aimed to identify what household characteristics contribute to vulnerability and resilience to droughts and flooding. The following

TABLE 13.1 *Overview of the case studies: research area, climate stressors and sample size*

Country	District/Region	Climate-related stressor	Sample size
Burkina Faso	Sahel	Drought	465
Ethiopia	Gambella	Flooding	431
Gambia	North Bank	Drought	373
Kenya	Budalangi	Flooding	400
Mozambique	South & Central	Drought and floods	304

sections dealt with extreme weather events and slow-onset processes with a focus on impact, coping, adaptation and residual loss and damage. Open questions were combined with closed questions to optimize the balance between listening to the voices of rural people and being able to quantify how widespread different impacts and coping and adaptation responses were.

The selection of study sites was based on criteria such as the presence of climatic stressors, their predominantly agricultural livelihoods and access to communities through existing contacts. Within the study sites, the survey households were selected randomly using different sampling techniques based on the local situation. Households in Burkina Faso, The Gambia and Mozambique were selected randomly from a list of all households provided by local authorities and spatial sampling methods were used in Kenya and Ethiopia based on survey maps and community maps.

The research objective and methods have several limitations. Firstly, the attribution of local climatic changes and extreme events to global warming[2] is beyond the scope of this research. Climate systems in Africa have always been characterized by occasional droughts and floods are not just caused by heavy rainfall but also by local human drivers, upstream land conversion and deforestation. While this research on climate-related loss and damage contributes to knowledge-building on the impact of climate change and limits to adaptation, the climatic stressors examined could also have occurred without anthropogenic global warming. Secondly, no attempt was made to estimate total losses and damages in monetary terms at local, national or global scale. And finally, the local case studies are not representative of entire countries. These studies should be treated as a point of departure for further research on loss and damage in vulnerable communities.

2 See Huggel *et al.* (2013) and Wrathall *et al.* (in press) for a discussion of 'the attribution problem'.

Overview of Survey Findings

This section considers the quantitative findings from the questionnaire survey and the following one discusses three 'loss and damage pathways', which are illustrated with examples from the field.

The survey findings reveal that the impacts of drought and flooding are widely experienced in the African study sites. The proportion of households reporting adverse effects on their household economy was very high (>97%, see Table 13.2). The questionnaire asked specifically about impacts in eight different domains: crop cultivation, livestock keeping, fishing, trees, trade, food prices, houses/properties and 'other'. The three areas that were most affected in each study site are listed in Table 13.2. In all five study sites, the main domain impacted was crop cultivation, with over 90% of households affected in each site. In four out of the five research areas, food prices came second. Most respondents were small-scale farmers. Although over 50% of the households also engaged in non-farm activities, this mostly involved low-yielding activities such as petty trade. Given their low cash incomes, households that experienced harvest failure often had difficulty accessing food, especially as food prices rise in the aftermath of a drought or floods, which further aggravates the problem.

In the study areas that had experienced droughts, severe impacts on livestock were common and many respondents reported damage to physical assets, particularly houses, in the flood-affected research areas. The Ethiopian

TABLE 13.2 *Impacts of climate-related stressors (% of households)*

Country	Climate stressor	Experienced impact (%)	Impact sector 1 (%)	Impact sector 2 (%)	Impact sector 3 (%)
Burkina Faso	Drought	97	Crops (96)	Food prices (90)	Livestock (87)
Ethiopia	Flood	100	Crops (94)	Health (92)	Assets (79)
Gambia	Drought	100	Crops (98)	Food prices (89)	Livestock (75)
Kenya	Flood	98	Crops (98)	Food prices (95)	Physical assets (66)
Mozambique	Both	99	Crops (100)	Food prices (83)	Livestock (35)

Note: The percentages for 'impact sector' (Columns 4, 5 & 6) were calculated for the households that experienced impacts (Column 3).

case study, which looked at floods, added an impact domain to the questionnaire, namely health. This was a useful addition as 92% of households affected reported health impacts, particularly water-borne diseases (Haile *et al.* 2013).

For each impact domain, respondents were asked – in an open question – to specify how droughts or floods had affected their households. In the case of agriculture, the impact process differed between droughts and floods but the outcome was the same: loss of harvest as a first-order impact, followed by second-order impacts such as food insecurity, low income, a lack of seeds, impoverishment and indebtedness. With regard to the biophysical impact process, crops were mostly swept away, with farmers often losing their entire harvest in flood-affected study areas. Less commonly, farmers reported that field crops rotted after standing in flood water for too long. One farmer in Budalangi District in Kenya complained because he had, for the first time ever, invested in expensive maize seeds and fertilizer for his farm. He did not therefore only lose his harvest but also a major investment for which he had taken out a loan. With no income from his farm and needing to buy food for his family against soaring prices, this farmer had clearly entered a cycle of indebtedness and impoverishment.

In the case of drought, total harvest loss was less common. Participants in focus-group discussions in The Gambia explained that total crop failures were extremely rare there because of risk spreading in traditional subsistence agriculture. When crops planted in upland areas fail because of long dry spells, lowland fields may still produce a harvest. Farmers vary not only the locations of their fields but also crop types, sowing dates and seedbed types. These measures help reduce the impact of drought on agricultural production but are not enough to avoid very poor yields and associated second-order impacts.

The majority of survey respondents across the study sites (93%) indicated that they adopted coping measures to deal with the impact of droughts or floods (see Table 13.3). In the flood-affected study areas in Ethiopia and Kenya, many households tried to cope by relying on aid and support from the government and NGOs, while households in areas affected by drought in Burkina Faso and The Gambia tried to cope by selling livestock so they could buy food when their harvests failed. Across all the study areas, households tried to cope by relying on other sources of (mostly non-farm) income to buy food when their primary, more climate-sensitive livelihood activity was affected by drought or floods. Reliance on social networks was also a key element in people's portfolios of coping options. Migrants were often the people who provided financial support for food and other essentials when their relatives at home were hit by a drought or floods. Using migration as a coping measure was least common, but still in 23% of the sampled households at least one person migrated to cope with the impact of climate-related stressors. They moved to safer areas or places that provided opportunities to earn an income so they could buy food when their own harvests failed. In the Kenyan case study, a much higher proportion of households (62%) migrated, mostly short distances to camps or to relatives living on higher ground, in the aftermath of floods.

In the questionnaire, household members were also asked whether they had modified their food consumption pattern in response to drought or flood impacts. When designing the questionnaire, this was considered a coping strategy but one could question whether reducing one's number of meals per day or portion sizes is really a 'coping' strategy or a sign that all the other available options have failed. The number of households that indicated they had to modify food consumption was very high (78.5%), which indicates that other coping strategies were not effective enough to avoid food insecurity.

Among those who adopted coping measures, most indicated that they were not successful in preventing residual impacts (72%, see Table 13.3). This was either because the measures were insufficient or because they had costs or other negative consequences attached. In four of the five study sites, the figure was between 66% and 72%. The situation in Ethiopia was even more severe: up to 96% of households reported that their coping measures in response to flooding had not prevented residual losses and damage.

TABLE 13.3 *Coping measures and residual impacts (% of households)*

	Adopted measures (%)	Relied on other income (%)	Relied on aid (%)	Sold assets (%)	Relied on social network (%)	Household or members migrated (%)	Had residual impact (%)
Burkina Faso	79	33	51	79	29	41	72
Ethiopia	98	58	76	42	50	18	96
Gambia	97	57	49	56	48	23	67
Kenya	98	40	80	20	34	62	73
Mozambique	93	67	45	34	31	12	69
Median	93	57	51	42	34	23	72

Notes: Percentages for specific coping measures (Columns 3, 4, 5, 6 & 7) were calculated for households that adopted at least one coping measure (Column 2). The last column ('residual impact') shows the percentage of households whose coping measures did not prevent loss and damage.

Figure 13.1 compares the ability of adopters and non-adopters of different coping strategies to avoid residual loss and damage. Only the data for Kenya and The Gambia are shown.[3] The diagram should not be interpreted as an evaluation of the effectiveness of each of the coping strategies because respondents were asked only if the combination of coping strategies they adopted had helped them to avoid residual impacts. It does, however, show that certain coping strategies are associated with success in avoiding loss and damage and others with failure. For example, households that relied on food aid and other types of relief were more likely to incur residual loss and damage. A negative interpretation would be that government and/or NGO support to affected households did not work well or had adverse side-effects. On the other hand, it could equally be that food aid and other types of relief were used in the most dire cases where the stressor and impacts were most severe. Households that migrated or that sent individual household members away were also less likely to avoid loss and damage, but the difference was quite small and not significant statistically.

Households that tried to cope with drought and flood impacts on their main source of food and income (mostly crop cultivation) by relying on other sources of income were most likely to avoid loss and damage (though still not more than about 33%). Selling assets, mostly livestock, was also positively associated with success in avoiding loss and damage As highlighted earlier, only findings from Kenya and The Gambia are presented in figure 13.1 and Table 13.4.

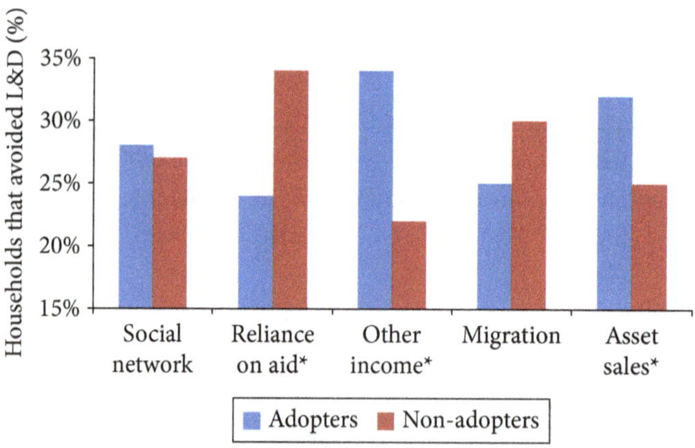

FIGURE 13.1 *Ability to avoid loss and damage in The Gambia and Kenya (by coping measure).*
Note: *indicates significant differences ($p < 0.05$) between adopters and non-adopters; Figure based on data from The Gambia and Kenya only.

3 The three other case studies – in Burkina Faso, Ethiopia and Mozambique – were funded by the African Climate Policy Centre and the authors of this chapter had only limited rights to use the data generated by these studies.

TABLE 13.4 *Uptake of adaptation measures in The Gambia and Kenya*

	Any adaptation (%)	Agricultural change (%)	Livelihood diversification (%)	Migration (%)	Other (%)
Gambia	77	55	47	23	8
Kenya	77	42	38	47	8

Besides *ex-post* coping measures, the study also inquired into *ex-ante* preventive measures and adaptation to (perceived) changes in drought and flood risk. Only findings from Kenya and The Gambia are presented here. In both study sites, 77% of the surveyed households adopted such measures. Most adaptation and *ex-ante* risk management occurred in agriculture, followed by livelihood diversification into non-farm activities and migration. Agricultural adaptations primarily involved planting early maturing, drought-resistant or water-tolerant crop cultivars, shifting to horticulture in small irrigated gardens and going into fruit-tree cultivation. Respondents in Kenya also mentioned that they shifted the location of their farms to reduce the risk of harvest losses.

Figure 13.2 compares the ability of adopters and non-adopters of different adaptation measures to avoid residual loss and damage. Separate figures for The

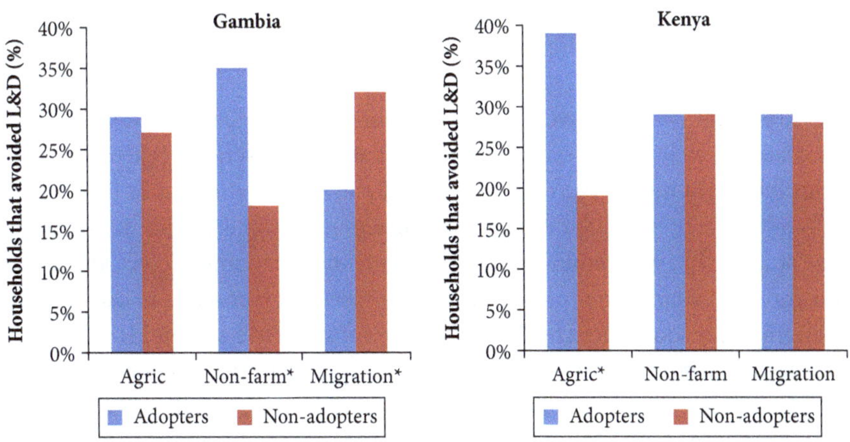

FIGURE 13.2 *Ability to avoid loss and damage in The Gambia and Kenya (by type of adaptation measure).*
Note: *indicates significant differences ($p < 0.05$) between adopters and non-adopters.

Gambia and Kenya are shown because the findings differ markedly in the two case-study areas. Households in The Gambia that were most likely to avoid loss and damage from drought adapted by diversifying into non-farm activities. By contrast, households that relied on migration were less likely to avoid loss and damage. No significant ($p < 0.05$) differences in ability to avoid loss and damage were found between adopters and non-adopters of agricultural adaptation measures. In the Kenyan case study, households that adapted by implementing agricultural changes were most 'successful' in avoiding loss and damage.

The findings on adaptation highlight the importance of local assessments of household strategies in dealing with climatic stressors. What works in one place does not necessarily work in another. The next section presents more qualitative and site-specific findings on how households incur loss and damage.

Pathways to Loss and Damage

The community-based case studies in the Loss and Damage in Vulnerable Country Initiative revealed four pathways where vulnerable people incur loss and damage (Warner & van der Geest 2013). Despite efforts to adapt and cope with the impact of drought and floods, many households across the research sites were not able to avoid adverse effects. Residual impacts occurred when:

- measures to cope or adapt were *insufficient* to avoid loss and damage
- measures involved economic, social or cultural *costs* that were not regained
- measures were *erosive* in the longer term and made people more vulnerable
- *no measures* were adopted at all.

The first three pathways were most common in the African case studies. The fourth pathway was less common as almost all the households (93%) adopted at least one coping measure to deal with the impact of drought or floods and 77% adopted preventive or adaptive measures. In some of the non-African case studies that were conducted in the context of the same project, the focus was more on adaptation to slow-onset processes, such as a rise in sea level or changing monsoon patterns. In these locations, more households indicated that they had not adopted any coping or adaptation measures because they simply did not know what to do or because they lacked the resources to respond adequately (Rabbani *et al.* 2013; Monnereau & Abraham 2013; Kusters & Wangdi 2013).

Measures are Not Enough to Avoid Loss and Damage

The vast majority of households across the five study sites adopted coping and adaptation measures to prevent or deal with the impact of drought and floods. However, these measures were often not enough to avoid loss and damage to household economies, livelihoods, health, and cultural assets. Examples from the five case studies are given below to demonstrate how the loss and damage pathways play out in real life.[4]

The case study in The Gambia (Yaffa 2013) looked at the drought-prone area of the North Bank Region. Long-term rainfall data from the area follow a similar pattern as in other parts of the Sahel that suffered severe consecutive droughts in the 1970s and 1980s, followed by a partial recovery with high variability and occasional years of drought (Hulme 2001; Dietz *et al.* 2004). The North Bank Region experienced a severe drought in 2011 with farmers seeing their yields decimated. Of the households surveyed, 98% were affected by drought despite the preventive measures they had adopted, such as planting drought-resistant crop cultivars and taking measures to spread the risk of crop failure, for example by growing different crops, spreading sowing dates over a longer period of time and farming upland as well as lowland fields. The most common way to cope with harvest failure was to 'look for money' to buy food, either through local non-farm activities or by sending household members to urban centres to engage in informal economic activities. However, due to scarce supplies, food prices increased in the aftermath of the drought and, with many households in the same situation, money was scarce, which made it more difficult to cope. About half of the households surveyed received some food aid and many sold livestock to buy grain. For most households, these measures were only partly successful: 64% of respondents reported that despite all their efforts to cope with the drought and harvest losses, they could not access enough food to feed their households. Buying cheaper, less nutritious food was one way of dealing with this situation but most households had to reduce their portion sizes or the number of meals they had as well (Yaffa 2013).

Haile *et al.* (2013) studied loss and damage due to flooding in the Gambella Region in Ethiopia. The research area floods very frequently but the degree of flooding is usually moderate. This case study paid special attention to the measures households adopted to prevent flood impacts and assessed their effectiveness in preventing loss and damage. Most commonly, households dug ditches or erected boundary walls around their houses and farms to keep the

4 Some of these examples were used in Warner *et al.* (2012b, 2013).

water out. Others tried to prevent any impact by moving household property, and especially livestock, to higher grounds. The study found that these preventive measures minimized flood impacts in most years when floods were moderate. However, when the area was hit by extremely high water levels in 2007, virtually none of the households were able to avoid losses and damage to their farms, houses, property, livestock and granaries (Haile *et al.* 2013).

Measures have Costs that are Not Regained

The measures that households adopt to cope with the impacts of extreme weather events often have costs themselves, which can be either monetary or non-monetary.

In the study areas in Mozambique, households had to deal with the double threat of drought and floods (Brida *et al.* 2013). Many of the households surveyed had been resettled from lowland areas to higher ground after severe flooding in 2001 and 2007. This adaptation made them less vulnerable to rising water levels but it came at significant cost. In the upland areas, many people are unable to produce enough food for their families because the upland soils are less fertile and crop yields are thus significantly lower. Some households moved their fields back to more fertile lowland areas, while still living in upland areas as this gives them better yields, but valuable time and energy are then wasted commuting the large distances between the upland and lowland areas. Moreover, when a flood hits, households still risk losing their entire harvest from their lowland farms (Brida *et al.* 2013).

The study area in Burkina Faso showed clear signs of cultural losses. Many former pastoralists have lost their herds in recurrent droughts and have since had to take up other occupations. Besides any material losses, the loss of their cattle is shameful for many of these former herdsmen and ceasing to be a herder brings a loss of cultural identity and lifestyle. Their experience and skills are highly specialized and learning to make a living in another profession and building up a new life and livelihood in an urban centre can often be a challenge for them (Traore *et al.* 2013).

Erosive Coping: Short-Term Merits, Adverse Long-Term Effects

Coping measures are 'erosive' when they jeopardize future food and livelihood sustainability (de Waal 1989; van der Geest & Dietz 2004). Households that have to deal with the impacts of climatic stressors will prefer to first adopt 'non-erosive' coping measures, such as drawing on buffers. When non-erosive coping options are not – or are no longer – available, people are forced to adopt erosive measures, such as selling productive assets, consuming seeds that have been reserved for the next agricultural cycle or taking their children out of school so they can find work to support the household. Across the five African case study sites, researchers found that many households were being forced to employ erosive coping strategies that would enable them to survive in the short term but that ultimately make them more vulnerable in the longer term.

Opondo (2013) examined the coping strategies that households adopted after the River Nzoia in western Kenya broke its banks in December 2011 and caused havoc in Budalangi Division, a low-lying area on the shores of Lake Victoria. Floods have become more frequent and intense in recent decades but those in December 2011 led to people and livestock drowning, crops being washed away and severe damage to houses, and they also spawned an outbreak of water-borne diseases. Some of the coping strategies that people adopted to gain access to food or to reconstruct their houses have had severe implications for their future livelihood security. For example, some households were forced to sell their bullocks to buy food after the floods washed away their crops. The following season, these bullocks were then not available to plough their fields and the households' situation became even more precarious. And able-bodied household members had to invest a great deal of their time in non-farm activities or migration to urban centres to resolve their urgent food needs. This meant that they had much less time to spend on their household farm, which undermined food security in the next cultivation cycle. In addition, some households withdrew their children from school so they could help generate income, mostly through non-farm activities.

The case study in The Gambia, which focused on the impact of droughts, yielded similar findings (Yaffa 2013) and in Ethiopia it was found that when preventive measures are not enough to avoid severe impacts, households often rely on their social networks for survival. However, repeated floods erode this social capital, overburden social networks and leave affected households in a more vulnerable position with each subsequent flood (Haile *et al.* 2013).

Conclusions and Policy Implications

This study defined loss and damage as the 'negative effects of climate variability and climate change that people have not been able to cope with or adapt to'. The findings presented in this chapter show that loss and damage from climate-related stressors is already a reality for rural households in Africa. Almost all the households that were interviewed for this research (N = 1973) had adopted coping strategies to deal with the impact of drought and floods. In over 70% of these households, their efforts were, however, not sufficiently effective to avoid loss and damage.

The case-study findings suggest that some of the most notable current impacts are on household food production and livelihoods. A minority of households in these study areas have been able to cope with the impacts of droughts and floods and to adapt to climatic changes in ways that enhance their resilience. Of concern, however, are the numerous vulnerable households that have not been able to cope and adapt successfully. To survive when a drought or flood strikes, they have often been forced to use erosive coping measures that can trap them in a downward spiral of declining well-being and security. Such households are experiencing diminishing coping and adaptive capacity, and increasing losses and damages.

The chapter went a step beyond earlier research by evaluating which household coping and adaptation measures are associated with success and failure when it comes to avoiding loss and damage. Levels of success were generally low, but households that adapted through agricultural adaptations, such as planting drought- or water-tolerant crop varieties or by diversifying their livelihoods with non-farm activities, were significantly more likely to avoid loss and damage. Migration, on the other hand, was associated with failure to avoid loss and damage.

Vulnerable countries, like those studied in this research, are at the frontline of loss and damage realities. This chapter originated from a first generation of local-level research that is systematically assessing loss and damage from climate-related stressors. The case studies serve as a point of departure for further research to understand how climate variability and climate change are affecting vulnerable households today, and the consequences of adaptation shortfalls for livelihood security, sustainable development and human well-being in the future.

References

Adger, W.N., S. Huq, K. Brown, D. Conway & M. Hulme (2003), Adaptation to Climate Change in the Developing World. *Progress in Development Studies* 3(3): 179–195.

Adger, W.N., I. Lorenzoni & K. O'Brien (eds) (2009), *Adapting to Climate Change: Thresholds, Values, Governance*. Cambridge: Cambridge University Press.

Berrang-Ford, L., J.D. Ford & J. Paterson (2011), Are We Adapting to Climate Change? *Global Environmental Change* 21: 25–33.

Birkmann, J. (2011), First- and Second-Order Adaptation to Natural Hazards and Extreme Events in the Context of Climate Change. *Natural Hazards* 58(2): 811–840.

Boko, M., I. Niang, A. Nyong, C. Vogel, A. Githeko, M. Medany, B. Osman-Elasha, R. Tabo & P. Yanda (2007), Africa. In: M.L. Parry, O.F. Canziani, J.P. Palutikof, P.J. van der Linden & C.E. Hanson (eds), *Climate Change 2007: Impacts, Adaptation and Vulnerability. Contribution of Working Group II to the Fourth Assessment Report of the Intergovernmental Panel on Climate Change*. Cambridge: Cambridge University Press, pp. 433–467.

Brida, A.B., T. Owiyo & Y. Sokona (2013), Loss and Damage from the Double Blow of Flood and Drought in Mozambique. *International Journal of Global Warming* 5(4): 514–531.

Burton, I. (2009), Climate Change and the Adaptation Deficit. In: E.L.F. Schipper & I. Burton (eds), *Earthscan Reader on Adaptation to Climate Change*. London: Earthscan, pp. 89–95.

de Waal, A. (1989), Is Famine Relief Irrelevant to Rural People?. *IDS Bulletin* 20(2): 63–67.

Di Baldassarre, G., A. Montanari, H. Lins, D. Koutsoyiannis, L. Brandimarte & G. Blöschl (2010), Flood Fatalities in Africa: From Diagnosis to Mitigation. *Geophysical Research Letters* 37(22): 402.

Dietz, T., R. Rueben & J. Verhagen (eds) (2004), *The Impact of Climate Change on Drylands*. Dordrecht: Kluwer.

Foresight (2011), 'Migration and Global Environmental Change, Final Project Report'. London: The Government Office for Science.

Fung, R., A. Lopez & M. New (2011), Water Availability in +2°C and +4°C Worlds. *Philosophical Transactions of the Royal Society* 369(1934): 99–116.

Haile, A.T., N. Wagesho & K. Kusters (2013), Loss and Damage from Flooding in the Gambela Region, Ethiopia. *International Journal of Global Warming* 5(4): 483–497.

Hertel, T., M. Burke & D. Lobell (2010), The Poverty Implications of Climate-induced Crop Yield Changes by 2030. *Global Environmental Change* 20: 575–585.

Huggel, C., D. Stone, M. Auffhammer & G. Hansen (2013), Loss and Damage Attribution, *Nature Climate Change* 3(8): 694–696.

Hulme, M. (2001), Climatic Perspectives on Sahelian Desiccation: 1973–1998. *Global Environmental Change* 11(1): 19–29.

Jones, P.G. & P.K. Thornton (2009), Croppers to Livestock Keepers: Livelihood Transitions to 2050 in Africa Due to Climate Change. *Environmental Science & Policy* 12(4): 427–437.

Kates, R.W., W.R. Travis & T.J. Wilbanks (2012), Transformational Adaptation When Incremental Adaptations to Climate Change are Insufficient. *PNAS* 109(19): 7156–7161.

Kusters, K. & N. Wangdi (2013), The Costs of Adaptation: Changes in Water Availability and Farmers' Responses in Punakha District, Bhutan. *International Journal of Global Warming* 5(4): 387–399.

Li, Y.P., W. Ye, M. Wang & X.D. Yan (2009), Climate Change and Drought: A Risk Assessment of Crop-Yield Impacts. *Climate Research* 39: 31–46.

Meinshausen, M., N. Meinshausen, W. Hare, S.C.B. Raper, K. Frieler, R. Knutti, D.J. Frame & M.R. Allen (2009), Greenhouse-gas Emission Targets for Limiting Global Warming to 2°C. *Nature* 458(7242): 1158–1163.

Monnereau, I. & S. Abraham (2013), Limits to Autonomous Adaptation in Response to Coastal Erosion in Kosrae, Micronesia. *International Journal of Global Warming* 5(4): 416–432.

Morrissey, J. & A. Oliver-Smith (2013), Perspectives on Non-economic Loss and Damage: Understanding Values at Risk from Climate Change. http://www.lossanddamage.net/4933. Accessed 8 September 2014.

Moser, S.C. & J.A. Ekstrom (2010), A Framework to Diagnose Barriers to Climate Change Adaptation. *PNAS* 107(51): 22026–22031.

Müller, C., W. Cramer, W. Hare & H. Lotze-Kampen (2011), Climate Change Risks for African Agriculture. *PNAS* 108(11): 4313–4315.

O'Brien, K. & J.A. Wolf (2010), A Values-Based Approach to Vulnerability and Adaptation to Climate Change. *WIREs Climate Change* 1(2): 232–242.

Opondo, D. (2013), Erosive Coping after the 2011 Floods in Kenya. *International Journal of Global Warming* 5(4): 452–466.

Preston, B.L., K. Dow & F. Berkhout (2013), The Climate Adaptation Frontier. *Sustainability* 5(3): 1011–1035.

Rabbani, G., A. Rahman & K. Mainuddin (2013), Salinity Induced Loss and Damage to Farming Households in Coastal Bangladesh. *International Journal of Global Warming* 5(4): 400–415.

Roudier, P., B. Sultan, P. Quirion & A. Berg (2011), The Impact of Future Climate Change on West African Crop Yields: What Does the Recent Literature Say? *Global Environmental Change* 21: 1073–1083.

Thornton, P.K., P.G. Jones, P.J. Ericksen & A.J. Challinor (2011), Agriculture and Food Systems in Sub-Saharan Africa in a 4°C World. *Philosophical Transactions of the Royal Society* 369(1934): 117–136.

Traore, S., T. Owiyo & Y. Sokona (2013), Dirty Drought Causing Loss and Damage in Northern Burkina Faso. *International Journal of Global Warming* 5(4): 498–513.

Tschakert, P., R. Sagoe, G. Ofori-Darko & S.N. Codjoe (2010), Floods in the Sahel: An Analysis of Anomalies, Memory, and Anticipatory Learning. *Climatic Change* 10(3–4): 471–502.

van der Geest, K. (2004), 'We're Managing!' Climate Change and Livelihood Vulnerability in Northwest Ghana. Leiden: African Studies Centre.

van der Geest, K. & T. Dietz (2004), A Literature Survey about Risk and Vulnerability in Drylands, with a Focus on the Sahel. In: T. Dietz, R. Rueben & J. Verhagen (eds), *The Impact of Climate Change on Drylands*. Dordrecht: Kluwer, pp. 117–146.

van Vliet, J., M. van der Berg, M. Schaeffer, D.P. van Vuuren, M. den Elzen, A.F. Hof, A.M. Beltran & M. Meinshausen (2012), Copenhagen Accord Pledges Imply Higher Costs for Staying below 2°C Warming. *Climatic Change* 113(2): 551–561.

Warner, K., S. Kreft, M. Zissener, P. Höppe, C. Bals, T. Loster, J. Linnerooth-Bayer, S. Tschudi, U. Gurenko, A. Haas, S. Young, P. Kovacs, A. Dlugolecki & A. Oxley (2012a), Insurance Solutions in the Context of Climate Change-related Loss and Damage. UNU-EHS Policy Brief No. 6. Bonn: United Nations University Institute for Environment and Human Security.

Warner, K. & K. van der Geest (2013), Loss and Damage from Climate Change: Local-level Evidence from Nine Vulnerable Countries. *International Journal of Global Warming* 5(4): 367–386.

Warner, K., K. van der Geest & S. Kreft (2013), Pushed to the Limits: Evidence of Climate Change-related Loss and Damage when People Face Constraints and Limits to Adaptation. UNU-EHS) Report No. 11. Bonn: United Nations University Institute for Environment and Human Security.

Warner, K., K. van der Geest, S. Kreft, S. Huq, K. Kusters & A. de Sherbinin (2012b), Evidence from the Frontlines of Climate Change: Loss and Damage to Communities Despite Coping and Adaptation. Loss and Damage in Vulnerable Countries Initiative. UNU-EHS Policy Report No. 9. Bonn: United Nations University Institute for Environment and Human Security.

Wrathall, D., A. Oliver-Smith, A. Fekete, E. Gencer, M. Lepana & P. Sakdapolrak (in press), Problematising Loss and Damage, *International Journal of Global Warming*.

Yaffa, S. (2013), Coping Measures Not Enough to Avoid Loss and Damage from Drought in the North Bank Region of the Gambia. *International Journal of Global Warming* 5(4): 467–682.

Zaal, F., T. Dietz, J. Brons, K. van der Geest & E. Ofori Sarpong (2004), Sahelian Livelihoods on the Rebound: A Critical Analysis of Rainfall, Drought Index and Yields in Sahelian Agriculture. In: T. Dietz, R. Rueben & J. Verhagen (eds), *The Impact of Climate Change on Drylands*. Dordrecht: Kluwer, pp. 61–78.

CHAPTER 14

Agriculture and Nutrition

Linkages and Complementarities

Inge D. Brouwer

Malnutrition remains a pervasive public-health concern in many low- and middle-income countries and has long-lasting and often irreversible consequences for health, cognitive ability, school performance and future earnings. To further reduce malnutrition, traditional nutrition-specific interventions should be aligned with so-called nutrition-sensitive activities in food security, agriculture, water and sanitation, and social protection/safety nets. Scientifically, there is still a large gap of evidence of efficacy of nutrition-sensitive agriculture, which emphasizes the need for well-designed, rigorous studies to test pathways and measure impact using valid, preferably simple indicators of food and nutrient intake.

Poor nutritional status and limited financial resources often compromise individual welfare in the developing world. In combination with a high concurrent disease load, a self-perpetuating cycle of poverty, malnutrition and mortality often arises. Adequate nutrition is a human right that underpins progress towards most of the UN's Millennium Development Goals (MDG) in an attempt to reduce child mortality, improve maternal health and decrease the burden of malaria, HIV/AIDS and other diseases by 2015. It will also affect the achievement of future Sustainable Development Goals beyond 2015 if it is not addressed properly. Although the proportion of undernourished people in developing countries is decreasing, the reduction is modest and more than 800 m people are still undernourished (FAO 2013).

Maternal and child malnutrition remains a pervasive and damaging public-health concern in many low- and middle-income countries. To date, it is estimated that globally over 165 m children under five years of age are stunted (too short for their age) and about 52 m children are too thin (wasted) and thus require special treatment (Black et al. 2013). The prevalence of low BMI among adult women (<18.5 kg/m^2), which is indicative of maternal undernutrition, remains higher than 10% in Africa and Asia (Black et al. 2013). Over the past fifteen years, rates of malnutrition have decreased but this improvement is stagnating, especially in Sub-Saharan Africa where the absolute number of undernourished people continues to rise (Stevens et al. 2012).

 | DOI 10.1163/9789004282698_015

In addition to undernourishment, stunting and wasting, there is rising concern about the quality of diets in terms of adequate micronutrients in developing countries. Hidden hunger, which involves a lack of micronutrients, affects a far greater percentage of the world's population than overt hunger due to a mere lack of food. Worldwide, the most common micronutrient deficiencies are of iron, zinc and vitamin A, but deficiencies of iodine, folate and B12 and other B vitamins are also widespread. Over a quarter of the world's population is anaemic, including 293 m children younger than five years of age (47%) and 486 m non-pregnant women (30%) (Balarajan et al. 2011; McLean et al. 2008). Africa has the highest prevalence of anaemia: 65% and 44% of preschool children and non-pregnant women are anaemic (McLean et al. 2008). Iron deficiency is considered the major cause of anaemia, accounting for around 50% of cases (Ezzati et al. 2004) and has adverse health effects on the outcome of pregnancies, infant growth, cognitive performance, psychomotor development, immune status and work capacity. It also results in substantial economic costs related to impaired school performance and decreased productivity (Zimmermann & Hurrell 2007). Anaemia is a contributing factor in over 20% of all maternal deaths among pregnant women.

As iron and zinc are found in many of the same foods, high rates of iron deficiency in Sub-Saharan Africa suggest widespread cases of zinc deficiency in these populations. Zinc is required by about 50 enzymes in the body and many metabolic functions are affected by a deficiency in zinc, including physical growth, immune competence, reproductive function and neural development (Hess et al. 2009). Several supplementation trials among high-risk young children, including some carried out in Africa, have shown that an increased zinc intake reduces the burden of diarrhoea and acute lower respiratory infections (Zinc Investigators' Collaborative Group et al. 1999, 2000) although the evidence from among populations in malaria-endemic areas in Africa has been inconsistent (Sazawal et al. 2007). Estimates suggest that 17% of the world's population is at risk of zinc deficiency on the basis of an analysis of national diets. The figure ranges from 7.5% in high-income countries to 26% and 30% in Sub-Saharan Africa and Asia respectively (Wessels & Brown 2012). Young children in Sub-Saharan Africa in particular suffer from an inadequate intake of bioavailable zinc, which leads to zinc deficiency as assessed by plasma or serum zinc concentrations. A vitamin A deficiency can be seen in an estimated 190 m preschool-aged children (33%) and 19 m pregnant women (15%) worldwide (WHO 2009). Almost every third child – or about 56 m children – with a vitamin A deficiency lives in Sub-Saharan Africa. The health consequences of a vitamin A deficiency include mild to severe systemic effects on innate and acquired immunity, higher death rates from infectious diseases, xerophthalmia, blindness and increased mortality in general (Sommer and Davidson, 2002). Although clinical signs of vitamin A deficiency are showing a downward trend, subclinical vitamin A deficiency (defined as serum retinol concentration <0.70 µmol/L) that is associated with higher maternal and infant mortality rates seems to be increasing.

At the same time, the prevalence of adult overweight (BMI ≥ 25 kg/m²) and obesity (BMI ≥ 30 kg/m²) has been rising in all regions and has reached more than 40% in Africa (Stevens et al. 2012), while prevalence of overweight in children younger than five years of age increased from 4% in 1990 to 7% in 2011 in Africa and is expected to reach 11% in 2015 (Black et al. 2013). Worldwide, the risk factors to disease burden shifted away from communicable diseases in children associated with undernutrition to non-communicable diseases in adults due to overnutrition. However, in Africa, undernutrition among children associated with poverty remains the dominant contributor to the disease burden that is responsible for over 45% of the total number of deaths among children under the age of five (Lim et al. 2012; Black et al. 2013). Improving nutritional status especially in Sub-Saharan Africa is therefore critical to breaking the present cycle of malnutrition, poverty and mortality.

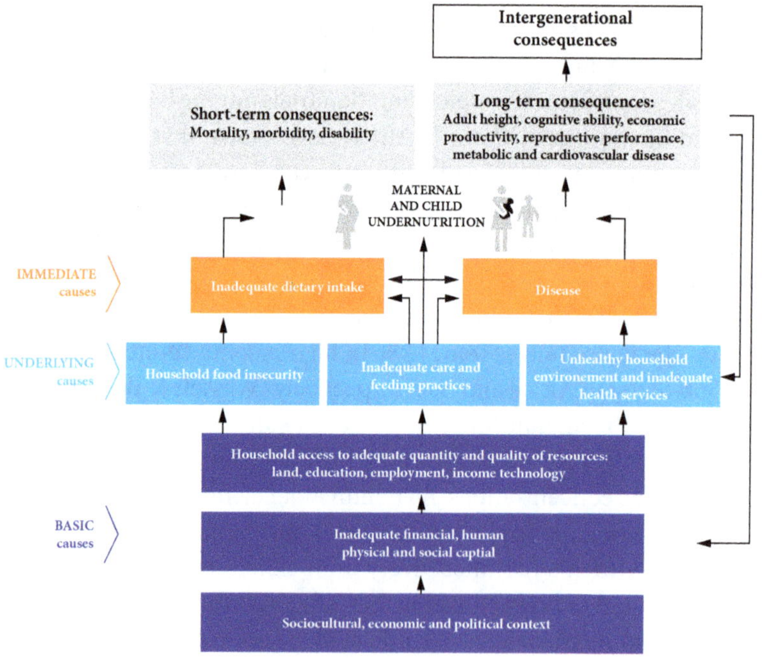

FIGURE 14.1 *Conceptual framework of the determinants of child malnutrition*
SOURCE: UNICEF (2013)

The most vulnerable period is during the first 1,000 days of a child's life, i.e. during pregnancy and into the second year of life (Black et al. 2008, 2013). It is in this period that a child has increased nutritional needs to support his/her rapid growth and development, is more susceptible to infections and has heightened sensitivity to biological programming. In addition, an optimal nutritional status prior to pregnancy is essential among women. This often depends on nutrition and growth during adolescence (Black et al. 2013) and indicates the importance of care before conception occurs (Dean et al. 2013). Undernourished girls have a greater likelihood of becoming undernourished mothers who, in turn, have a greater chance of giving birth to low birth-weight babies (Özaltin et al. 2010). This perpetuates an intergenerational cycle. Sadly, women generally only access health and nutrition-promoting services at the end of the second or third trimester of their pregnancy. Malnutrition during adolescence and in the first 1,000 days of life has a damaging effect at an early age on morbidity leading to mortality and disability (Black et al. 2008, 2013), with long-lasting harmful and often irreversible consequences for cognitive ability, school performance and future earnings that can affect the economic development of whole regions (Victora et al. 2008). As such, undernutrition can trap children, families, communities and even nations in an intergenerational cycle of poor nutrition, illness and poverty.

BOX 14.1. *Nutrition-Specific Interventions and Programmes*

The programmes that can achieve optimum fetal and child nutrition and development include:

- adolescent health and preconception nutrition
- maternal dietary supplementation
- micronutrient supplementation and fortification
- breastfeeding and complementary feeding
- dietary supplementation for children
- dietary diversification
- feeding behaviour and stimulation
- treatment of severe acute malnutrition
- disease prevention and management
- emergency nutrition interventions

Source: Black et al. (2013)

BOX 14.2. *Food and Nutrition Security*

Food security is recognized as essential, but not sufficient, to ensure nutrition and prevent malnutrition. Food security and nutrition security are related concepts but the distinction between the concepts is important because it affects what can be expected from investments aimed at boosting agricultural productivity and promoting global food security worldwide (Ruel 2013). Food security was defined, at the World Food Summit in 1996 and adapted in the 2009 Declaration of the World Summit on Food Security as being 'when people, at all times, have physical, social and economic access to sufficient, safe and nutritious food to meet their dietary needs and food preferences for an active and healthy life'.

The four pillars of food security are availability, access, stability and utilization. This definition entails that food should be available and accessible in sufficient quantity as well as being of sufficient quality, and it should be culturally acceptable and available at all times throughout the year. Building on UNICEF's conceptual framework, good nutrition also needs children to have access to adequate care-giving practices and a safe and clean environment that allows them to stay healthy and utilize the food they eat effectively (Ibid.).

Different terms are used to combine elements of food security and nutrition security, such as 'food and nutrition security', 'food security and nutrition', 'nutrition security' or 'food security'. The lack of consistency in the use of these terms across different disciplines and languages has prevented meaningful discussions on how to best achieve food security and improved nutrition. In 2012 and based on an in-depth review and discussions of the meaning and different uses of the terms, the Committee on World Food Security recommended using the term 'food and nutrition security' as it best reflects the conceptual linkages between food security and nutrition security, while also expressing a single integrated development goal to help guide policy and programmes effectively. The Committee stated that 'food and nutrition security exists when all people at all times have physical, social and economic access to food, which is safe and consumed in sufficient quantity and quality to meet their dietary needs and food preferences, and is supported by an environment of adequate sanitation, health services and care, allowing for a healthy and active life' (CFS 2012).

UNICEF's conceptual framework shows the multifactorial causality of undernutrition (Figure 14.1). At the immediate level, malnutrition is caused by a lack of appropriate food, nutrient intake and disease, whose interaction can create a lethal cycle of worsening illness and deteriorating nutritional status (Black et al. 2008). These are influenced by three broad factors: food, health and care (UNICEF 2013). An optimal nutritional status results when children have access to affordable, diverse and nutrient-rich food; appropriate maternal and child-care practices; adequate health services; and a healthy environment including safe water, sanitation and good hygiene practices. Food, health and care determinants are shaped by social, economic and political conditions, the combination and importance of which differ by country, depending on national and global contexts, resources and governance.

BOX 14.3. *Pathways Where Agriculture Can Affect Nutrition*

1. As a source of food: This is the most direct pathway where household agricultural production translates into consumption (via crops cultivated by the household).
2. As a source of income: either by agricultural workers' wages or by the marketed sales of food produced and how it is actually spent on food.
3. The link between agricultural policy and food prices, involving supply and demand that affects the prices of various marketed food and non-food crops. This, in turn, affects the incomes of net sellers and the ability to ensure household food security (including the quality of the diet) of net buyers.
4. As a source of income for non-food expenditures: the extent to which these are allocated to nutrition-relevant activities, for example, expenditures on health, education and social welfare.
5. Women's socioeconomic status and their ability to influence household decision making and intra-household allocations of food, health and care.
6. Women's ability to manage the care, feeding and health of young children.
7. Women's own nutritional status: if their work-related energy expenditure exceeds their intakes, their dietary diversity is compromised or agricultural practices may be hazardous to their health and, in turn, may affect their nutritional status.

Source: Gillespie et al. (2012) & Ruel et al. (2013)

Progress has been made and many interventions have been implemented at scale. Evidence supporting the effectiveness of nutrition interventions and delivery strategies has grown since The Lancet Series on Maternal and Child Undernutrition in 2008 (Bhutta et al. 2008). These involve mainly nutrition-specific interventions and programmes (see Box 14.1) that address the immediate determinants of fetal and child nutrition and development, focusing on adequate food and nutrient intakes, feeding, care-giving and parenting practices and a low burden of infectious diseases (Ruel et al. 2013). The 2013 Lancet Series on Maternal and Child Nutrition indicated that the current number of deaths among children younger than five years of age could be reduced by 15% and the current stunting rates could drop by 20% if populations could access ten evidence-based nutrition-specific interventions at 90% coverage (Bhutta et al. 2013). The interventions considered were: periconceptional folic acid supplementation or fortification; maternal balanced energy protein supplementation; maternal calcium supplementation; multimicronutrient supplementation in pregnancy; the promotion of breastfeeding; appropriate complementary feeding; vitamin A and preventive zinc supplementation of children between 6 and 59 months of age; the management of severe acute malnutrition and the management of moderate acute malnutrition. Enhancing the coverage and effectiveness of these interventions requires effective nutrition-sensitive programmes that address the key underlying determinants of nutrition in the areas of agriculture and food security, care-giving resources, health services and safe and hygienic environment by incorporating specific nutrition goals and actions (Ruel et al. 2013).

Nutrition-Sensitive Agriculture

As people depend on food for most of their micronutrient needs and agriculture is often the only realistic way of obtaining food and/or an income to provide the required nutrition, investment in agriculture is widely seen as important in an attempt to reduce malnutrition. Agricultural interventions aimed at improving nutrition have been implemented by governments and development agencies since the 1960s but their focus and the types of interventions have changed considerably over the years as knowledge and the theoretical understanding of the causes of hunger and malnutrition have developed (Massett et al. 2011). Early interventions focused mainly on increasing food production (mostly cereals) as undernutrition was seen to be the result of a lack of a (staple) food. However, it was soon realized that increasing food production alone was not going to be sufficient to eradicate malnutrition unless the poorest were given access to the food produced. Malnutrition was thus linked to food security rather than food availability, resulting in projects focusing on increasing incomes and livelihoods rather than food availability alone. But empirical studies showed that changes in income did not immediately translate into changes in consumption and these highlighted the limitations of increasing income alone in an attempt to improve nutritional status. Since the 1990s, interventions have also been aimed at ensuring that diets include more nutritious foods in terms of their caloric, protein or micronutrient contents. These interventions have had the explicit objective of reducing malnutrition as this would increase the chance of improving the nutritional status of children and women (Herforth et al. 2012). Projects have included, for example, production diversification (home gardens) and biofortification (breeding crops with an increased nutritional value). Likewise, the definition of food security has evolved and the term 'food and nutrition security' is used more often (Box 14.2).

The bidirectional linkages between agriculture and nutrition would seem to be obvious. A well-nourished and healthy population is needed to provide the necessary labour productivity in agriculture, and agriculture produces the food and incomes required to ensure a supply of the energy and nutrients required. Evidence shows that when farmers are malnourished, they are less productive (Sahn 2010; Haas et al. 1995; McNamara et al. 2010). Adult height is positively associated with income and analysis by Haddad & Bouis (1991) concluded that every 1% increase in height is associated with a 4% increase in agricultural wages. The effects of malnutrition on agricultural productivity are more profound when it is combined with disease. Malnourished people are more susceptible to infection and illness, and illness results in a loss of labour, time diverted to caring for the sick and a loss of savings and assets due to health-related expenditures (Asenso-Okyere et al. 2011; Chuma et al. 2006). As labour demand in agriculture is seasonal, a brief period of illness that delays planting, hampers weeding or coincides with harvesting can have a severe economic effect. Unfortunately, most cases of malaria in Africa coincide with critical periods in the farming agenda (Girardin et al. 2004) and result in a significant loss of labour (Malaney et al. 2004).

The linkages between agriculture and nutrition are theorized rather than empirically based as evidence for the causal pathways is still weak and mechanisms are poorly understood. Several theorized pathways have been developed to guide the search for evidence on agriculture-nutrition interactions. Most recognize that agriculture can influence nutrition through multiple direct and indirect pathways occurring at household or at macroeconomic level (Webb 2013; World Bank 2007). Education on nutrition is often identified as an important factor that modifies the effect of household-level determinants (Herforth et al. 2012). Using the conceptual framework in Figure 14.1, it can be seen that agriculture affects food security through the availability of and access to foods to be consumed in diets, and it also impacts on income. It influences women's decision-making powers, income, time and knowledge, each of which affects maternal and childcare practices. Agriculture can also have a huge impact on health, with exposure to agrochemicals causing health risks and mosquito-borne diseases based on agronomic practices depending on stagnant water. Agricultural development can also affect the basic causes of malnutrition, such as poverty and environmental resources (Ibid.).

Ruel et al. (2013) emphasize that the number of pathways is not fixed, that evidence for some is stronger than for others and that there are additional ways through more distal mechanisms that serve as a programming platform for other sectors. Webb (2013), referring to Hawkes et al (2012), indicates that agriculture-nutrition linkages are not simplistic associations whereby agricultural improvements translate directly into better nutrition but they are a chain of associations. Massett et al. (2011) present such a chain of causal inferences from household participation in agriculture through the direct and indirect impacts on food and income to nutritional status (Figure 14.2). This (simple) chain approach is suggested as crucial in studying the nutritional impact of agricultural interventions (Hawkes and Ruel 2011).

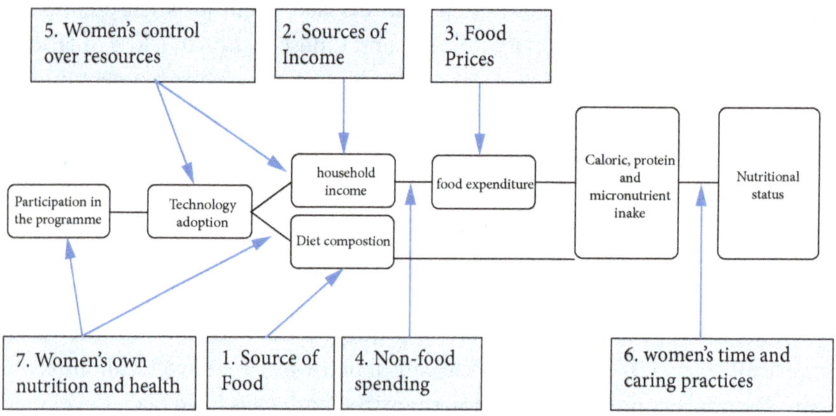

FIGURE 14.2 *Chain approach to assess the impact of agricultural interventions on nutrition*
SOURCE: MASSETT ET AL. (2011), WITH THE SEVEN KEY PATHWAYS ADDED BY WEBB (2013)

Improving Dietary Diversity

The most direct impact of agriculture or food systems on nutrition is the improvement in the delivery of foods that have a high nutritional quality, which leads to increased dietary diversification. Dietary diversity is essential to nutrient adequacy as there is no single food, other than breast milk, that contains all of the nutrients required to maintain good health and nutritional status. Consuming a variety of foods among and within food groups is a recommended strategy to ensure an adequate intake of micronutrients (FAO/WHO 2004). Dietary diversity is, therefore, a key element of high-quality diets and the recommendation to consume a variety of foods is included in many nutritional guidelines. In Africa, however, monotonous diets based mainly on grains, roots and tubers with few vegetables and hardly any animal-source foods are often consumed and are generally expensive for the rural poor (Gittelsohn and Vastine 2003). In Mali, for example, the diets of women in Bamako were based on a starchy staple of refined white rice, refined wheat flour or millet accompanied by a sauce typically made from vegetables (tomato, onion, dried okra or cabbage) and beef or fish in very small quantities (Kennedy et al. 2010). Studies in different age groups have shown that an increase in dietary diversity results in the increased nutrient adequacy of the diet. Dietary diversity has been positively correlated with the adequate micronutrient density of complementary foods for infants and young children (FANTA 2006) and the macronutrient and micronutrient adequacy of the diet for non-breastfed children (Hatloy et al. 1998; Ruel et al. 2004; Steyn et al. 2006; Kennedy et al. 2007), adolescents (Mirmiran et al. 2004) and adults (Ogle et al. 2001; Foote et al. 2004; Arimond et al. 2010).

Agriculture, as part of the food system that produces, markets, stores, trades, processes and retails food for consumers, can promote more nutritious and sustainable diets. Cultivating and promoting foods that stay as closely as possible to local dietary patterns are the most sustainable and easiest to implement. The question remains whether local diets can be optimized in such a way that they fulfil the requirements of all the nutrients needed. Recently, linear programming has been used to develop optimized diets based on existing dietary patterns and food habits (Ferguson et al. 2006; Maillot et al. 2008; Santika et al. 2009; Daelmans et al. 2013; Skau et al. 2014). Most of these studies indicate that optimized diets improve the fulfilment of many nutrients but, for some nutrients, alternative foods are needed to ensure an adequate nutrient intake by vulnerable groups. These so-called problem nutrients are iron, zinc, folic acid and calcium, and sometimes also vitamin A, B12 and other B vitamins. Diversification of production systems may offer alternatives, with one approach being increased vegetable and fruit production, for example in fields or home gardens. Farmers in Africa traditionally grow vegetables between their cereal crops, utilizing open spaces but also cultivate then in or near their compounds in well-kept home gardens. A large variety of vegetables and fruits are produced such as spinach, pumpkins, bitter leaf, amaranth, drumstick tree (*Moringa Oleifera*), papaya, banana, and mango (Grubben et al. 2014). Although most studies on the effect of home gardens on dietary diversity indicate a higher consumption of green leafy vegetables, orange roots and fruits, which leads to an increased micronutrient intake (especially of vitamin A, iron and zinc), there is little evidence of the effect on micronutrient status (Massett et al. 2012). Most evidence is available on the effect on people's vitamin A status, measured as serum retinol concentration. A meta-analysis of four studies shows a summary impact of an overall difference of 2.4 microgram/dl serum retinol between project and control areas, providing some support to the hypothesis that agricultural interventions improve the vitamin A status of children (Massett et al. 2012). The low bioavailability of micronutrients in vegetables and fruit may explain the weak association with micronutrient status.

Bioavailability is the extent to which micronutrients are absorbed and used for normal body functions (Hurrell & Egli 2010). In Africa, dietary carotenoids in green leafy vegetables, in yellow or orange-coloured vegetables and fruits are major sources of vitamin A and can provide between 70% and 90% of a person's daily vitamin A intake (FAO/WHO 1988; Amoussa-Hounkpatin et al. 2012). The bioavailability of β carotene from a mixed diet is ~8% but can drop to ~4% if low quantities of fruit are included, as is often the case in Sub-Saharan Africa. It is known that the food matrix plays an important role in the bioavailability of β carotene from plant sources (Castenmiller et al. 1999). The bioavailability of β-carotene is also enhanced by the presence of fat, cellulose and pectin in a composite meal (van het Hof et al. 2000; de Pee et al. 1995). In addition, variations in the presence and intensity of intestinal parasites affect absorption (Stephensen 2001). Concerning iron, the bioavailability of plant-based iron is particularly affected by the presence of phytic acid and polyphenols in the diet, which can be exacerbated by chronic parasitemia, such as malaria infections, while ascorbic acid, other organic acids and the so-called 'meat factor' enhance iron absorption (Hurrell & Egli 2010; Cercamondi et al. 2010). A stable iron isotope study in Benin assessing the iron absorption from traditional and iron-improved leafy amaranth and jute sauces indicated that vegetable sauces can significantly contribute to iron intake but that the absorption from the improved sauces was lower compared to traditional sauces resulting in a comparable quantity of total iron absorbed (Cermondi et al. 2014). A study in Kenya showed that enriching maize porridge with iron-rich amaranth grain did not improve the iron status of 12-59 month-old children (Macharia-Mutie et al. 2012). Further measures, such as adding iron absorption enhancers to sauces and porridges and decreasing the concentration of inhibiting phenolic compounds in sauces to increase iron absorption, should be investigated.

Another promising approach is biofortification, either by adopting natural breeding techniques or by using genetically modified organisms. For some crops, wild varieties with specific nutrient-producing traits can be used for breeding. HarvestPlus, part of the CGIAR Research Program on Agriculture for Nutrition, promotes biofortification through natural breeding processes by focusing on (i) increasing the micronutrient density of staple crops, such as cassava, sweet potato, rice, wheat, beans and millet; (ii) providing evidence of crops' ability and acceptability to improve the micronutrient status of vulnerable groups under controlled conditions; and (iii) the delivery and promotion of the crop to increase farmer adoption and consumption by communities (Bouis et al. 2011). Examples of biofortified staple crops are high-zinc rice and wheat; pro-vitamin-A-rich maize, cassava (see Photo 14.1), and orange-fleshed sweet potato; and high-iron beans and pearl millet (Saltzman et al. 2013). One of the successful biofortified crops that has already been introduced in Africa is the orange-fleshed sweet potato. Varieties with a high concentration of carotenoids (5023-11,278 μg total β carotene per 100 g, Hotz et al. 2012a) have been proven to increase the vitamin A status in children in South Africa (Jaarsveld et al. 2005), Mozambique (Low et al. 2007) and Uganda (Hotz et al. 2012b). Effectiveness trials in Mozambique and Uganda showed that the crop's adoption by both farmers and consumers was excellent given that 77% of consumers were cultivating it themselves and this resulted in a 42%-169% increase in intake of vitamin A (Hotz et al. 2012a; 2012b). Orange-fleshed sweet potato is currently available in many African countries and is contributing to reducing the prevalence of vitamin A deficiency. Although evidence for efficacy under 'controlled' conditions and effectiveness in 'real life' situations for orange-fleshed sweet potato is virtually complete, evidence is still lacking for most of the other newly developed biofortified crops.

These traits have not (yet) been identified for other crops. For example, in the EU FP7-supported INSTAPA research programme in Mali, research focused on identifying iron-biofortified varieties of sorghum, a staple food for a large part of the population. Screening on station and on-farm trials indicated, however, that genetic diversity in iron and zinc concentrations of sorghum exists but at present the potential contribution of biofortified sorghum varieties to iron and zinc intake is not sufficient to solve the problem of iron and zinc deficiency (Weltzien 2014, personal communication). Moreover, the improved varieties

PHOTO 14.1 *Consumption of biofortified yellow cassava improves the vitamin A intake of schoolchildren in Kenya*
PHOTO: ELISE TALSMA

with the highest yields were not the varieties with higher iron and zinc content. On the contrary, concentrations were often below those of the traditional varieties. From a nutrition perspective, the release of new varieties with a lower content of iron and zinc content should be prevented. As a result of INSTAPA, it was decided to only release or recommend the adoption of sorghum and millet varieties that have satisfactory levels of Fe and Zn (i.e. as good as or better contents of Fe and Zn than local varieties). This measure would ensure that varieties that are more productive would not only improve food security but also make a contribution to nutrition security as well (Weltzien 2014, personal communication). Alternative ways of changing the iron and zinc profile of sorghum is by agronomic approaches and applying, for example, zinc fertilizer (Rengel et al. 1999) or by genetic engineering (Welch & Graham 2004).

Evidence for Agriculture Leveraging for Nutrition

Many reviews have been carried out in the search for evidence into whether and how agricultural interventions improve nutrition. Some have focused on micronutrient outcomes, some on animal-source food and some on dietary diversification strategies, while others have looked more generally at a range of productivity-enhancing interventions. Despite differences in methodology, the types of agricultural activity considered and the analytical techniques, most reviews were consistent and concluded that the current state of knowledge about nutritional improvements through agriculture-based interventions is weak and the understanding of mechanisms of change is poor (Webb 2013; Massett et al. 2011; Girard et al. 2012). It is important to emphasize that the current lack of evidence does not mean that agriculture does not support gains in nutrition and health but rather that the evidence of positive impacts is still weak. This is mainly due to methodological weaknesses in existing studies and the lack of rigorous and well-designed studies (Massett et al. 2011).

Strengthening studies considering the often-complex linkages between agriculture and nutrition using a causal-chain approach influenced by a myriad of factors often has challenges and constrain any claims of attribution of impact to specific interventions. A rigorous design is defined as one that minimizes selection and information bias, controls for confounding factors and attempts to rule out chance and secular trends as contributors to changes observed over time (Habicht et al. 1999; Victora et al. 2004). This is best done by comparing changes over time between groups or individuals who are exposed to the interventions and groups or individuals that are not (double difference), assuming that both intervention and non-intervention groups are similar in all other aspects. In principle, randomized controlled trials (RCT) are the best method of proving a causal relationship between an intervention and various outcomes and attributing changes in these outcomes (solely) to the implementation of the intervention. However, in practice, the conditions of an ideal RCT almost never hold. The randomization of food systems is also not always possible or acceptable in the development setting. And interventions in the food system tend to be (re)shaped continuously to adapt to changing conditions, which limits the possibility of assessing their impact over time (Ton 2012). Withholding programme interventions in non-intervention groups for the duration of an entire project may be considered unethical. One solution could be to use a stepped-wedge approach or to set up comparison groups with different intensities of programme delivery (Menon et al. 2013). Where randomization is not feasible, options for establishing valid comparison groups include matching by design or by analysis using econometric methods such as propensity score matching (PSM) or nearest neighbour matching (NNM) (Ibid.). To strengthen overall methodological rigour, Ton (2012) proposes mixing methods instead of simple one-method designs combining quantitative and qualitative methods in a 'nested' design to study the same mechanisms, processes and outcomes to strengthen the validity of conclusions.

The nutrition indicators used depend on the project's aims and context. However, priority should be given to measuring the determinants of nutrition most likely to be affected by agriculture, such as household food security, diet and dietary diversity, women's workloads, childcare practices, women's control of income, their health status, the seasonality of income and the efficiency of water use (Herforth et al. 2012). Dietary diversity has gained increased interest as a proxy indicator of adequacy of intake of micronutrients (Arimond et al. 2010; Kennedy et al. 2011). There are a number of issues that call for further refinement of the dietary diversity score: correlated measurement errors may have falsely inflated the correlations between indicators of dietary diversity and nutrient adequacy in many validation studies; the need for agreement on the number of food groups to be used in the score and whether the association would be improved by assigning weights to food groups or by including specific nutrient foods; questions on the robustness of the associations between dietary diversity score and nutrient adequacy throughout different seasons and on the sensitivity of the score to interventions; and the need for a generally agreed cut-off point for a dichotomous score indicating inadequacy of intake. If child nutritional status is the target, it is important to consider the optimal ages at which assessment should be made to detect the greatest difference between intervention and non-intervention groups. The earlier and the longer children are exposed to nutrition interventions during their first two years of life, the more they are likely to benefit in terms of linear growth (Menon et al. 2013). Although the interventions should reach infants and young children during their first two years of life (Black et al. 2013), the impact of interventions should ideally be measured once this period of greatest potential benefit has ended (Ruel et al. 2008). Therefore, measuring the impact on height (and stunting) should be assessed among children older than 24 months who have benefitted from the intervention during their first two years of life. Measuring impact on stunting at an earlier age (for example, under 12 months) could underestimate the real impact of the intervention (Menon et al. 2013). Even if agricultural interventions fail to show improvements in child nutritional status, they will not necessarily have failed to affect nutrition. If they show a positive impact on underlying determinants of nutrition, as shown in Figure 14.1, then they will have succeeded in creating some of the conditions necessary (but not always sufficient) for good nutrition (Herforth et al. 2012).

The Global Context and Future Developments

National and international interest in nutrition has increased dramatically in the past ten years due to the inclusion of underweight as an indicator of the Millennium Development Goals, the publication of the Lancet Series on maternal and child nutrition in 2008 and 2013, prioritization by top economists of the need to reduce micronutrient malnutrition (Horton et al. 2008) and the global leadership offered by the Scaling up Nutrition (SUN) movement. In the coming years, reducing micronutrient deficiencies will remain a high priority internationally and there will be a shift in emphasis from underweight towards reducing stunting. The private sector has recently become more actively engaged in finding solutions to malnutrition by complementing the efforts of the public sector and NGO nutrition programmes. In addition, there is growing awareness that traditional nutrition-specific interventions should be aligned with so-called nutrition-sensitive activities in food security, agriculture, water and sanitation, and social protection/safety nets. Especially within the food security and agriculture sector, interest in addressing nutrition has increased and has resulted in a shift from the concept of food security to the concept of food and nutrition security. An increased interest in nutrition in the Netherlands too is mainly embedded in the food-security strategy triggered by the Dutch Parliament's decision to include nutrition and nutritional value in its food-security policy. As a result of this renewed focus, more funding is being channelled through the agricultural sector for research with the aim of encouraging agricultural policies and programmes to become nutrition sensitive or, more specifically, to make agriculture work for nutrition.

Scientifically, there is still a gap in the evidence regarding the efficacy of nutrition-sensitive agriculture. Underutilized foods or wild foods are often promoted for the improvement they offer in nutrition but the evidence base for such claims is very weak. In addition, efficacy does not guarantee the effectiveness of the approach. Very few strategies have been tested in terms of their feasibility to achieve improved nutrition in practice. The evidence base for the theoretical linkages and causal pathways is weak and it is generally unclear which innovative approaches could be used to improve agriculture leveraging for nutrition. This calls for specific well-designed studies to test efficacy and pathways, and rigorous evaluation studies to measure the impact of food-based approaches, agricultural programmes and projects on nutrition.

Monitoring and evaluation of food-based and agricultural programmes is becoming increasingly important. There is a need for scientifically sound and valid but also (preferably) simple indicators of nutrient intake. The dietary diversity score is such an indicator. There is ample evidence from developed countries to show that dietary diversity is indeed strongly associated with nutrient adequacy and is thus an essential element of diet quality (Foote et al. 2004). There is less evidence of this in developing countries where monotonous diets, mostly relying on a few plant-based staple foods, are typical. Further evaluation and validation of dietary diversity scores is required.

References

Amoussa-Hounkpatin, W., C. Mouquet-Rivier, R.A.M. Dossa et al. (2012), Contribution of Plant-Based Sauces to the Vitamin A Intake of Young Children in Benin. *Food Chemistry* 131: 948–955.

Arimond, M., D. Wiesmann, E. Becquey et al. (2010), Simple Food Group Diversity Indicators Predict Micronutrient Adequacy of Women's Diets in Five Diverse, Resource-Poor Settings. *Journal of Nutrition* 140: 2059S–2069S.

Asenso-Okyere, K., C. Chiang, P. Thangata et al. (2011), Understanding the Interaction between Farm Labour Productivity, and Health and Nutrition: A Survey of the Evidence. *Journal of Development and Agricultural Economics* 3: 80–90.

Balarajan, Y., U. Ramakrishnan, E. Özaltin et al. (2011), Anaemia in Low-Income and Middle-Income Countries. *The Lancet* 378: 2123–2135.

Bhutta, Z.A., T. Ahmed, R.E. Black et al. (2008), What Works? Interventions for Maternal and Child Undernutrition and Survival. *The Lancet* 371: 417–440.

Bhutta, Z.A., J.K. Das, A. Rizvi et al. (2013), Evidence-Based Interventions for Improvement of Maternal and Child Nutrition: What Can Be Done and at What Cost? *The Lancet* 371: 243–260.

Black, R.E., L.H. Allen, Z. Bhutta et al. (2008), Maternal and Child Undernutrition: Global and Regional Exposures and Health Consequences. *The Lancet* 371: 243–260.

Black, R.E., C.G. Victora, S.P. Walker et al. (2013), Maternal and Child Nutrition and Overweight in Low-Income and Middle-Income Countries. *The Lancet* 382(9890): 427–451.

Bouis, H.E., C. Hotz, B. McClafferty et al. (2011), Biofortification: A New Tool to Reduce Micronutrient Malnutrition. *Food and Nutrition Bulletin* 32: 531–540.

Castenmiller, J.J., C.E. West, J.P. Linssen, K.H. van het Hof & A.G. Voragen (1999), The Food Matrix of Spinach Is a Limiting Factor in Determining the Bioavailability of Beta-Carotene and to a Lesser Extent of Lutein in Humans. *Journal of Nutrition* 129(2): 349–355.

Cercamondi, C.I., I.M. Egli, E. Ahouandjinou et al. (2010), Afebrile *Plasmodium Falcipa*rum Parasitemia Decreases Absorption of Fortification Iron but Does Not Affect Systemic Iron Utilization: A Double Stable-Isotope Study in Young Beninese Women. *American Journal of Clinical Nutrition* 92: 1385–1392.

Cercamondi, C.I., C. Icard-Veniere, I.M. Egli IM, et al. (2014), A Higher Proportion of Iron-rich Leafy Vegetables in a Typical Burkinabe Maize Meal does not increase the Amount of Iron Absorped in Young Women. *Journal of Nutrition* 144: 1394–1400.

Chuma, J., M. Thiede & C. Molyneux (2006), Rethinking the Economic Costs of Malaria at Household Level: Evidence from Applying a New Analytical Framework in Rural Kenya. *Malaria Journal* 5(1): 76.

Committee on World Food Security (2012), Coming to Terms with Terminology. 39th session. CFS 39/4. Rome: FAO. http://www.fao.org/docrep/meeting/026/MD776E.pdf.

Daelmans, B., E. Ferguson, C.K. Lutter et al. (2013), Designing Appropriate Complementary Feeding Recommendations: Tools for Programmatic Action. *Maternal & Child Nutrition* 9: 116–130.

de Pee, S., C.E. West, Muhilal et al. (1995), Lack of Improvement in Vitamin A Status with Increased Consumption of Dark-Green Leafy Vegetables. *The Lancet* 346(8967): 75–81.

Dean, S.V., A.M. Imam, Z.S. Lassi & Z.A. Bhutta (2013), Importance of Intervening in the Preconception Period to Impact Pregnancy Outcomes. In: J. Bhatia, Z.A. Bhutta & S.C. Kalhan (eds), Maternal and Child Nutrition: The First 1000 Days. Nestlé Nutrition Institute Workshop Series 74: 63–73.

Ezzati, M., A.D. Lopez, A.A. Rodgers et al. (2004), *Comparative Quantification of Health Risks: Global and Regional Burden of Disease Attributable to Selected Major Risk Factors*. Geneva: WHO.

FANTA (2006), Developing and Validating Simple Indicators of Dietary Quality and Energy Intake of Infants and Young Children in Developing Countries: Summary of Findings from Analysis of 10 Data Sets. Working Group on Infant and Young Child Feeding Indicators. Food and Nutrition Technical Assistance (FANTA) Project, Academy for Educational Development (AED), Washington DC.

FAO (2013), *The State of Food Insecurity in the World: The Multiple Dimensions of Food Security*. Rome: FAO.

FAO/WHO (1988), Requirements of Vitamin A, Iron, Folate, and Vitamin B12. Report of Joint FAO/WHO Expert Consultation 23. Rome: FAO.

FAO/WHO (2004), *Vitamin and Mineral Requirements in Human Nutrition*. Geneva: FAO/WHO.

Ferguson, E., N. Darmon, U. Fahmida et al. (2006), Design of Optimal Food-Based Complementary Feeding Recommendations and Identification of Key 'Problem Nutrients' Using Goal Programming. *Journal of Nutrition* 136: 2399–2404.

Foote, J.A., S.P. Murphy, L.R. Wilkens et al. (2004), Dietary Variety Increases the Probability of Nutrient Adequacy among Adults. *Journal of Nutrition* 134: 1779–1785.

Gillespie, S., L. Harris & S. Kadiyala (2012), The Agriculture-Nutrition Disconnect in India, What Do We Know? International Food Policy Research Institute Discussion Paper 01187. Washington DC: IFPRI.

Girard, A.W., J.L. Self & C. McAuliffe (2012), The Effect of Household Food Production Strategies on the Health and Nutrition Outcomes of Women and Young Children: A Systematic Review. *Paediatric and Perinatal Epidemiology* 26: 205–222.

Girardin, O., D. Dao, B.G. Koudou et al. (2004), Opportunities and Limiting Factors of Intensive Vegetable Farming in Malaria Endemic Côte d'Ivoire. *Acta Tropica* 89: 109–123.

Gittelsohn, J. & A.E. Vastine (2003), Sociocultural and Household Factors Impacting on the Selection, Allocation and Consumption of Animal Source Foods: Current Knowledge and Application. *Journal of Nutrition* 133: 4036S–4041S.

Grubben, G., W. Klaver, R. Non-Womdim et al. (2014), Vegetables to Combat the Hidden Hunger in Africa. *Chronica Horticulturae* 54(1): 24–32.

Haas, J.D., E.J. Martinez, S. Murdoch et al. (1995), Nutrition Supplementation during Pre-school Years and Physical Work Capacity in Adolescents and Young Adult Guatemalans. *Journal of Nutrition* 125: 1078S–1089S.

Habicht, J.P., C.G. Victora & J.P. Vaughan (1999), Evaluation Designs for Adequacy, Plausibility and Probability of Public Health Programme Performance and Impact. *International Journal of Epidemiology* 28(1): 10–18.

Haddad, L.J. & H.E. Bouis (1991), The Impact of Nutritional Status on Agricultural Productivity: Wage Evidence from the Philippines. *Oxford Bulletin of Economics and Statistics* 53(1): 45–68.

Hatloy, A., L. Torheim & A. Oshaug (1998), Food Variety: A Good Indicator of Nutritional Adequacy of the Diet? A Case Study from an Urban Area in Mali, West Africa. *European Journal of Clinical Nutrition* 52(12): 891–898.

Hawkes, C. & M.T. Ruel (2011), Value Chains for Nutrition. 2020 Conference Paper 4. *IFPRI 2020 International Conference Leveraging Agriculture for Improving Nutrition and Health*. New Delhi: IFPRI.

Hawkes, C., R. Turner and J. Waage (2012), Current and Planned Research on Agriculture for Improved Nutrition: A Mapping and a Gap Analysis. Report for the Department of International Development (DFID). London: Leverhulme Centre for Integrative Research on Agriculture and Health/University of Aberdeen/Center for Sustainable International Development.

Herforth, A., A. Jones & P. Pinstrup-Anderson (2012), Prioritizing Nutrition in Agriculture and Rural Development: Guiding Principles for Operational Investments. Health, Nutrition and Population Family (HNP) Discussion Paper. Washington DC: World Bank.

Hess, S.Y., B. Lönnerdal, C. Hotz et al. (2009), Recent Advances in Knowledge of Zinc Nutrition and Human Health. *Food and Nutrition Bulletin* 30(1): S5–S11.

Horton, S., H. Alderman & J.A. Rivera (2008), Hunger and Malnutrition: Copenhagen Consensus 2008. Challenge Paper. Copenhagen Consensus Centre.

Hotz, C., C. Loechl, A. de Brauw et al. (2012a), A Large-Scale Intervention to Introduce Orange Sweet Potato in Rural Mozambique Increases Vitamin A Intakes among Children and Women. *British Journal of Nutrition* 108:163–176.

Hotz, C., C. Loechl, A. Lobowa et al. (2012b), Introduction of Beta Carotene Rich Orange Sweet Potato in Rural Uganda Results in Increased Vitamin A Intakes among Children and Women and Improved Vitamin A Status among Children. *Journal of Nutrition* 142: 1871–1880.

Hurrell, R. & I. Egli (2010), Iron Bioavailability and Dietary Reference Values. *American Journal of Clinical Nutrition* 91; 1461S–1467S.

Jaarsveld, P., A. Carriquiry & J.V. Meenakshi (2005), Beta Carotene Rich Orange-Fleshed Sweet Potato Improves the Vitamin A Status of Primary School Children Assessed with the Modified-Relative-Doseresponse Test. *American Journal of Clinical Nutrition* 81: 1080–1087.

Kennedy, G., M.R. Pedro, C. Seghieri et al. (2007), Dietary Diversity Score Is a Useful Indicator of Micronutrient Intake in Non-Breast-Feeding Filipino Children. *Journal of Nutrition* 137: 1–6.

Kennedy, G., N. Fanou-Fogny, C. Seghieri et al. (2010), Food Groups Associated with a Composite Measure of Probability of Adequate Intake of 11 Micronutrients in the Diets of Women in Urban Mali. *Journal of Nutrition* 140: 2070S–2078S.

Kennedy, G., T. Ballard & M. Claude-Dop (2011), *Guidelines for Measuring Household and Individual Dietary Diversity*. Rome: FAO.

Lim, S.S., T. Vos, A.D. Flaxman et al. (2012), A Comparative Risk Assessment of Burden of Disease and Injury Attributable to 67 Risk Factors and Risk Factor Clusters in 21 Regions, 1990–2010: A Systematic Analysis for the Global Burden of Disease Study 2010. *The Lancet* 380: 2224–2260.

Low, J.W., M. Arimond, N. Osman et al. (2007), A Food-Based Approach Introducing Orange Fleshed Sweet Potato Increased Vitamin A Intake and Serum Retinol Concentrations in Young Children in Rural Mozambique. *Journal of Nutrition* 137: 1320–1327.

Macharia-Mutie, C.W., D. Moretti, N. van der Briel et al. (2012), Maize Porridge Enriched with a Micronutrient Powder Containing Low-Dose Iron as NaFeEDTA But Not Amaranth Grain Flour Reduces Anemia and Iron Deficiency in Kenyan Preschool Children. *Journal of Nutrition* 142(9): 1756–1763.

Maillot, M., E. Ferguson, A. Drewnowski et al. (2008), Nutrient Profiling Can Help Identify Foods of Good Nutritional Quality for Their Price: A Validation Study with Linear Programming. *Journal of Nutrition* 138: 1107–1113.

Malaney, P., A. Spielman & J. Sacks (2004), The Malaria Gap. *American Journal of Tropical Medicine and Hygiene* 71: 141–146.

Massett, E., L. Haddada, A. Cornelius et al. (2011), *A Systematic Review of Agricultural Interventions that Aim to Improve Nutritional Status of Children.* London: EPPI Centre, Social Science Research Unit, Institute of Education, University of London.

Massett, R., L. Haddad, A. Cornelius et al. (2012), Effectiveness of Agricultural Interventions That Aim to Improve Nutritional Status of Children: Systematic Review. *British Medical Journal* 344: d8222 doi: 10.1136/bmj.d8222.

McLean, E., M. Cogswell, I. Egli et al. (2008), Worldwide Prevalence of Anaemia, WHO Vitamin and Mineral Nutrition Information System, 1993–2005. *Public Health Nutrition* 12(4): 444–454.

McNamara, P.E., J.M. Ulimwengu & K.L. Leonard (2010), *Do Health Investments Improve Agricultural Productivity?* International Food Policy Research Institute Discussion Paper. Washington DC.

Menon, P., R. Rawat & M. Ruel (2013), Bringing Rigor to Evaluations of Large-Scale Programs to Improve Infant and Young Child Feeding and Nutrition: The Evaluation Designs for the Alive & Thrive Initiative. *Food and Nutrition Bulletin* (UNU) 34: S195–S211.

Mirmiran, P., L. Azadbakht, A. Esmaillzadeh et al. (2004), Dietary Diversity Score in Adolescents: A Good Indicator of the Nutritional Adequacy of Diets: Tehran Lipid and Glucose Study. *Asia Pacific Journal of Clinical Nutrition* 13(1): 56–60.

Ogle, B.M., P.H. Hung & H.T. Tuyet (2001), Significance of Wild Vegetables in Micronutrient Intakes of Women in Vietnam: An Analysis of Food Variety. *Asia Pacific Journal of Clinical Nutrition* 10: 21–30.

Özaltin, E., K. Hill & S.V. Subramanian (2010), Association of Maternal Stature with Offspring Mortality, Underweight, and Stunting in Low- to Middle-Income Countries. *JAMA* 303(15): 1507–1516.

Rengel, Z., G.D. Batten & D.E. Crowley (1999), Agronomic Approaches for Improving the Micronutrient Density of Edible Portions of Field Crops. *Field Crops Research* 60: 27–40.

Ruel, M. (2013), Food Security and Nutrition: Linkages and Complementarities. In: M. Eggersdorfer, K. Kraemer, M. Ruel et al. (eds), *The Road to Good Nutrition*. Basel: Karger.

Ruel, M., J. Graham, S. Murphy et al. (2004), Validating Simple Indicators of Dietary Diversity and Animal Source Food Intake that Accurately Reflect Nutrient Adequacy in Developing Countries. Report submitted to GL-CRSP.

Ruel, M.T., P. Menon, J.P. Habicht et al. (2008), Age-Based Preventive Targeting of Food Assistance and Behaviour Change and Communication for Reduction of Childhood Malnutrition in Haiti: A Cluster Randomized Trial. *The Lancet* 371(9612): 588–595.

Ruel, M.T., H. Alderman et al. (2013), Nutrition-Sensitive Interventions and Programmes: How Can They Help to Accelerate Progress in Improving Maternal and Child Nutrition? *The Lancet* 382(9890): 536–551.

Sahn, D.E. (2010), The Impact of Poor Health and Nutrition on Labor Productivity, Poverty and Economic Growth in Sub-Saharan Africa. In: P. Pinstrup-Anderson (ed.), *The African Food System and Its Interaction with Human Health and Nutrition*. Ithaca, NY: Cornell University Press.

Saltzman, A., B. Birol, H.E. Bouis et al. (2013), Biofortification: Progress toward a More Nourishing Future. *Global Food Security* 2: 9–17.

Santika, O., U. Fahmida & E. Ferguson (2009), Development of Food-Based Complementary Feeding Recommendations for 9 to 11 Month Old Peri-urban Indonesian Infants Using Linear Programming. *Journal of Nutrition* 139: 135–141.

Sazawal, S., R.E. Black, M. Ramsam et al. (2007), Effect of Zinc Supplementation on Mortality in Children Aged 1–48 Months: A Community-Based Randomised Placebo-Controlled Trial. *The Lancet* 369: 927–934.

Skau, J.K., T. Bunthang, C. Chamnan et al. (2014), The Use of Linear Programming to Determine Whether a Formulated Complementary Food Product Can Ensure Adequate Nutrients for 6–11 Month Old Cambodian Infants. *American Journal of Clinician Nutrition* 99: 130–138.

Sommer, A. & F.R. Davidson (2002), Assessment and Control of Vitamin A Deficiency: The Annecy Records. *Journal of Nutrition* 132: 2845S–28505S.

Stephensen, C.B. (2001), Vitamin A, Infection, and Immune Function. *Annual Review of Nutrition* 21: 167–192.

Stevens, G.A., M.M. Finucane, C.J. Paciorek et al. (2012), Trends in Mild, Moderate, and Severe Stunting and Underweight, and Progress towards MDG1 in 141 Developing Countries: A Systematic Analysis of Population Representative Data. *The Lancet* 380: 824–834.

Steyn, N.P., J.H. Nel, G. Nantel et al. (2006), Food Variety and Dietary Diversity Scores in Children: Are They Good Indicators of Dietary Adequacy? *Public Health Nutrition* 9(5): 644–650.

Ton, G. (2012), The Mixing of Methods: A Three-Step Process for Improving Rigour in Impact Evaluations. *Evaluation* 18: 5–25.

UNICEF (2013), *Improving Child Nutrition. The Achievable Imperative for Global Progress.* New York: UNICEF.

van het Hof, K.H., C.E. West, J.A. Weststrate & J.G. Hautvast (2000), Dietary Factors that Affect the Bioavailability of Carotenoids. *Journal of Nutrition* 130(3): 503–506.

Victora, C.G., L. Adair, C. Fall et al. (2008), Maternal and Child Undernutrition: Consequences for Adult Health and Human Capital. *The Lancet* 371: 349–357.

Victora, C.G., J.P. Habicht & J. Bryce (2004), Evidence-Based Public Health: Moving beyond Randomized Trials. *American Journal of Public Health* 94(3): 400–405.

Webb, P. (2013), *Impact Pathways from Agricultural Research to Improved Nutrition and Health: Literature Analysis and Research Priorities.* Geneva: FAO/WHO.

Welch, R.M. & R.D. Graham (2004), Breeding Micronutrients in Staple Food Crops from a Human Nutrition Perspective. *Journal of Experimental Botany* 55(396): 353–364.

Wessels, K.R. & K.H. Brown (2012), Estimating the Global Prevalence of Zinc Deficiency: Results Based on Zinc Bioavailability in National Food Supplies and the Prevalence of Stunting. *PLoS ONE* 7(11): e50568. doi:10.1371/journal.pone.0050568.

WHO (2009), Global Prevalence of Vitamin A Deficiency in Populations at Risk 1995–2005. WHO Global Database on Vitamin A Deficiency. Geneva: WHO.

World Bank (2007), *From Agriculture to Nutrition: Pathways, Synergies and Outcomes.* Washington DC: World Bank.

Zimmermann, M.B. & R.F. Hurrell (2007), Nutritional Iron Deficiency. *The Lancet* 370: 511–520.

Zinc Investigators' Collaborative Group, Z.A. Bhutta, R.E. Black et al. (1999), Prevention of Diarrhea and Pneumonia by Zinc Supplementation in Children in Developing Countries: Pooled Analysis of Randomized Controlled Trials. *Journal of Pediatrics* 135: 689–697.

Zinc Investigators' Collaborative Group, Z.A. Bhutta, S.M. Bird et al. (2000), Therapeutic Effects of Oral Zinc in Acute and Persistent Diarrhea in Children in Developing Countries: Pooled Analysis of Randomized Controlled Trials. *American Journal of Clinical Nutrition* 72: 1516–1522.

Appendix A

Selected Statistics of Major World Regions and Selected Countries in Sub-Saharan Africa

Source: FAO Food Security Statistics, 20 December 2013: Second release for 2013: http://www.fao.org/fileadmin/templates/ess/foodsecurity/Food_Security_Indicators.xlsx. Accessed 30 March 2014.

In the run-up to the Second International Conference on Nutrition 21 years later (ICN+21) that is being held in Rome from 19–21 November 2014, the FAO has made available an international core set ('suite') of indicators of food insecurity that aims to capture various aspects of food insecurity across regions and over time (FAO 2013). Indicators are classified along the four dimensions of food security: availability, access, utilization and stability. To facilitate their interpretation, they are also classified as indicators of determinants and outcomes of food insecurity. Determinants, which can be either static or dynamic, refer to structural conditions that worsen food insecurity in the absence of adequate policy interventions, including emergency assistance. Outcome indicators capture results in terms of inadequate food consumption or anthropometric failures.

This appendix presents a subset of the suite of indicators for different world regions, in particular for the four case-study countries selected for analysis in Chapters 3 and 4 in this volume. The tables below present snapshots for three selected years, while the graphs show the full time series found in the FAO worksheets.

List of Tables in Appendix A	FAO worksheet[1]	Type[2]	Dimension of food security
Table A.1: Food production value and variability			
Average Value of Food Production (Int$ per caput)	V_1.2	SD	Availability
Per Capita Food Production Variability (constant 2004–2006 Int$ per capita)	V_6.3	DD	Stability
Table A.2: Cereal imports and food losses at retail level			
Cereal Import Dependency Ratio (%)	V_5.1	DD	Vulnerability
Share of Dietary Energy Supply derived from cereals, roots and tubers (%)	V_1.3	SD	Availability (diversity)

1 Worksheet identifier as per the FAO file Food_Security_Indicators.xlsx. Note: Calculated = calculated by the authors.

2 Type of indicator: SD = static determinant; DD = dynamic determinant; O = outcome; A = additional statistic needed to estimate adequacy.

 DOI 10.1163/9789004282698_016

(*Table continued*)

List of Tables in Appendix A	FAO worksheet	Type	Dimension of food security
Table A.3: Food availability and stability of food supply			
Dietary Energy Supply (kcal/caput/day)	A_9	SD	Availability
Per Capita Food Supply Variability (kcal/caput/day)	V_6.4	DD	Stability
Table A.4: Adequacy of food supply compared to requirements			
Average Dietary Energy Supply Adequacy (%)	V_1.1	SD	Availability
Average Dietary Energy Requirement (ADER) (kcal/caput/day)	A_4	A	Adequacy
Table A.5: Food price level and volatility			
Domestic Food Price Level Index (index)	V_3.1	SD	Economic access
Domestic Food Price Volatility Index (index)	V_6.2	DD	Stability
Table A.6: Energy requirement levels: Cut-off value for chronic food deprivation, for Food Inadequacy and average requirements			
Minimum Dietary Energy Requirement (MDER) – PAL 1.55 (kcal/caput/day)	A_3	A	Adequacy
Less conservative Minimum Dietary Energy Requirement (MDER) – pal 1.75 (kcal/caput/day)	A_5	A	Adequacy
Average Dietary Energy Requirement (ADER) – PAL 1.85 (kcal/caput/day)	A_4	A	Adequacy
Table A.7: Total population and no. of people who are hungry (undernourished), i.e. consuming less than the minimum dietary requirement			
Total population (millions)	A_1	A	Adequacy
No. of people undernourished (millions)	A_2	O	Access
Table A.8: % of people who are hungry (undernourished), i.e. consuming less than the minimum dietary requirement			
Prevalence of Undernourishment (PoU) (%)	V_7.1	O	Access
Minimum Dietary Energy Requirement (MDER) (kcal/caput/day)	A_3	A	Adequacy
Table A.9: Depth of food deficit			
Depth of food deficit (kcal/caput/day)	V_7.3	O	Access
Average dietary energy supply of the chronically undernourished (food-deprived) population	Calculated	O	Access

List of Tables in Appendix A	FAO worksheet	Type	Dimension of food security
Average dietary energy supply of those not chronically undernourished (food-deprived)	Calculated	O	Access
Table A.10: No. and % of people with inadequate food (new indicator of access)			
No. of people with Food Inadequacy (millions)	Calculated	O	Access
Prevalence of Food Inadequacy (%)	V_7.4	O	Access
Minimum Dietary Energy Requirement (MDER) – PAL 1.75 (kcal/caput/day)	A_5	A	Adequacy
Tables A.11a–c: Prevalence of under-fives anthropometric failure			
Table A.11a: Prevalence of acute undernutrition (wasting) among under-fives			
% of children aged <5 years affected by wasting in total children <5 population (%)	V_8.1	O	Utilization
Table A.11b: Prevalence of chronic undernutrition (stunting) among under-fives			
% of children aged <5 years stunted in total children <5 population (%)	V_8.2	O	Utilization
Table A.11c: Prevalence of underweight among under-fives			
% of children aged <5 years underweight in total children <5 population (%)	V_8.3	O	Utilization
Tables A.12 & A.13: Prevalence of micronutrient deficiency among young children			
Table A.12a: Prevalence of anaemia among under-fives			
% of children aged <5 years affected by anaemia in total children <5 population (%)	V_8.5	O	Utilization
Table A.12b: Prevalence of Vitamin A deficiency among under-fives			
% of children aged <5 years affected by Vitamin A deficiency in total children <5 population (%)	V_8.6	O	Utilization
Table A.13a: Prevalence of iodine deficiency among children 6–12 years			
% of children aged 6–12 years affected by iodine deficiency	V_8.7	O	Utilization
Table A.13b: Prevalence of anaemia among pregnant women			
% of pregnant women affected by anaemia (%)	V_8.8	O	Utilization

TABLE A.1 *Food production value and variability*

	Average value of food production (Int$ per caput)[a]			Per capita food production variability (constant 2004–2006 Int$ per capita)[b]		
	1991	**2000**	**2010**	**1995**	**2000**	**2010**
Developed regions	460	460	480	8.7	7.5	6.8
Latin America and the Caribbean	313	368	453	3.4	4.9	9.4
Asia	166	206	256	1.6	0.9	1.6
North Africa	175	197	245	6.8	11.7	6.0
Sub-Saharan Africa	144	151	166	2.3	1.8	3.4
Nigeria	170	204	195	5.3	1.7	14.4
Tanzania	139	125	154	5.4	4.2	7.1
Kenya	150	134	160	3.9	6.6	6.3
Uganda	178	171	155	5.6	7.5	2.6

[a] The total value of Annual Food Production, as estimated by the FAO and published by FAOSTAT in International Dollars (I $) divided by the total population. The indicator is calculated on three-year averages to reduce the errors due to the difficulties in recording annual stock changes.

SOURCE: FAO WORKSHEET V_1.2.

[b] per capita food production variability corresponds to the variability of the net food production value in constant 2004–2006 1,000 international $ (net food PIN) divided by the population as taken from UN 2010 estimates. variability is based on the trend of the net food PIN per capita from 1985 to 2010 and corresponds with the standard deviation of the deviation from the trend over a period of five years. the time series starts in 1995. note that the variability at continental level is less than at country level due to the effect of aggregation.

SOURCE: FAO WORKSHEET V_6.3.

Time series graphs with Table A.1 Average Value of Food Production

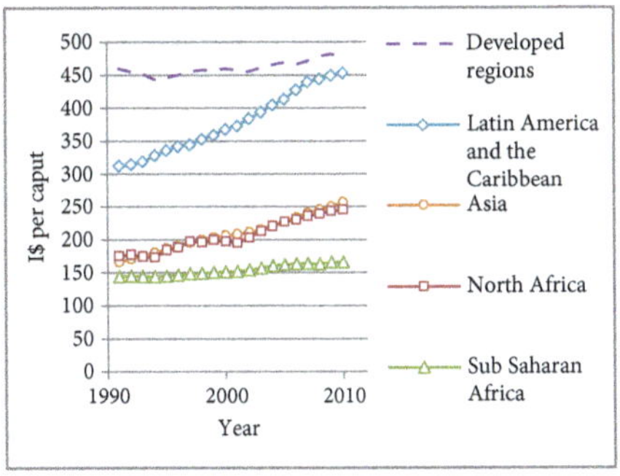

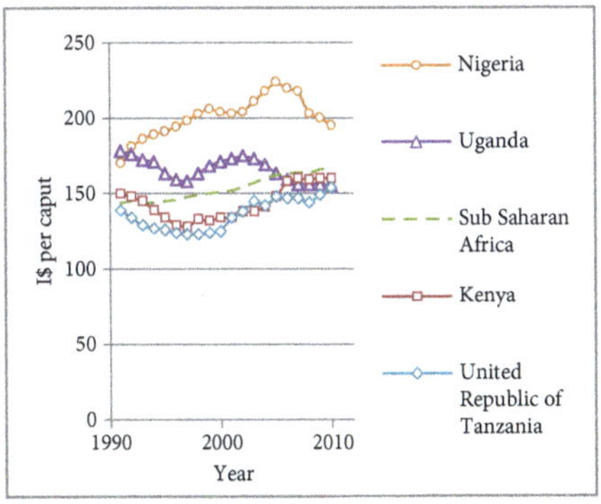

Time series graphs with Table A.1 Per Capita Food Production Variability

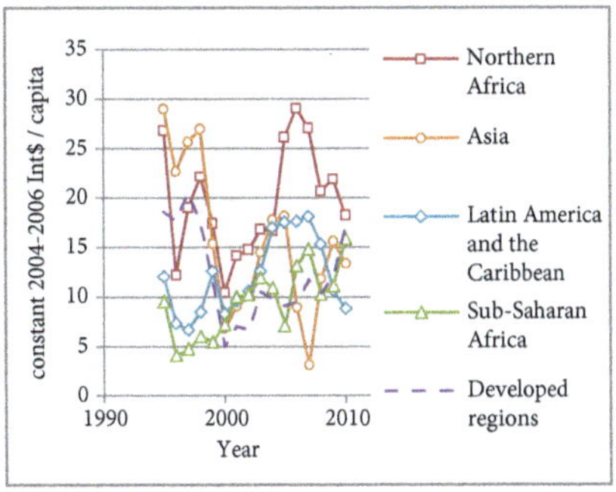

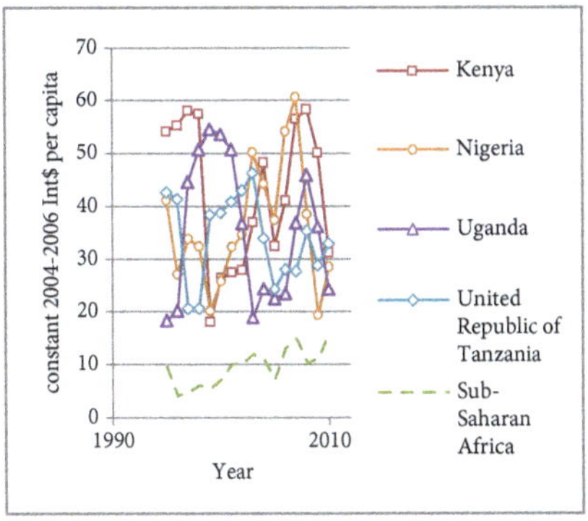

TABLE A.2 *Cereal imports and food losses at retail level*

	Cereal Import Dependency Ratio (%)[a]			Share of Dietary Energy Supply derived from cereals, roots and tubers (%)[b,c]		
	1991	2000	2008	1990	2000	2008
Developed regions	16.7	15.2	16.0	34	33	32
Latin America and the Caribbean	22.5	30.3	29.9	43	41	40
Asia	9.7	10.0	10.0	67	63	58
North Africa	43.2	52.8	49.9	64	63	62
Sub-Saharan Africa	18.0	18.3	21.5	64	65	63
Nigeria	4.4	13.7	14.5	67	65	64

(cont.)

	Cereal Import Dependency Ratio (%)[a]			Share of Dietary Energy Supply derived from cereals, roots and tubers (%)[b,c]		
	1991	**2000**	**2008**	**1990**	**2000**	**2008**
Tanzania	4.0	11.9	13.8	70	65	59
Kenya	14.3	26.4	36.1	55	56	56
Uganda	1.8	6.8	17.7	45	45	45

[a] The ratio is calculated as cereal imports/(cereal production + cereal import-cereal export).

SOURCE: FAO WORKSHEET V_5.1.

[b] energy supply (in kcal/caput/day) provided by cereals, roots and tubers divided by total dietary energy supply (DES) (in kcal/caput/day) calculated from the FAOSTAT food balance sheets. the share is calculated on three-year averages, to reduce the errors due to the difficulties in recording annual stock changes.

SOURCE: FAO WORKSHEET V_1.3.

Note: this indicator does not include plantains, which are an important staple food in uganda.

Time series graphs with Table A.2 Cereal Import Dependency Ratio

Time series graphs with Table A.2 Share of Dietary Energy Supply derived from cereals, roots and tubers

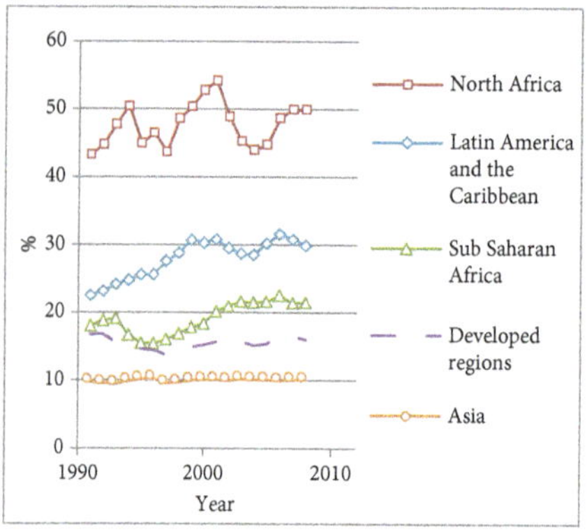

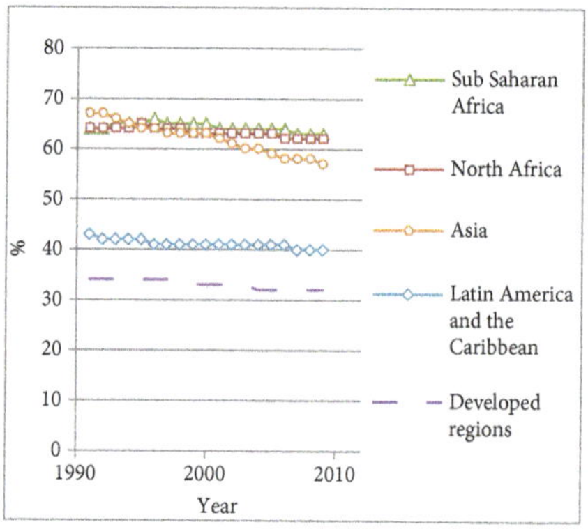

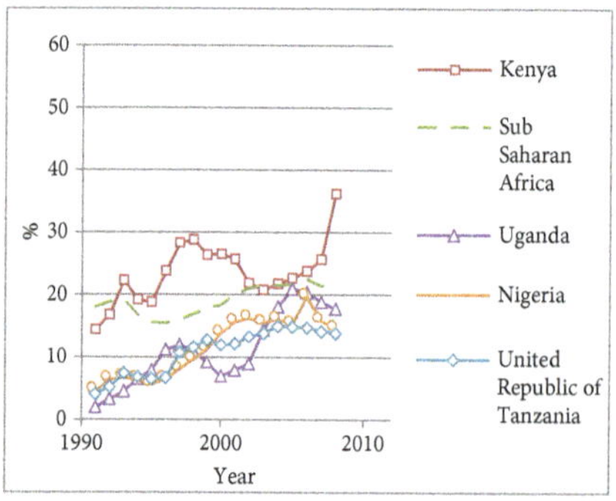

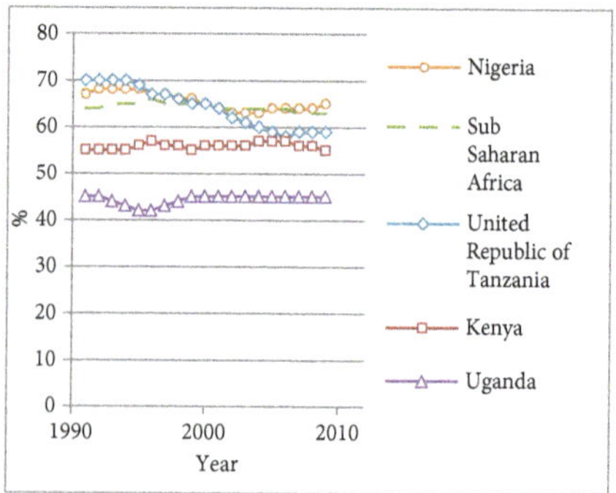

TABLE A.3 *Food availability and stability of food supply*

	Dietary Energy Supply (kcal/caput/day)[a]			Per Capita Food Supply Variability (kcal/caput/day)[b]		
	1991	**2000**	**2010**	**1995**	**2000**	**2010**
Developed regions	3,270	3,340	3,370	18.6	5.0	17.2
Latin America and the Caribbean	2,670	2,800	2,970	12.1	8.3	8.8
Asia	2,410	2,550	2,720	29.0	6.8	13.3
North Africa	3,060	3,190	3,310	26.8	10.4	18.2
Sub-Saharan Africa	2,150	2,220	2,380	9.5	7.2	15.9
Nigeria	2,280	2,610	2,750	41.0	25.7	28.5
Tanzania	2,150	2,030	2,160	42.6	38.7	32.9
Kenya	1,980	2,060	2,170	54.1	26.3	31.2
Uganda	2,250	2,270	2,230	18.3	53.5	24.2

[a] Based on data from Food Balance Sheets, calculated on three-year averages to reduce the impact of errors in recording of annual stock variations.

SOURCE: FAO WORKSHEET A_9.

[b] the variability is obtained as the standard deviation over five years of the deviation from the trend of per capita food supply observed during the period 1990 to 2010.

SOURCE: FAO WORKSHEET V_6.4. The time series starts in 1995.

Note that the variability at continental level is less than at country level due to the effect of aggregation.

Time series graphs with Table A.3
Dietary Energy Supply

Time series graphs with Table A.3
Per Capita Food Supply Variability

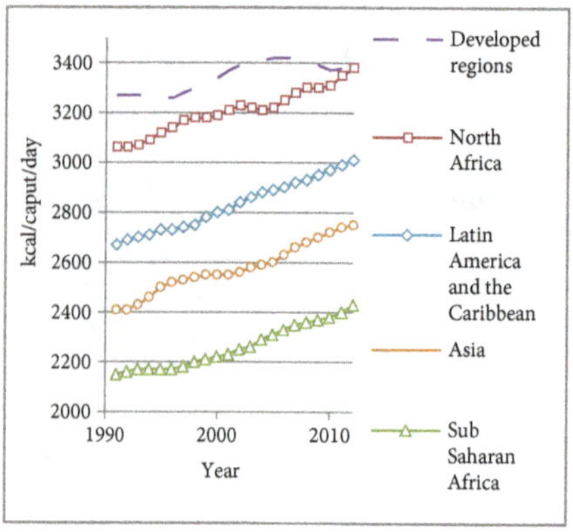

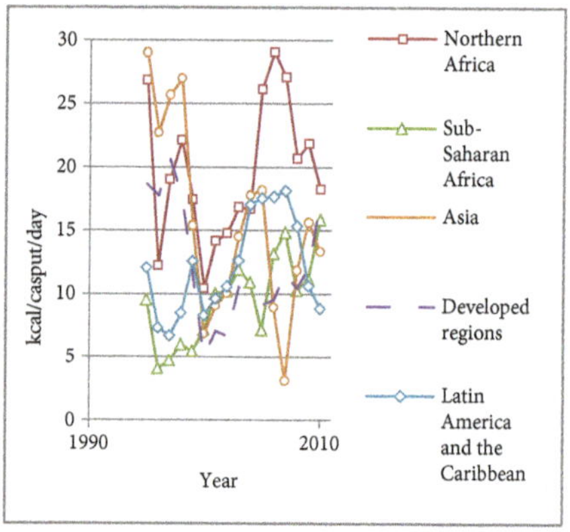

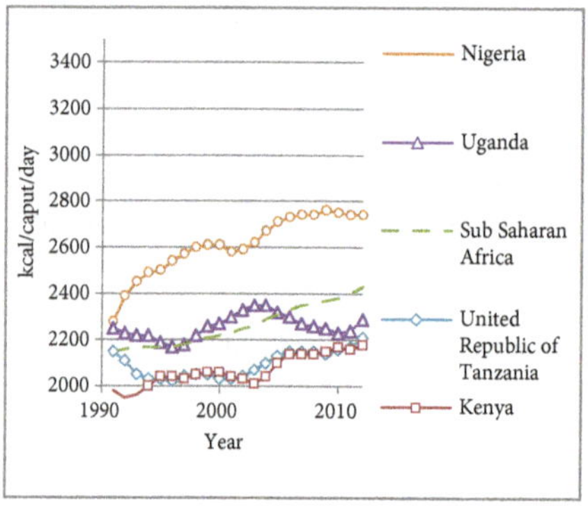

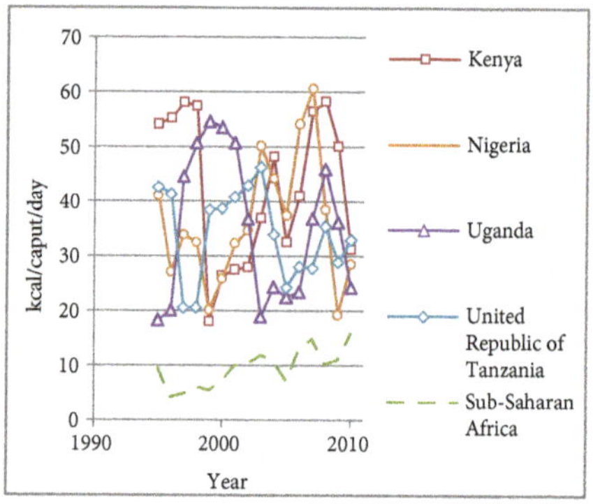

TABLE A.4 *Adequacy of food supply compared to requirements*

	Average Dietary Energy Supply Adequacy (%)[a]			Average Dietary Energy Requirement (ADER) (kcal/caput/day)[b]		
	1991	**2000**	**2010**	**1991**	**2000**	**2010**
Developed regions	131	133	134	2,490	2,510	2,510
Latin America and the Caribbean	117	121	126	2,280	2,320	2,360
Asia	107	112	117	2,240	2,290	2,330
North Africa	138	139	141	2,210	2,290	2,340
Sub-Saharan Africa	100	103	110	2,140	2,160	2,180
Nigeria	107	121	127	2,140	2,160	2,160
Tanzania	103	96	103	2,090	2,110	2,100

	Average Dietary Energy Supply Adequacy (%)[a]			Average Dietary Energy Requirement (ADER) (kcal/caput/day)[b]		
	1991	2000	2010	1991	2000	2010
Kenya	95	96	100	2,080	2,150	2,160
Uganda	108	109	107	2,080	2,080	2,090

[a] Calculated as Dietary Energy Supply (DES – see Table A.3) as % of the Average Dietary Energy Requirement (ADER).

SOURCE: FAO WORKSHEET V_1.1.

[b] ADER is a proper normative reference for adequate nutrition in the population, assuming a moderate activity pattern (set at 1.85 times basal metabolic rate). as it is an average for all age groups and both sexes, the form of the population pyramid matters: the younger the population, the lower the requirements. the requirements are also based on normal body weight for actual height; so stunting among the population leads to a lower ADER.

SOURCE: FAO WORKSHEET A_4.

Time series graphs with Table A.4
Average Dietary Energy Supply Adequacy

Time series graphs with Table A.4
Average Dietary Energy Requirement (ADER)

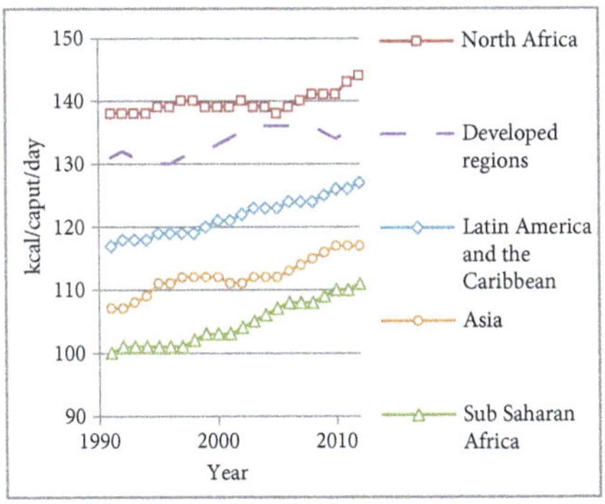

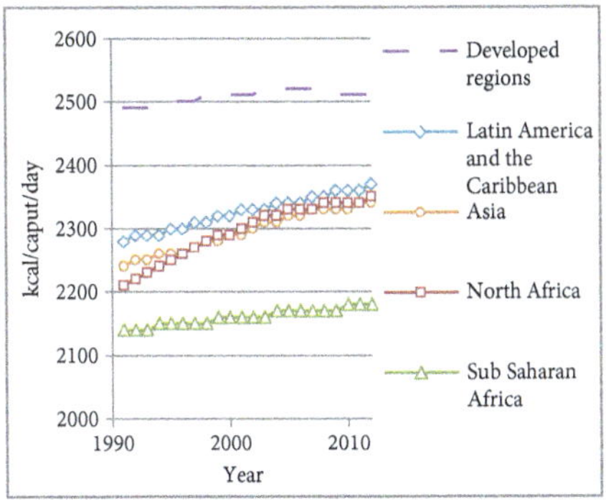

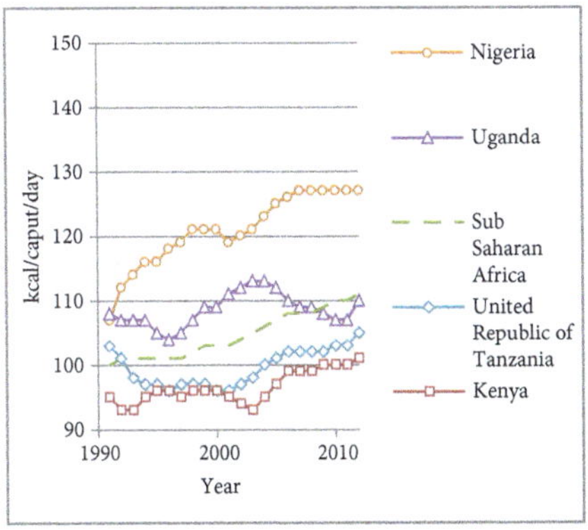

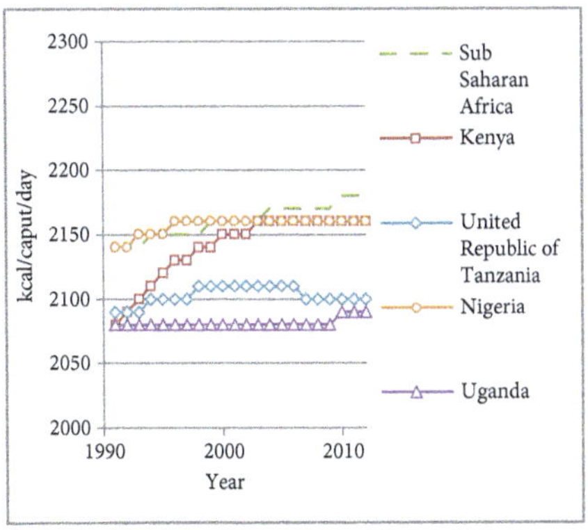

TABLE A.5 *Food price level and volatility*

	Domestic Food Price Level Index (index)[a]			Domestic Food Price Volatility Index (index)[b]		
	1990	2000	2010	1995	2000	2010
Developed regions	1.24	1.21	1.20	19.4	11.6	7.4
Latin America and the Caribbean	1.42	1.28	1.40	33.7	33.4	13.7
Asia	1.51	1.59	1.79	24.2	22.2	36.4
North Africa	1.85	1.84	1.79	41.6	10.2	48.9
Sub-Saharan Africa	1.75	1.79	1.97	78.0	40.8	15.9
Nigeria	2.75	2.37	2.42	80.2	64.6	45.6
Tanzania	1.85	1.91	2.27	23.6	12.3	46.4
Kenya	1.61	1.72	2.24	73.4	26.9	42.7
Uganda	1.56	1.66	1.88	57.0	57.7	59.2

[a] Calculated by dividing the Food Purchasing Power Parity (FPPP) by the General PPP, thus providing an index of the price of food in the country relative to the price of the generic consumption basket.

SOURCE: FAO WORKSHEET V_3.1.

[b] calculated as the standard deviation (SD) of the deviations from the trend of the domestic food price level index over the previous five years.

SOURCE: FAO WORKSHEET V_6.2.

Time series graphs with Table A.5
Domestic Food Price Level Index

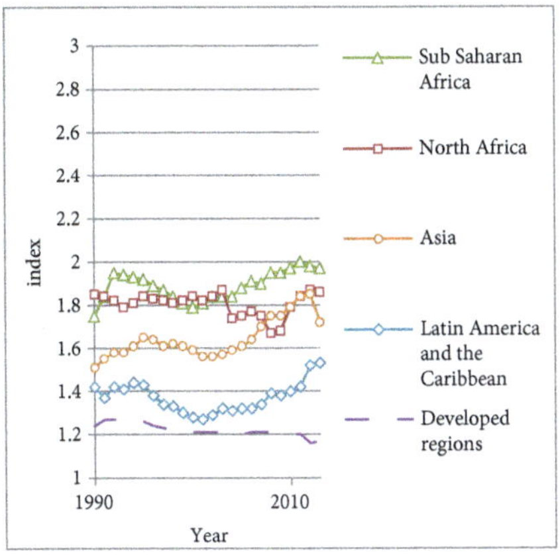

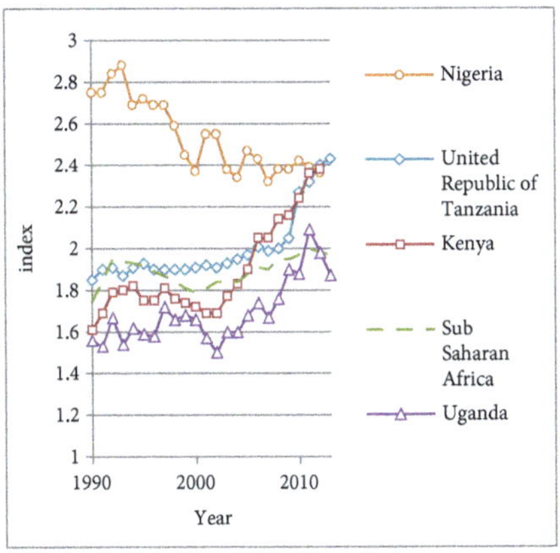

Time series graphs with Table A.5
Domestic Food Price Volatility Index

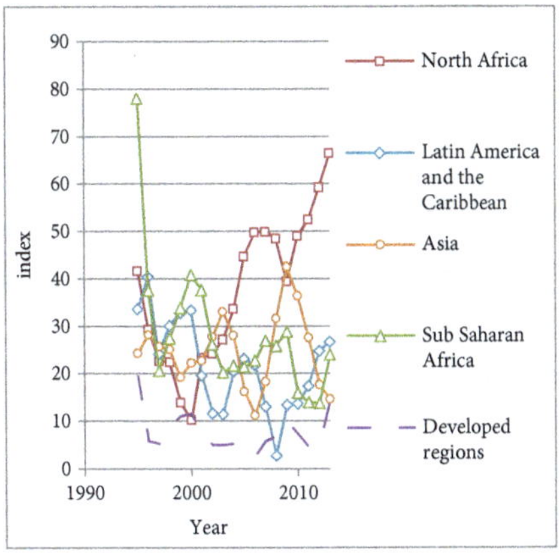

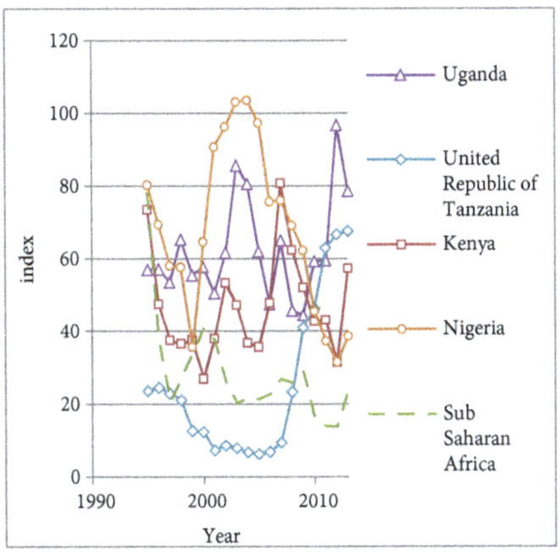

TABLE A.6 *Energy requirement levels: Cut-off value for chronic food deprivation, for Food Inadequacy and average requirements*

	Minimum Dietary Energy Requirement (MDER) – PAL 1.55 (kcal/caput/day)[a]			Less conservative Minimum Dietary Energy Requirement (MDER) – PAL 1.75 (kcal/caput/day)[b]			Average Dietary Energy Requirement (ADER) – PAL 1.85 (kcal/caput/day[c]		
	1991	2000	2010	1991	2000	2010	1991	2000	2010
Developed regions	1,930	1,950	1,950	2,130	2,150	2,160	2,490	2,510	2,510
Latin America & the Caribbean	1,810	1,830	1,860	1,960	1,990	2,030	2,280	2,320	2,360
Asia	1,780	1,800	1,830	1,930	1,960	2,010	2,240	2,290	2,330
North Africa	1,760	1,810	1,840	1,900	1,960	2,010	2,210	2,290	2,340
Sub-Saharan Africa	1,720	1,730	1,740	1,840	1,860	1,870	2,140	2,160	2,180
Nigeria	1,720	1,730	1,730	1,850	1,860	1,860	2,140	2,160	2,160
Tanzania	1,680	1,690	1,690	1,810	1,820	1,820	2,090	2,110	2,100
Kenya	1,680	1,720	1,730	1,800	1,840	1,860	2,080	2,150	2,160
Uganda	1,680	1,680	1,680	1,800	1,790	1,800	2,080	2,080	2,090

[a] Amount of energy needed for minimum acceptable weight for attained-height and light physical activity level (PAL), i.e. 1.55 times Basal Metabolic Rate). Cut-off value for (chronically) undernourished.

SOURCE: FAO WORKSHEET A_3.

[b] Amount of energy needed for minimum acceptable weight for attained-height and moderate PAL (1.75*BMR). Cut-off value for people who are likely to be in their economic activity due to insufficient food.

SOURCE: FAO WORKSHEET A_5.

[c] normative reference for adequate nutrition in the population, assuming a moderate activity pattern (PAL = 1.85). as it is an average for all age groups and both sexes, the form of the population pyramid matters: the younger the population, the lower the requirements. the requirements are also based on normal body weight for attained height; so stunting among the population leads to a lower ADER.

SOURCE: FAO WORKSHEET A_4.

Time series graphs with Table A.6
Energy requirement levels by country

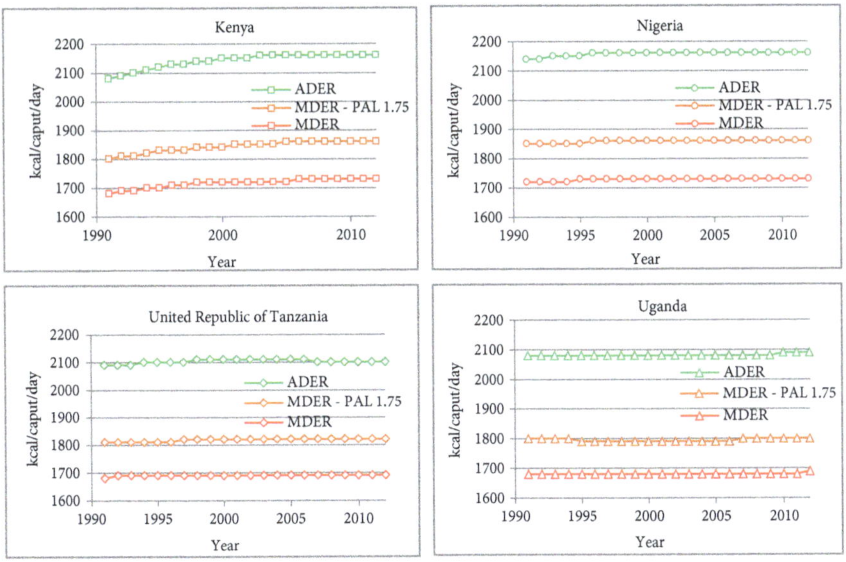

TABLE A.7 *Total population and no. of people who are hungry (undernourished), i.e. consuming less than the minimum dietary requirement (MDER, see Table A.6)*

	Total population (millions)[a]			**No. of people undernourished (millions)**[b]		
	1991	**2000**	**2010**	**1991**	**2000**	**2010**
Developed regions	1,155.1	1,195.6	1,244.1	19.8	20.2	15.9
Latin America and the Caribbean	451.0	521.4	590.1	65.7	61.5	49.1
Asia	3,127.4	3,586.0	4,029.1	751.3	646.7	575.2
North Africa	122.0	142.0	165.9	4.6	4.9	4.3
Sub-Saharan Africa	529.9	669.3	856.5	173.1	205.6	224.5
Nigeria	100.0	123.7	158.5	21.3	12.6	11.1
Tanzania	26.3	34.1	44.9	7.6	13.8	15.9
Kenya	24.2	31.3	40.5	8.4	10.1	10.7
Uganda	18.3	24.2	33.4	5.0	6.5	11.0

[a] UN Population Division, revision 2,010, calculated on three-year averages.
SOURCE: FAO WORKSHEET A_1.

[b] calculated by applying the estimated prevalence of undernourishment (see table A.8) to the total population in each period.
SOURCE: FAO WORKSHEET A_2.

Time series graphs with Table A.7
Total population

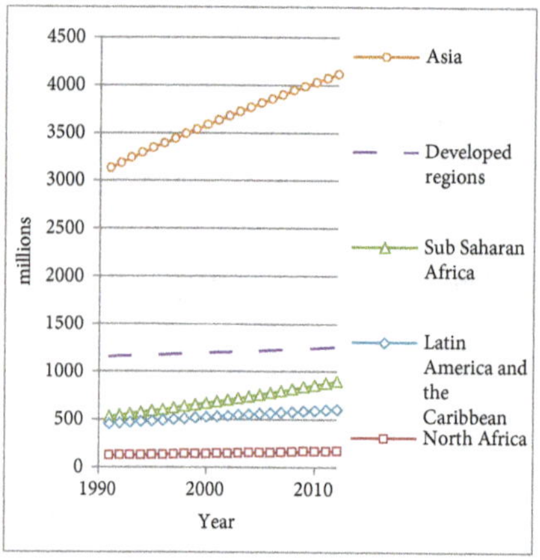

Time series graphs with Table A.7
No. of people undernourished

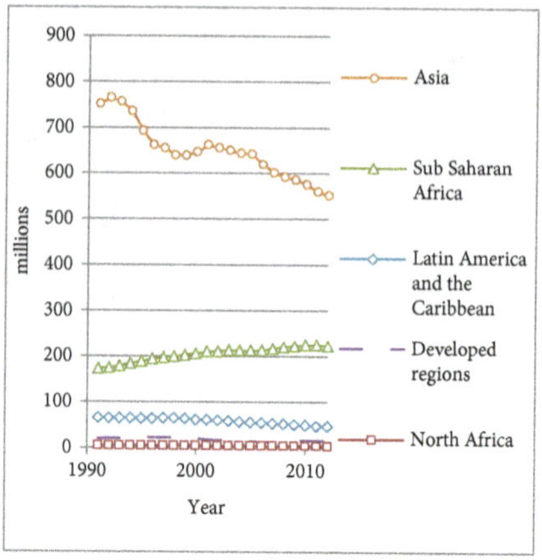

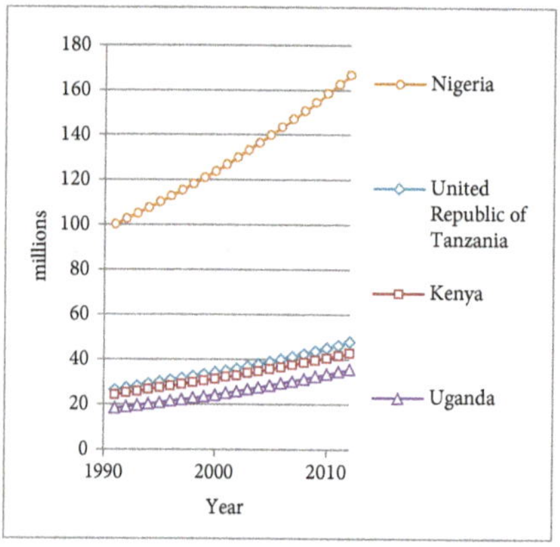

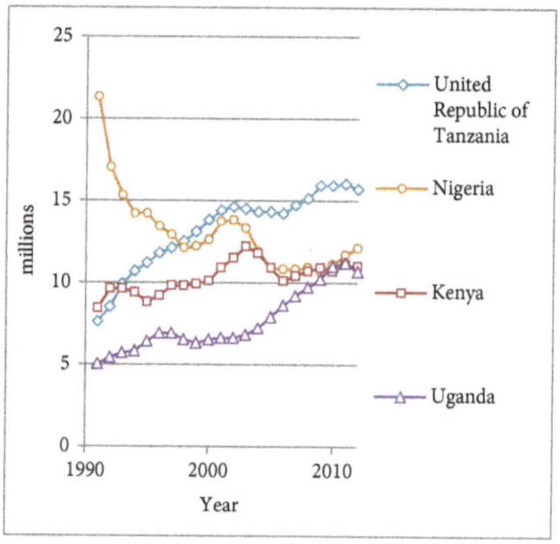

TABLE A.8 *% of people who are hungry (undernourished), i.e. consuming less than the minimum dietary requirement*

	Prevalence of Undernourishment (PoU) (%)[a]			Minimum Dietary Energy Requirement (MDER) (kcal/caput/day)[b]		
	1991	2000	2010	1991	2000	2010
Developed regions	<5	<5	<5	1,930	1,950	1,950
Latin America and the Caribbean	14.7	11.9	8.4	1,810	1,830	1,860
Asia	24.1	18.1	14.3	1,780	1,800	1,830
North Africa	<5	<5	<5	1,760	1,810	1,840
Sub-Saharan Africa	32.7	30.8	26.3	1,720	1,730	1,740
Nigeria	21.3	10.2	7.0	1,720	1,730	1,730

	Prevalence of Undernourishment (PoU) (%)[a]			Minimum Dietary Energy Requirement (MDER) (kcal/caput/day)[b]		
	1991	2000	2010	1991	2000	2010
Tanzania	28.8	40.5	35.5	1,680	1,690	1,690
Kenya	34.8	32.2	26.3	1,680	1,720	1,730
Uganda	27.1	26.9	32.9	1,680	1,680	1,680

[a] The PoU expresses the probability that a randomly selected individual from the population consumes an amount of calories below the MDER, i.e. insufficient to cover her/his energy requirement for an active and healthy life.

SOURCE: FAO WORKSHEET V_7.1.

[b] the amount of energy needed for light activity and minimum acceptable weight for attained-height. for an entire population, the minimum energy requirement is the weighted average of the minimum energy requirements of the different gender-age groups in the population. it is used as a cut-off value, or threshold, to estimate the prevalence of the undernourished population in a country.

SOURCE: FAO WORKSHEET A_3.

Time series graphs with Table A.8
Prevalence of Undernourishment (PoU)

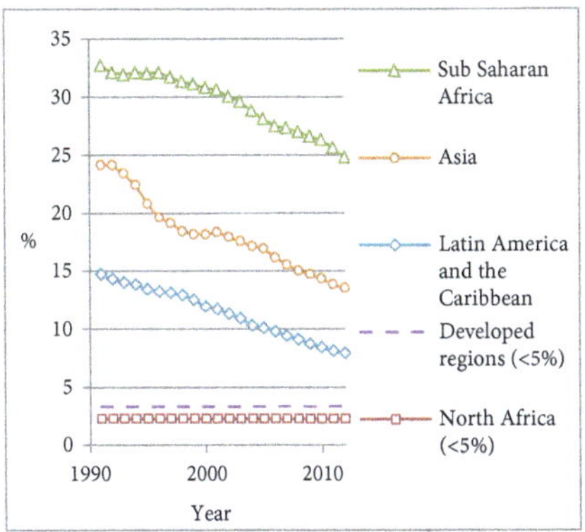

Time series graphs with Table A.8
Minimum Dietary Energy Requirement

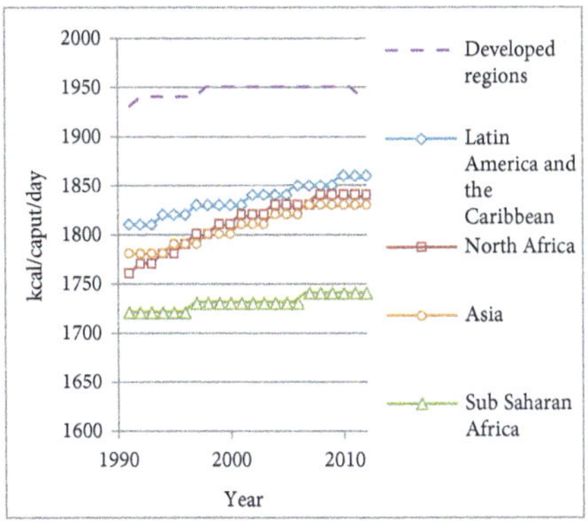

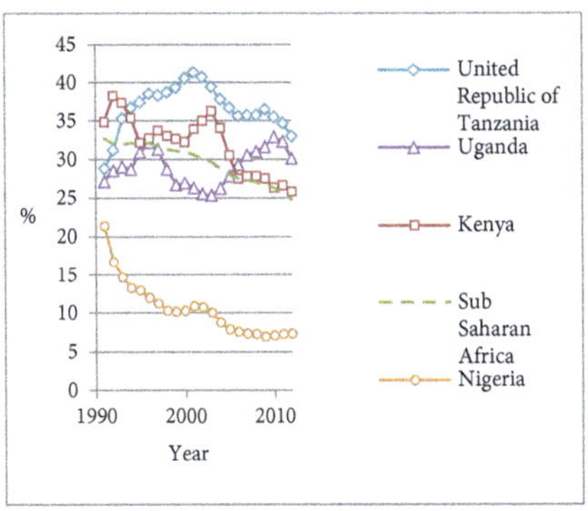

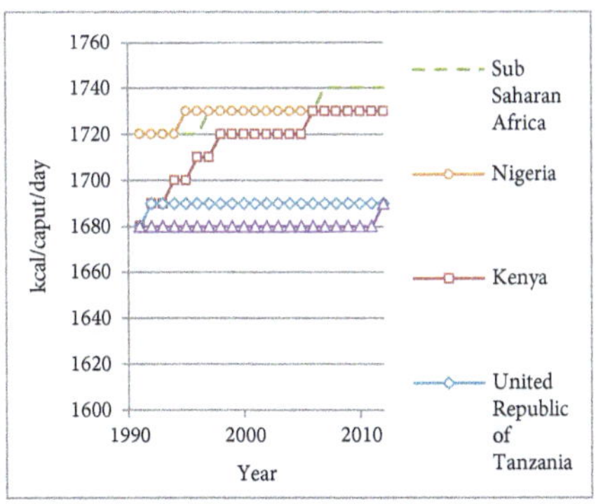

TABLE A.9 *Depth of food deficit*

	Depth of food deficit (kcal/caput/day)[a]			Calculated average dietary energy supply of the chronically undernourished (food-deprived) population[b]			Calculated average dietary energy supply of those not chronically undernourished (food-deprived)[c]		
	1991	2000	2010	1991	2000	2010	1991	2000	2010
Developed regions	12	12	9	1,918	1,938	1,941	2,834	2,941	3,077
Latin America without the Caribbean	94	77	58	1,716	1,753	1,802	2,662	2,743	2,885
Asia	165	123	99	1,615	1,677	1,731	3,111	3,241	3,350
North Africa	21	21	16	1,739	1,789	1,824	2,466	2,533	2,674
Sub-Saharan Africa	221	213	184	1,499	1,517	1,556	2,468	2,717	2,830
Nigeria	133	60	41	1,587	1,670	1,689	2,413	2,454	2,552
Tanzania	180	283	243	1,500	1,407	1,447	2,259	2,323	2,388
Kenya	222	213	170	1,458	1,507	1,560	2,524	2,548	2,605
Uganda	166	165	214	1,514	1,515	1,466	2,834	2,941	3,077

[a] The depth of the food deficit indicates how many calories would be needed to lift the undernourished from their food-deprived status to minimum requirement level (MDER), everything else being constant.

SOURCE: FAO WORKSHEET V_7.3.

[b] Calculated by the authors as MDER (see Table A.6) minus the depth of the deficit.

[c] Calculated by the authors by subtracting the part corresponding to the undernourished (b) from the total DES (see Table A.3) and dividing by (1-PoU/100).

Time series graphs with Table A.9
Depth of food deficit

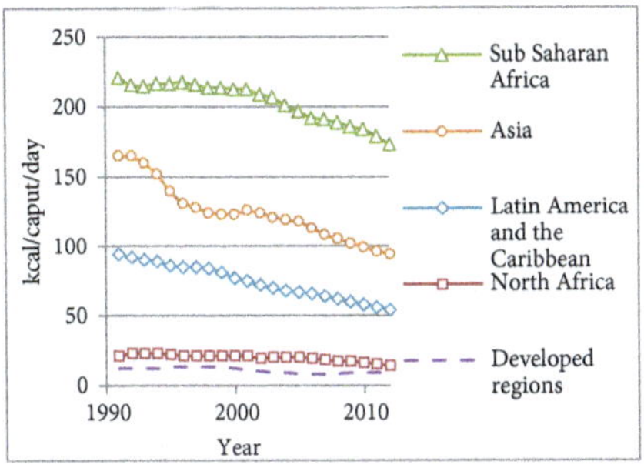

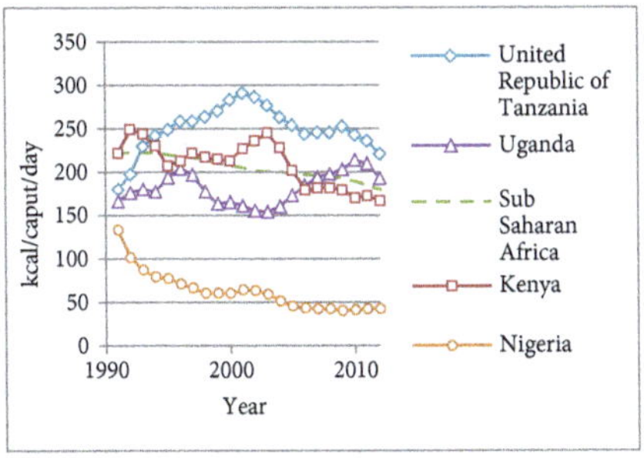

Time series graphs with Table A.9
Average DES of the non-hungry

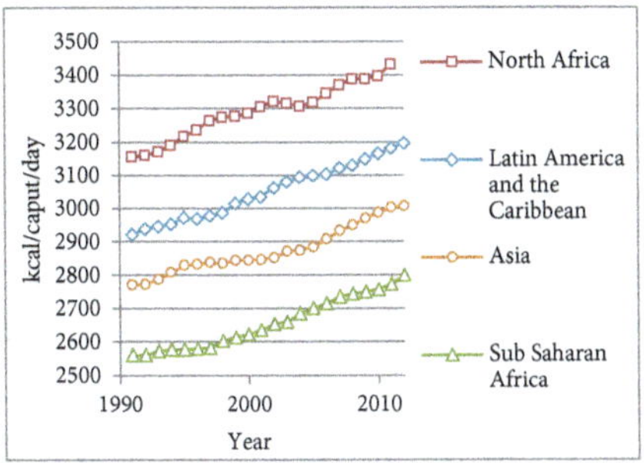

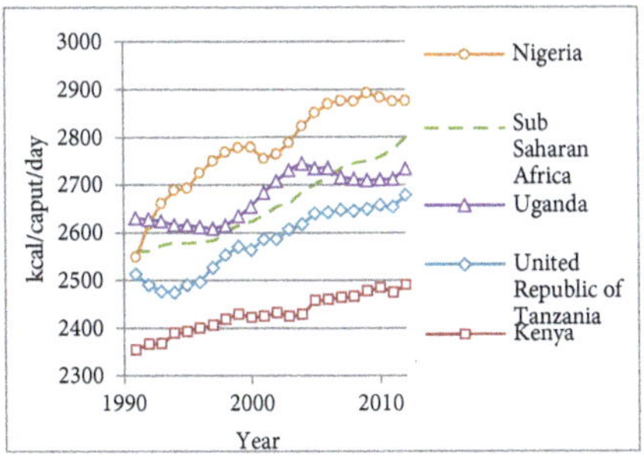

TABLE A.10 *No. and % of people with inadequate food (new outcome indicator of access)*

	No. of people with Food Inadequacy (millions)[a]			Prevalence of Food Inadequacy (%)[b]			Minimum Dietary Energy Requirement (MDER) – PAL 1.75 (kcal/caput/day)[c]		
	1991	**2000**	**2010**	**1991**	**2000**	**2010**	**1991**	**2000**	**2010**
Developed regions	<58	<60	<62	<5	<5	<5	2,130	2,150	2,160
Latin America & Caribbean	98	98	89	21.8	18.7	15	1,960	1,990	2,030
Asia	1,035	957	903	33.1	26.7	22.4	1,930	1,960	2,010
North Africa	9	9	9	7.0	6.6	5.7	1,900	1,960	2,010
Sub-Saharan Africa	216	256	280	40.8	38.3	32.7	1,840	1,860	1,870
Nigeria	30	20	18	29.7	15.8	11.5	1,850	1,860	1,860
Tanzania	10	17	19	38.1	48.6	43	1,810	1,820	1,820
Kenya	11	13	15	45	42.7	36.7	1,800	1,840	1,860
Uganda	7	9	13	35.8	35.3	40.2	1,800	1,790	1,800

[a] Calculated by the authors by multiplying the prevalence with the population size (see Table A.7).

[b] Food Inadequacy uses a less conservative cut-off value than the Prevalence of (chronically) Undernourished (PoU, see Table A. 8). It includes, in addition to the chronically undernourished, also those who are likely being conditioned in their economic activity by insufficient food.

SOURCE: FAO WORKSHEET V_7.4.

[c] see table a.6.

SOURCE: FAO WORKSHEET A_5

Time series graphs with Table A.10 No. of people with Food Inadequacy

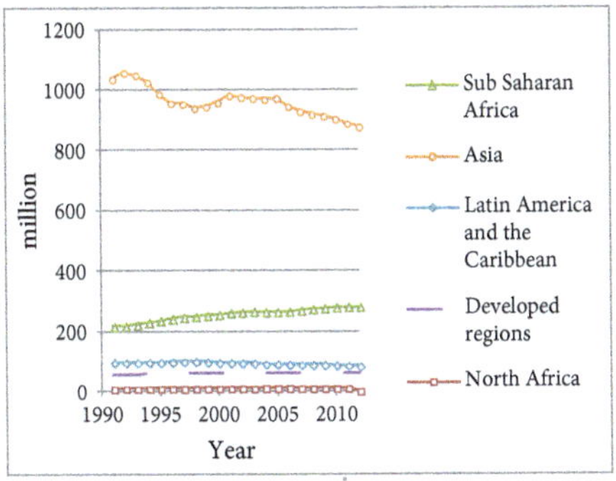

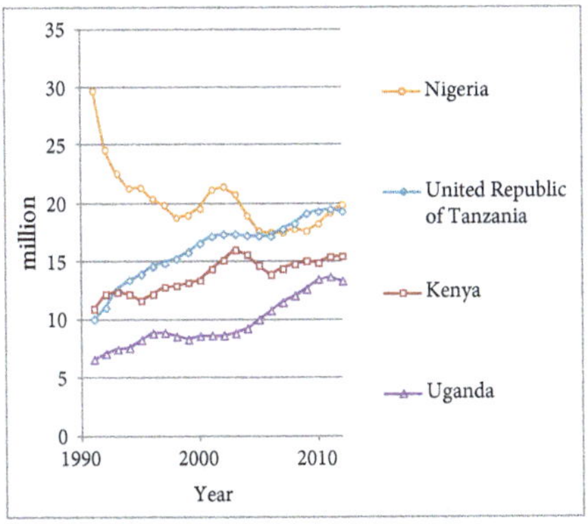

Time series graphs with Table A.10 Prevalence of Food Inadequacy

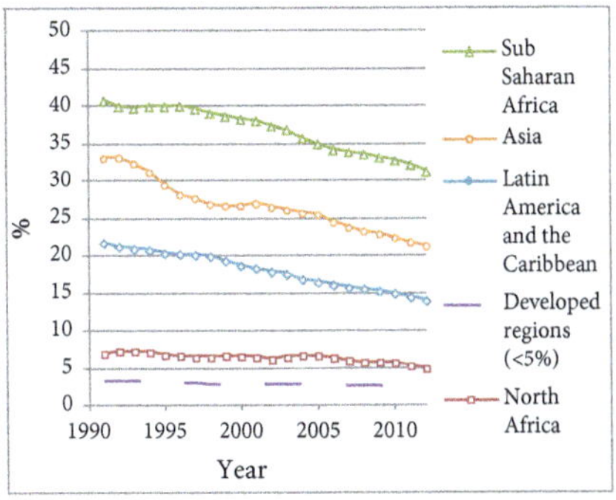

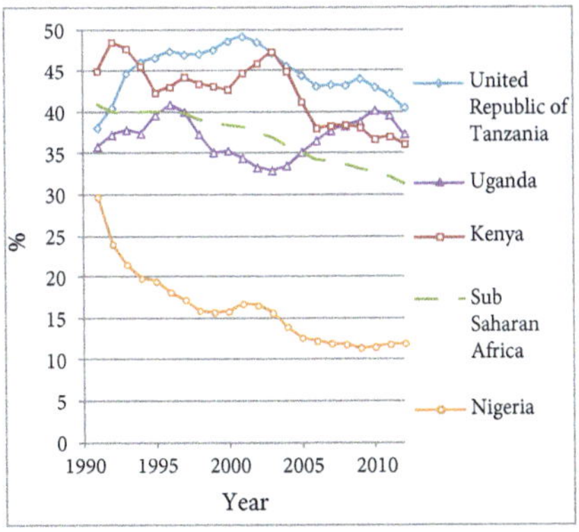

TABLES A.11A–A.11C: *Prevalence of under-fives with anthropometric failure*

TABLE A.11A *Prevalence of acute undernutrition (wasting) among under-fives*

	% of children aged <5 years affected by wasting[1] in total children <5 population (%)					
	Year closest to 1993		Year closest to 2000		Year closest to 2009	
Nigeria	1993	20.6	1999	17.6	2008	14.4
Tanzania	1991	7.9	1999	5.6	2010	4.9
Kenya	1993	7.1	2000	7.4	2009	7
Uganda	1995	7	2000	5	2006	6.3

[1] Having a weight-for-height more than 2 standard deviations below the WHO Child Growth Standards median among children aged 0–5 years.

SOURCE: FAO WORKSHEET V_8.1.

TABLE A.11B *Prevalence of chronic undernutrition (stunting) among under-fives*

	% of children aged <5 years stunted[2] in total children <5 population (%)					
	Year closest to 1993		Year closest to 2000		Year closest to 2009	
Nigeria	1993	43.8	1999	39.7	2008	41
Tanzania	1991	49.7	1999	48.3	2010	42.5
Kenya	1993	40.2	2000	41	2009	35.2
Uganda	1995	45	2000	44.8	2006	38.7

[2] Having a height-for-age more than 2 standard deviations below the WHO Child Growth Standards median among children aged 0–5 years.

SOURCE: FAO WORKSHEET V_8.2.

TABLE A.11C *Prevalence of underweight among under-fives*

	% of children aged <5 years underweight[3] in total children <5 population (%)					
	Year closest to 1993		Year closest to 2000		Year closest to 2009	
Nigeria	1993	35.1	1999	27.3	2008	26.7
Tanzania	1991	25.1	1999	25.3	2010	16.2

	% of children aged <5 years underweight in total children <5 population (%)					
	Year closest to 1993		Year closest to 2000		Year closest to 2009	
Kenya	1993	20.1	2000	17.5	2009	16.4
Uganda	1995	21.5	2000	19	2006	16.4

[3] Having a weight-for-age more than 2 standard deviations below the WHO Child Growth Standards median among children aged 0–5 years.

SOURCE: FAO WORKSHEET V_8.3.

Time series graph with Table A.11a
Prevalence of under-fives with acute undernutrition (wasting)

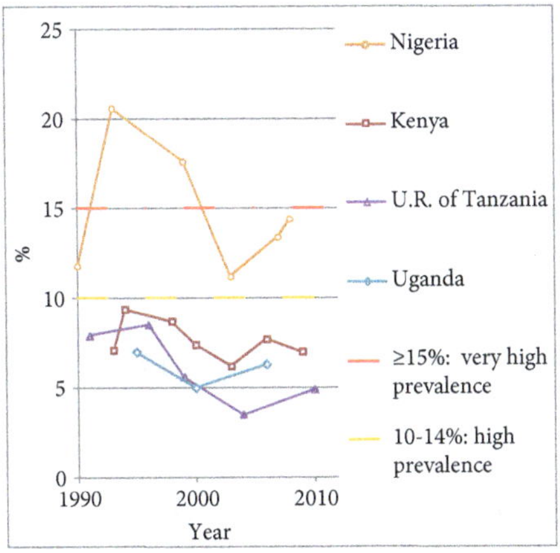

Time series graph with Table A.11b
Prevalence of under-fives with chronic undernutrition (stunting)

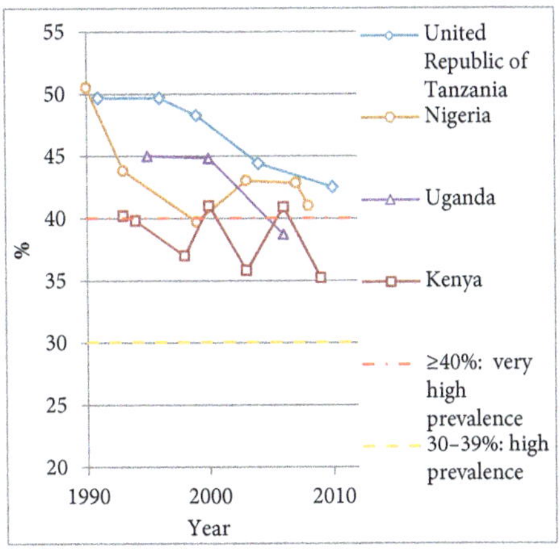

Time series graph with Tables A.11a & A.11b
% of under-fives wasted by % stunted from the early 1990s (open icon) till 2010 (largest icon)

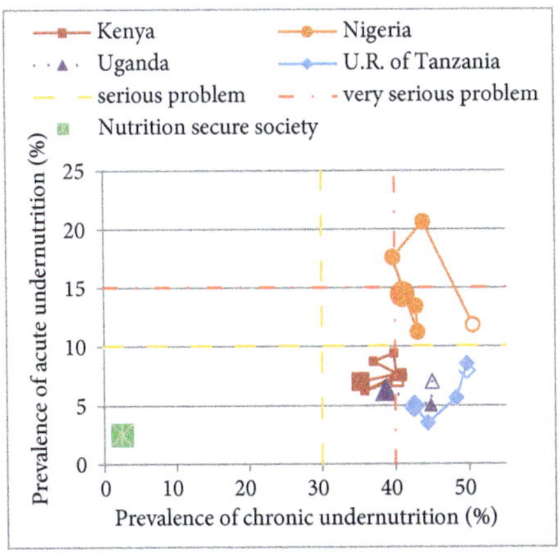

Time series graph with Table A.11c
Prevalence of under-fives with underweight (indicative of acute and/or chronic undernutrition)

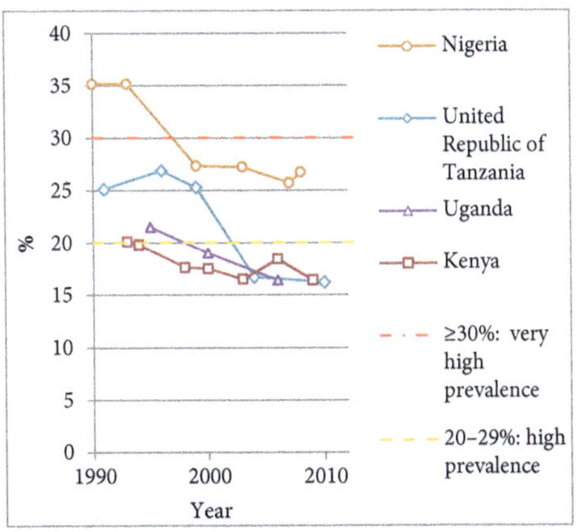

TABLES A.12 *Prevalence of micronutrient deficiency among young children* (1)

TABLE A.12A *Prevalence of anaemia among under-fives*

	% of children aged <5 years affected by anaemia[a] in total children <5 population (%)					
	Year closest to 1993		**Year closest to 2000**		**Year closest to 2009**	
Nigeria	1993	75.6				
Tanzania	1992	83.9	1996	80.6	2005	71.8
Kenya	1991	76.3	1999	69		
Uganda			2001	64.1		

[a] Less than 110 g/l of haemoglobin at sea level. The prevalence of anaemia is an important health indicator of nutritional imbalance. When used with other measurements of iron status, haemoglobin concentration can provide information about the severity of iron deficiency. A prevalence of anaemia equal to or higher than 40% signals a severe public-health problem.

SOURCE: FAO WORKSHEET V_8.5.

TABLE A.12B *Prevalence of Vitamin A deficiency among under-fives*

	% of children aged <5 years affected by Vitamin A deficiency[b] in total children <5 population (%)					
	Year closest to 1993		**Year closest to 2000**		**Year closest to 2009**	
Nigeria			2001	29.5		
Tanzania			1997	24.2		
Kenya	1994	40.6	1999	84.4		
Uganda			2001	27.9		

[b] Serum retinol equal or lower than 0.70 µmol/l. Deficiency of Vitamin A is associated with significant morbidity and mortality from common childhood infections and is the world's leading preventable cause of childhood blindness. A prevalence equal to or higher than 20% signals a severe public-health problem.

SOURCE: FAO WORKSHEET V_8.6.

Time series graph with Table A.12a Prevalence of anaemia among under-fives

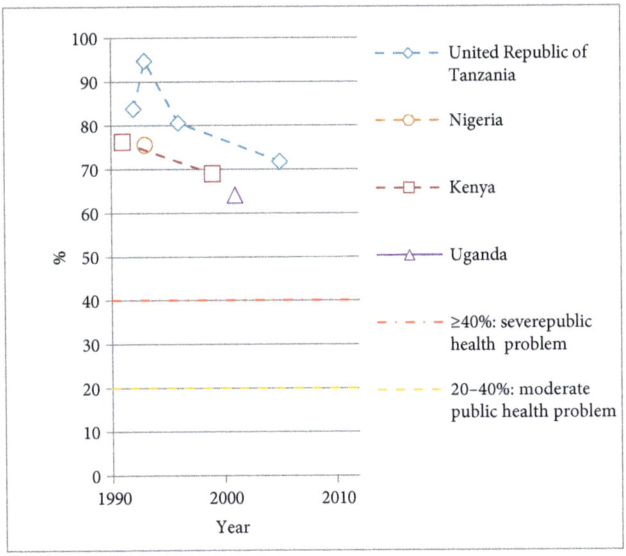

Time series graph with Table A.12b Prevalence of Vitamin A deficiency among under-fives

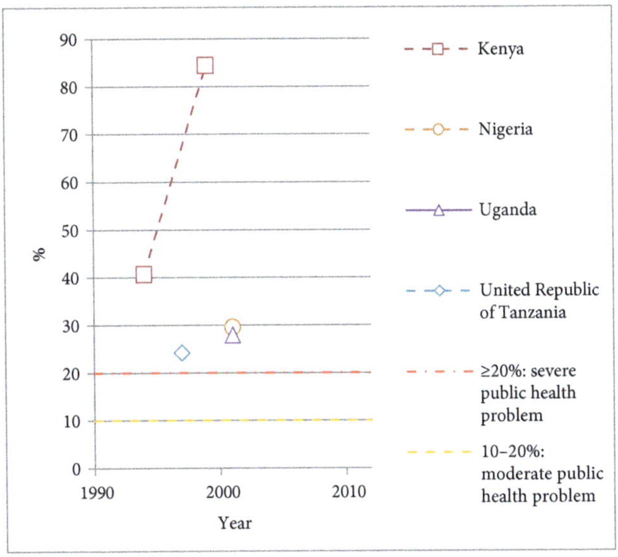

TABLES A.13: *Prevalence of micronutrient deficiency among young children* (2)

TABLE A.13A *Prevalence of iodine deficiency among children 6–12 years*

	% of children aged 6–12 years affected by iodine deficiency[a]					
	Year closest to 1993		**Year closest to 2000**		**Year closest to 2009**	
Nigeria			2001	27.5		
Tanzania	1996	37.7				
Kenya			2004	36.8		
Uganda					2005	3.9

[a] Urinary iodine equal to or lower than 100 μg/l. Iodine deficiency poses a threat throughout the lifecycle and has been associated with mental impairment, goitre, stillbirth and congenital anomalies. Inadequate iodine intake during pregnancy may lead to irreversible foetal brain damage. A prevalence of Iodine deficiency equal or higher than 20% signals a severe public-health problem.

SOURCE: FAO WORKSHEET V_8.7.

TABLE A.13B *Prevalence of anaemia among pregnant women*

	% of pregnant women affected by anaemia[b] (%)					
	Year closest to 1993		**Year closest to 2000**		**Year closest to 2009**	
Nigeria	1993	66.7				
Tanzania					2005	58.2
Kenya			1999	55.1		
Uganda			2001	41.2	2006	64.4

[b] Haemoglobin level below 110 grams per litre at sea level.

SOURCE: FAO WORKSHEET V_8.8.

Time series graph with Table A.13a
Prevalence of iodine deficiency among children aged 6–12 years

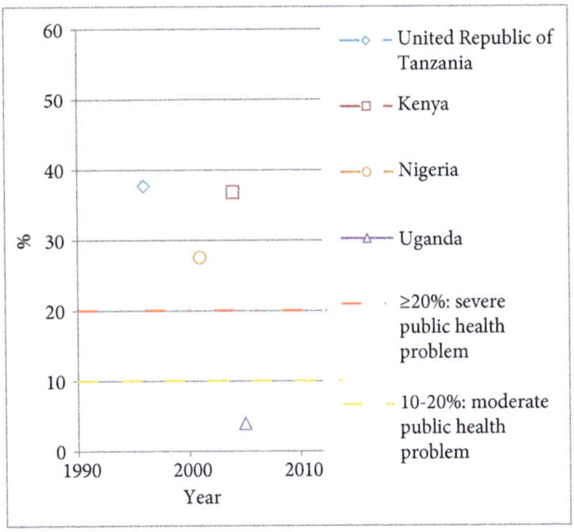

Time series graph with Table A.13b
Prevalence of anaemia among pregnant women

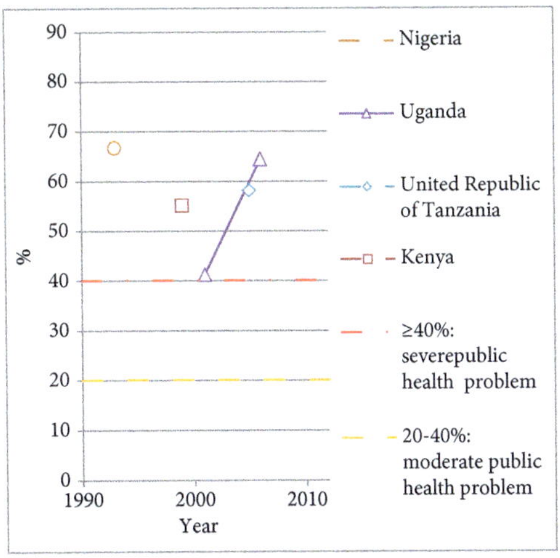

Appendix B

Fifty Years of Agricultural and Food Dynamics in Africa – Statistical Data

The database FAOSTAT contains time series data since 1961 on annual crop and animal production and the national food balance. While Chapter 1 contains the main findings, the tables and graphs in this Appendix give more detailed results[1]. Data on agricultural production as well as on commodity 'appearance' and 'disappearance' are given, which allow the indirect estimate of the amount available for human consumption: the so-called 'food balance'. Results for the year 2012 (or - when these were not available at the time of this analysis – 2011) are compared with the base year 1961. To standardize the comparison of changes, an index is given, which is calculated as the most recent value didived by the value in 1961 times 100. The indexes for results in terms of absolute quantities are compared with the index for population size, while the indexes for results in per capita terms are to be compared with the standard value 100.

TABLE B.1A *Africa's agricultural production volumes of staple food crops (in millions of metric tonnes), 1961 and 2012 (index compared to 1961=100)***

	1961	*2012*	index
Population	275m	*994m	*362
Cereals except beer	46.3	172.2	372
maize	16.1	69.6	431
wheat	5.1	24.7	483
sorghum	10.7	23.3	218
millet	6.6	16.1	244
Pulses	3.6	15.9	446
Roots and tubers	48.0	266.4	555
cassava	31.5	149.5	475
potatoes	2.1	28.2	1341
sweet potatoes	3.3	18.0	548
yam	7.6	56.5	745

* The population in 2011 was estimated to be 971 million. Compared to 1961, the index figure would then be 353. Growth was estimated at 2.36% between 2011 and 2012, the same as the FAO estimate for 2010–2011.

** These figures are taken from the FAO crop production statistics (that included the final figures for 2012).

1 All statistical data have been taken from FAOSTAT. http://faostat3.fao.org/faostat-gateway/go/to/home/E. Accessed March 2014.

 | DOI 10.1163/9789004282698_017

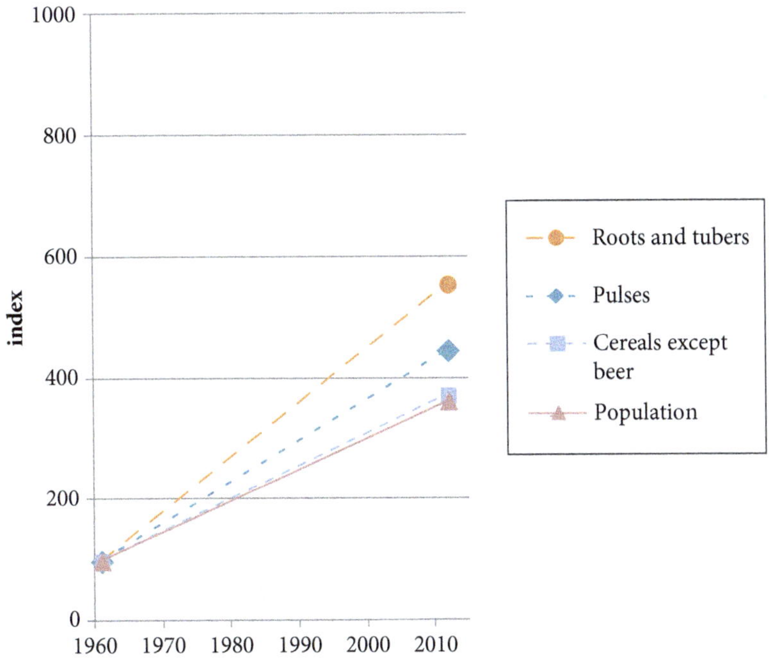

FIGURE B.1A *Africa's agricultural production volumes of staple food crops, 2012 (index compared to 1961=100)*

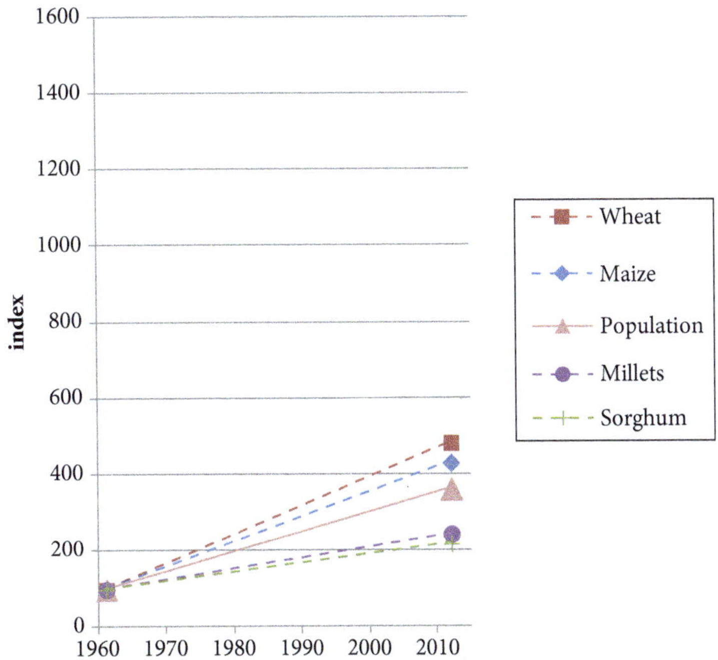

FIGURE B.1A.1 *Idem, different cereals*

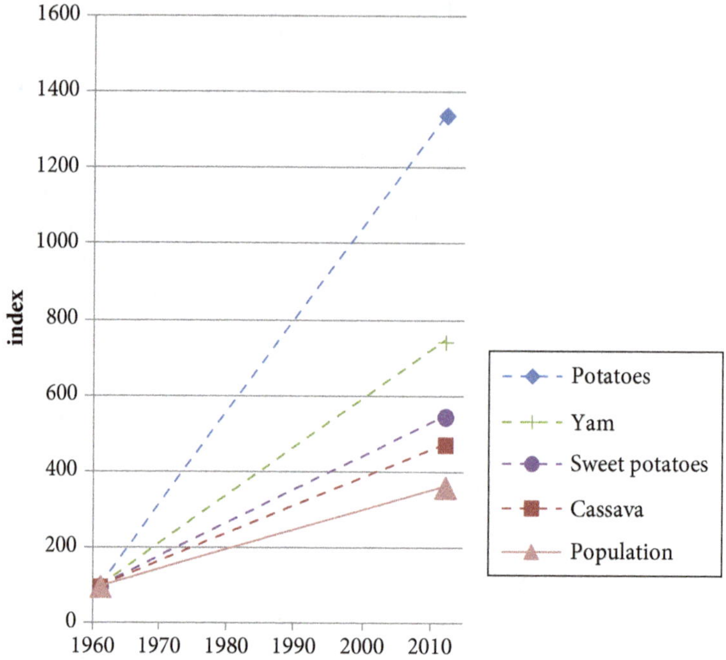

FIGURE B.1A.2 *Idem, different starchy foods*

TABLE B.1B *Africa's agricultural production volumes of other crops (in million metric tonnes), 1961 and 2012 (index compared to 1961=100)***

	1961	*2012*	index
Population	275m	*994m	*362
Fibre crops	1.2	1.8	149
Fruit, excl. melons	24.4	91.5	375
Oil crops	4.0	10.6	267
Tree nuts	0.2	2.3	1020
Vegetables and melons	13.6	74.2	546

* see note with Table B.1a
** These figures are taken from the FAO crop production statistics (that included the final figures for 2012).

Since 1961 all the major basic food categories – cereals, pulses and particularly roots and tubers – have seen greater increases than population growth (Table B.1a and Figure B.1a). Within these categories, there are major differences. For example, wheat and maize have experienced much faster growth than Africa's traditional grains (sorghum and millet – Table B.1a and Figure B.1a.1). And the roots and tubers category has seen

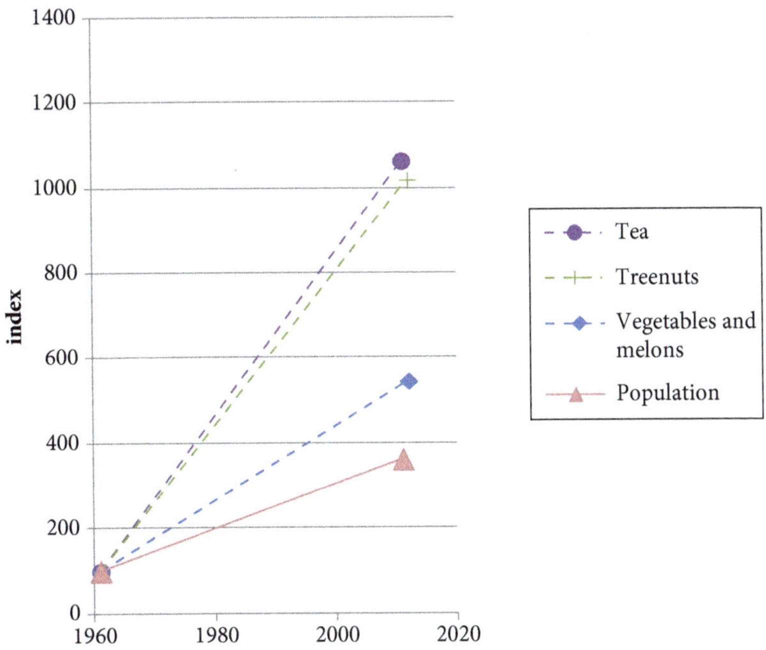

FIGURE B.1BI *Africa's agricultural production volumes of other crops, 2011 or 2012 (index compared to 1961=100): crops with higher than population growth*

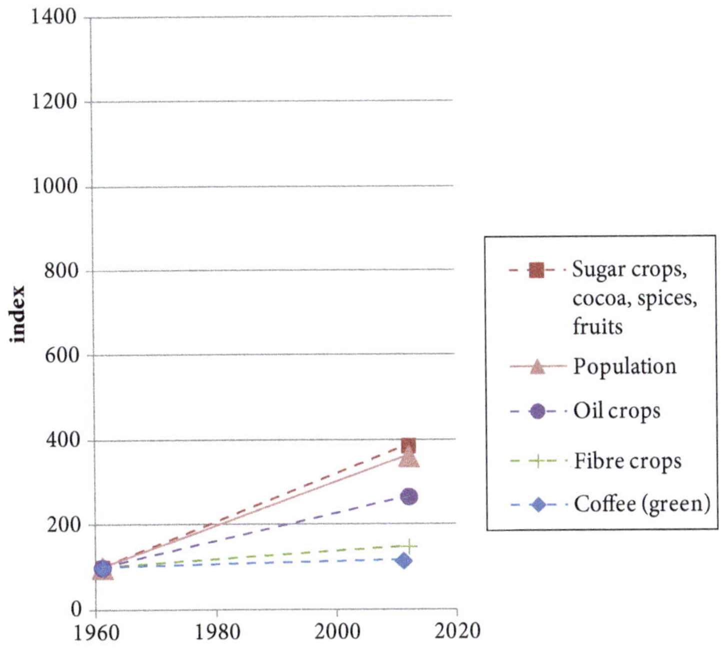

FIGURE B.1BII *Africa's agricultural production volumes of other crops, 2011 or 2012 (index compared to 1961=100): crops with modest or lower than population growth*

TABLE B.1C *Africa's agricultural production volumes of other crops and of animal products (in million metric tonnes), 1961 and 2011 (index compared to 1961=100)****

	1961	2011	
Population	275m	*971m	*353
Sugar crops	28.0	108.4	387
Stimulants	1.6	4.7	288
coffee (green)	0.9	**1.0	115
tea	0.1	**0.6	1064
cocoa	0.8	**3.3	394
Spices	0.2	0.9	383
Meat and offal	4.2	18.1	426
Animal fats	0.3	0.8	237
Milk and butter	11.0	44.2	404
Eggs	0.4	2.9	739
Fish, seafood, aquatic products	2.4	8.9	373

* The population in 2011 was estimated to be 971 million. Compared to 1961, the index figure would then be 353. Growth was estimated at 2.36% between 2011 and 2012, the same as the fao estimate for 2010–2011.

** 2012

*** These figures are taken from the FAO food balance sheets (that are available up to 2011 at the time of writing in June 2014), except for the different types of stimulants, which are taken from the FAO production statistics. For 1961, these two data sets show slight differences, partly based on different definitions of the product categories.

very high growth in potatoes and yams, followed by sweet potatoes and cassava (Table B.1a and Figure B.1a.2). Traditional export crops increased by less impressive percentages (Tables B.1b and c and Figure B.1ci); the exceptions are tea and treenuts (Tables B.1b and c and Figure B.1b). Among the animal products, the production of eggs increased twice as much as population growth, while meat increased 20% more and milk and butter 14% more than population growth (Table B.1c and Figure B.1cii).

There are differences in growth figures for agricultural produce within Africa. The data given here are for the increases in production volumes between 1961 and 2012 for the five macro-regions distinguished by the FAO (Table B.2).

By comparing the volumes of exports between 1961 and 2011 for all categories of agricultural products and also the percentages of the crops exported to foreign markets (including neighbouring countries), it can be noted that Africa's markets for agricultural produce have always been mainly internal, and are increasingly so (Table B.3).

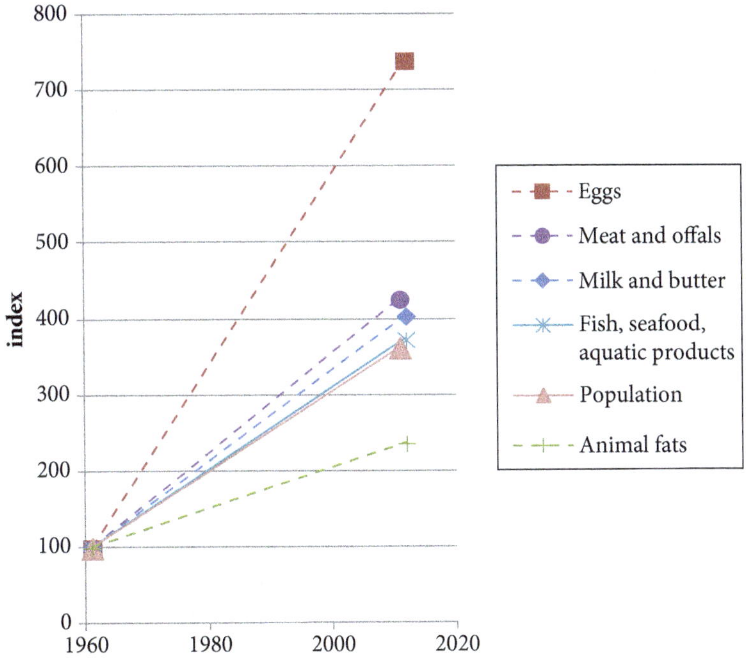

FIGURE B.1C *Africa's agricultural production volumes of animal products, 2011 (index compared to 1961=100)*

TABLE B.2 *Cereals and roots and tubers in five African regions (in millions of metric tonnes), 1961 and 2012 (index compared to 1961=100)*

Region	Cereals			Roots and tubers			Population		
	1961	2012	index	1961	2012	Index	1961	2012	index
North	9.8	39.5	402	1.3	12.3	938	70.4	212.3	302
West	13.7	57.5	421	20.0	146.1	732	86.5	322.3	373
Middle	2.6	8.9	343	13.6	40.4	296	16.7	65.7	392
Eastern	13.2	51.7	393	12.5	64.7	516	81.0	333.9	412
Southern	7.1	14.6	206	0.6	2.9	515	20.2	60.0	297
Africa as a whole	46.3	172.2	372	48.0	266.4	555	274.8	994.1	362

TABLE B.3 *Africa's agricultural exports (in millions of metric tonnes) and their relative importance as part of production volumes, 1961 and 2011 (index compared to 1961=100)*

Product categories	Export volumes in *millions of metric tonnes*			Exports as a % of total production		
	1961	2011	index	1961	2011	index
Cereals excl. beer	2.5	5.7	226	5.7	3.7	65
Pulses	0.3	0.7	274	8.0	5.0	63
Starchy roots	0.5	1.1	218	1.4	0.5	36
Sugar crops	1.1	3.5	305	4.1	3.2	78
Tree nuts	0.2	1.1	725	37.3	44.1	118
Oil crops and vegetable oils	3.4	3.3	97	32.5	10.4	32
Fruits excl. grapes	1.7	6.7	395	7.8	8.0	103
Stimulants	1.5	3.8	256	89.4	79.3	89
Spices	0.0	0.1	431	10.6	11.9	112
Alcoholic beverages	1.6	0.9	54	15.8	2.3	15
Meat and offal	0.1	0.2	158	2.6	1.0	38
Animal fat	0.1	0.1	140	26.3	15.5	59
Milk and butter	0.1	1.3	2133	0.5	2.9	580
Eggs	0.0	0.0	100	2.8	0.4	14
Fish, seafood and aquatic products	1.3	2.5	186	55.4	27.7	50

For most products, the relative importance of food imports as a percentage of available food has increased, as indicated by the index figure given in the final column of Table B.4.

Total volumes of feed increased eightfold from 12.9 million metric tonnes to 103.3 million metric tonnes. In 1961, 7% of Africa's agricultural production and import volumes were used as feed for the continent's animals. By 2011, this had increased to more than 12%. Starchy roots and cereals were and are the main products that 'disappear' into animal feed (Table B.5).

Table B.6 shows that across the board, the food situation has improved; from levels that were below WHO requirements for a healthy life to levels above these figures. The fact that all index figures in Table B.6, except for animal fats, are above 100, indicates that the availability of all these foods for human consumption increased more than population growth: the plant products on the whole by 31% and the animal products

TABLE B.4 *Africa's food imports (in millions of metric tonnes), 1961 and 2011 (index compared to 1961=100)*

Product category	Food imports in millions of metric tonnes			Food imports as a % of food production plus imports		
	1961	2011	index	1961	2011	index
Cereals excl. beer	5.0	73.1	1470	10.1	32.2	319
Pulses	0.1	1.4	1485	2.8	9.0	321
Starchy roots	0.4	1.1	287	1.0	0.4	40
Sugar crops	1.5	10.8	714	5.1	9.2	180
Tree nuts	0.0	0.1	342	9.2	5.4	59
Oil crops and vegetable oils	0.4	11.1	2793	3.7	26.0	703
Vegetables	0.2	3.2	1937	1.2	4.3	358
Fruits excl. wine	0.2	2.9	1394	1.0	3.3	330
Stimulants	0.2	0.9	536	9.5	16.4	173
Spices	0.0	0.1	971	5.4	12.6	233
Alcoholic beverages	0.4	1.4	339	3.8	3.5	92
Meat and offal	0.1	2.1	1870	2.6	11.8	454
Animal fats	0.1	0.4	246	32.0	32.8	103
Milk and butter	0.7	8.9	1316	5.8	16.7	288
Eggs	0.0	0.1	938	2.0	2.6	130
Fish, seafood and aquatic products	0.4	5.5	1478	13.5	38.2	283

as a whole by 39%. That the latter percentage is higher than the former, is again a sign of the increase in dietary diversification discussed in more detail in chapter 4.

Table B.7 highlights the construction of the food balance (food available at retail level for human consumption) of cereals and starchy roots for Africa in 2011 compared to 1961. The staple foods (cereals and starchy roots) have dropped from 66% of all kilocalories consumed to 62% and from 59% of all proteins consumed to 55%. Animal products have gone up from 8.1% of all kilocalories consumed to 8.6% and from 20.7% of all proteins consumed to 23.7%.

The total production of cereals in Africa has almost (for 99%) kept pace with population growth in the period 1961–2011. When comparing 2011 with 1961, domestic food supply increased 33% more than population growth thanks to massive imports of cereals (amounting to 33% of domestic food supply in 2011, versus 11% in 1961). The availability for human consumption per capita increased, but at a lower rate of 20% in

TABLE B.5 *Africa's food production used as feed (in millions of metric tonnes), 1961 and 2011 (index compared to 1961=100)*

Product categories	Feed volumes in millions of metric tonnes			Feed as part of food production plus food imports (%)		
	1961	2011	index	1961	2011	index
Cereals	4.8	39.3	823	10.3	17.3	168
Pulses	0.2	1.2	576	6.2	7.8	126
Starchy roots	4.4	54.4	1240	11.4	22.3	196
Sugar crops	0.6	1.5	268	2.0	1.3	65
Oil crops	0.0	0.2	2456	0.0	0.5	810
Fruits	1.0	2.5	267	4.4	2.9	66
Milk and butter	1.9	2.7	144	16.1	5.1	32
Fish etc.	0.1	1.3	1015	4.5	8.8	196

TABLE B.6 *Food quantities (expressed in terms of dietary energy and amount of protein) available per capita for Africa as a whole, 1961 and 2011 (index compared to 1961=100)*

Product category	kcal/capita/day			Protein grams/capita/day		
	1961	2011	index	1961	2011	index
Total	1990	2615	131	52.7	68.7	130
Plant products	1828	2391	131	41.7	52.4	126
cereals	1060	1279	121	28.6	33.7	118
pulses	75	101	135	5	6.7	134
starchy roots	247	337	136	2.6	3.9	150
sugars	104	162	156	-	-	-
tree nuts	9	14	156	0.3	0.4	133
oil crops	56	68	121	2.2	3.1	141
vegetable oils	124	217	175	-	-	-
vegetables	31	47	152	1.5	2.2	147
fruits	76	106	139	0.8	1.2	150
stimulants	1	4	400	0.2	0.3	150
spices	7	8	114	0.3	0.3	100

Product category	kcal/capita/day			Protein grams/capita/day		
	1961	2011	index	1961	2011	index
alcoholic beverages	38	46	121	0.3	0.4	133
Animal products	161	224	139	10.9	16.3	150
meat and offal	73	94	129	6.1	8.0	131
animal fats	19	19	100	-	-	-
milk and butter	55	83	151	2.9	4.5	155
eggs	4	9	225	0.3	0.7	233
fish, seafood, aquatic products	9	20	222	1.5	3.1	207

TABLE B.7 *Food balance sheet data for cereals and starchy roots in Africa, 1961 and 2011 (index compared to 1961=100)*

Food balance sheet (in million metric tonnes, unless otherwise indicated)	Cereals			Starchy roots		
	1961	2011	index	1961	2011	index
Production	44.2	153.8	348	37.9	242.9	640
Imports	5.0	73.0	1470	0.4	1.1	287
Stock variation	−0.1	−1.8	x	−0.1	−6.6	x
Exports	2.5	5.7	226	0.5	1.1	218
Total available	46.6	219.2	471	37.7	236.2	626
Used as seed	2.0	3.6	179	0.9	4.4	468
Used as feed	4.8	39.3	823	4.3	54.4	1240
Used for manufacturing and other uses	1.3	12.2	970	1.9	15.7	830
Wasted	3.8	17.9	470	4.1	34.7	851
Food balance = Amount available as primary food for human consumption in Africa	34.7	146.2	422	26.4	127.1	481
Amount of food (kg per capita per year)	126	151	120	96	131	136
Food energy (kcal per capita per day	1060	1279	121	247	337	136
Protein (grams per capita per day)	29	34	118	2.6	3.9	150
% of daily food energy	53	49	92	12	13	104

terms of cereal tonnage and 21% in terms of food energy (Table B.7). The differential is mainly due to a sizeable "disappearance" of cereals into animal feed (18% of domestic food supply in 2011 versus 10% in 1961) and manufacturing and other uses (6% versus 3% in 1961). Cereal losses due to waste were estimated at 8% in both years.

The total production of starchy roots in Africa increased 80% more than population growth in the period 1961-2011. When comparing 2011 with 1961, domestic food supply increased 77% more than population growth, a bit lower than production, inter alia due to almost 3% "disappearance" into stockage in 2011. The availability for human consumption per capita increased, but at a lower rate of 36% (Table B.7). The differential is mainly due to a sizeable "disappearance" of starchy roots into animal feed (23% of domestic food supply in 2011 versus 11% in 1961) and losses due to waste (15% in 2011 against 11% in 1961). Use of starchy roots for manufacturing and other uses was 5–6%.

It has to be noted that all the percentages mentioned above are the "weighted" result of different mixes of individual food items within the food group. For instance, the waste of the main cereal maize and its products increased from an estimated 8.1% in 1961 to 9.0% in 2011, while the waste of the main starchy root cassava and its products increased from 9.% in 1961 to 16.0% in 2011.

www.ingramcontent.com/pod-product-compliance
Lightning Source LLC
Chambersburg PA
CBHW071414230926
51846CB00001B/88